T4-AJT-242

MTP International Review of Science

Physiology
Series One

Consultant Editors

A. C. Guyton and
D. F. Horrobin

Publisher's Note

The MTP International Review of Science is an important new venture in scientific publishing, which is presented by Butterworths in association with MTP Medical and Technical Publishing Co. Ltd. and University Park Press, Baltimore. The basic concept of the Review is to provide regular authoritative reviews of entire disciplines. Chemistry was taken first as the problems of literature survey are probably more acute in this subject than in any other. Physiology and Biochemistry followed naturally. As a matter of policy, the authorship of the MTP Review of Science is international and distinguished, the subject coverage is extensive, systematic and critical, and most important of all, it is intended that new issues of the Review will be published at regular intervals.

In the MTP Review of Chemistry (Series One), Inorganic, Physical and Organic Chemistry are comprehensively reviewed in 33 text volumes and 3 index volumes. Physiology (Series One) consists of 8 volumes and Biochemistry (Series One) 12 volumes, each volume individually indexed. Details follow. In general, the Chemistry (Series One) reviews cover the period 1967 to 1971, and Physiology and Biochemistry (Series One) reviews up to 1972. It is planned to start in 1974 the MTP International Review of Science (Series Two), consisting of a similar set of volumes covering developments in a two year period.

The MTP International Review of Science has been conceived within a carefully organised editorial framework. The overall plan was drawn up, and the volume editors appointed by seven consultant editors. In turn, each volume editor planned the coverage of his field and appointed authors to write on subjects which were within the area of their own research experience. No geographical restriction was imposed. Hence the 500 or so contributions to the MTP Review of Science come from many countries of the world and provide an authoritative account of progress.

Butterworth & Co. (Publishers) Ltd.

PHYSIOLOGY SERIES ONE

Consultant Editors
A. C. Guyton,
Department of Physiology and Biophysics, University of Mississippi Medical Center and
D. F. Horrobin,
Department of Medical Physiology, University College of Nairobi

Volume titles and Editors

1 **CARDIOVASCULAR PHYSIOLOGY**
Professor A. C. Guyton and Dr. C. E. Jones, *University of Mississippi Medical Center*

2 **RESPIRATORY PHYSIOLOGY**
Professor J. G. Widdicombe, *St. George's Hospital, London*

3 **NEUROPHYSIOLOGY**
Professor C. C. Hunt, *Washington University School of Medicine, St. Louis*

4 **GASTROINTESTINAL PHYSIOLOGY**
Professor E. D. Jacobson and Dr. L. L. Shanbour, *University of Texas Medical School*

5 **ENDOCRINE PHYSIOLOGY**
Professor S. M. McCann, *University of Texas*

6 **KIDNEY AND URINARY TRACT PHYSIOLOGY**
Professor K. Thurau, *University of Munich*

7 **ENVIRONMENTAL PHYSIOLOGY**
Professor D. Robertshaw, *University of Nairobi*

8 **REPRODUCTIVE PHYSIOLOGY**
Professor R. O. Greep, *Harvard Medical School*

BIOCHEMISTRY SERIES ONE

Consultant Editors
H. L. Kornberg, F.R.S.
Department of Biochemistry University of Leicester and
D. C. Phillips, F.R.S., *Department of Zoology, University of Oxford*

Volume titles and Editors

1 **CHEMISTRY OF MACRO-MOLECULES**
Professor H. Gutfreund, *University of Bristol*

2 **BIOCHEMISTRY OF CELL WALLS AND MEMBRANES**
Dr. C. F. Fox, *University of California*

3 **ENERGY TRANSDUCING MECHANISMS**
Professor E. Racker, *Cornell University, New York*

4 **BIOCHEMISTRY OF LIPIDS**
Professor T. W. Goodwin, F.R.S., *University of Liverpool*

5 **BIOCHEMISTRY OF CARBO-HYDRATES**
Professor W. J. Whelan, *University of Miami*

6 **BIOCHEMISTRY OF NUCLEIC ACIDS**
Professor K. Burton, F.R.S., *University of Newcastle upon Tyne*

7 **SYNTHESIS OF AMINO ACIDS AND PROTEINS**
Professor H. R. V. Arnstein, *King's College, University of London*

8 **BIOCHEMISTRY OF HORMONES**
Professor H. V. Rickenberg, *National Jewish Hospital & Research Center, Colorado*

9 **BIOCHEMISTRY OF CELL DIFFER-ENTIATION**
Dr. J. Paul, *The Beatson Institute for Cancer Research, Glasgow*

10 **DEFENCE AND RECOGNITION**
Professor R. R. Porter, F.R.S., *University of Oxford*

11 **PLANT BIOCHEMISTRY**
Professor D. H. Northcote, F.R.S., *University of Cambridge*

12 **PHYSIOLOGICAL AND PHARMACO-LOGICAL BIOCHEMISTRY**
Dr. H. F. K. Blaschko, F.R.S., *University of Oxford*

INORGANIC CHEMISTRY SERIES ONE

Consultant Editor
H. J. Emeléus, F.R.S.
Department of Chemistry
University of Cambridge

Volume titles and Editors

1 **MAIN GROUP ELEMENTS—HYDROGEN AND GROUPS I-IV**
Professor M. F. Lappert, *University of Sussex*

2 **MAIN GROUP ELEMENTS—GROUPS V AND VI**
Professor C. C. Addison, F.R.S. and Dr. D. B. Sowerby, *University of Nottingham*

3 **MAIN GROUP ELEMENTS—GROUP VII AND NOBLE GASES**
Professor Viktor Gutmann, *Technical University of Vienna*

4 **ORGANOMETALLIC DERIVATIVES OF THE MAIN GROUP ELEMENTS**
Dr. B. J. Aylett, *Westfield College, University of London*

5 **TRANSITION METALS—PART 1**
Professor D. W. A. Sharp, *University of Glasgow*

6 **TRANSITION METALS—PART 2**
Dr. M. J. Mays, *University of Cambridge*

7 **LANTHANIDES AND ACTINIDES**
Professor K. W. Bagnall, *University of Manchester*

8 **RADIOCHEMISTRY**
Dr. A. G. Maddock, *University of Cambridge*

9 **REACTION MECHANISMS IN INORGANIC CHEMISTRY**
Professor M. L. Tobe, *University College, University of London*

10 **SOLID STATE CHEMISTRY**
Dr. L. E. J. Roberts, *Atomic Energy Research Establishment, Harwell*

INDEX VOLUME

PHYSICAL CHEMISTRY SERIES ONE

Consultant Editor
A. D. Buckingham
Department of Chemistry
University of Cambridge

Volume titles and Editors

1 **THEORETICAL CHEMISTRY**
Professor W. Byers Brown, *University of Manchester*

2 **MOLECULAR STRUCTURE AND PROPERTIES**
Professor G. Allen, *University of Manchester*

3 **SPECTROSCOPY**
Dr. D. A. Ramsay, F.R.S.C., *National Research Council of Canada*

4 **MAGNETIC RESONANCE**
Professor C. A. McDowell, F.R.S.C., *University of British Columbia*

5 **MASS SPECTROMETRY**
Professor A. Maccoll, *University College, University of London*

6 **ELECTROCHEMISTRY**
Professor J. O'M Bockris, *University of Pennsylvania*

7 **SURFACE CHEMISTRY AND COLLOIDS**
Professor M. Kerker, *Clarkson College of Technology, New York*

8 **MACROMOLECULAR SCIENCE**
Professor C. E. H. Bawn, F.R.S., *University of Liverpool*

9 **CHEMICAL KINETICS**
Professor J. C. Polanyi, F.R.S., *University of Toronto*

10 **THERMOCHEMISTRY AND THERMO-DYNAMICS**
Dr. H. A. Skinner, *University of Manchester*

11 **CHEMICAL CRYSTALLOGRAPHY**
Professor J. Monteath Robertson, F.R.S., *University of Glasgow*

12 **ANALYTICAL CHEMISTRY —PART 1**
Professor T. S. West, *Imperial College, University of London*

13 **ANALYTICAL CHEMISTRY —PART 2**
Professor T. S. West, *Imperial College, University of London*

INDEX VOLUME

ORGANIC CHEMISTRY SERIES ONE

Consultant Editor
D. H. Hey, F.R.S.,
Department of Chemistry
King's College, University of London

Volume titles and Editors

1 **STRUCTURE DETERMINATION IN ORGANIC CHEMISTRY**
Professor W. D. Ollis, F.R.S., *University of Sheffield*

2 **ALIPHATIC COMPOUNDS**
Professor N. B. Chapman, *Hull University*

3 **AROMATIC COMPOUNDS**
Professor H. Zollinger, *Swiss Federal Institute of Technology*

4 **HETEROCYCLIC COMPOUNDS**
Dr. K. Schofield, *University of Exeter*

5 **ALICYCLIC COMPOUNDS**
Professor W. Parker, *University of Stirling*

6 **AMINO ACIDS, PEPTIDES AND RELATED COMPOUNDS**
Professor D. H. Hey, F.R.S., and Dr. D. I. John, *King's College, University of London*

7 **CARBOHYDRATES**
Professor G. O. Aspinall, *Trent University, Ontario*

8 **STEROIDS**
Dr. W. F. Johns, *G. D. Searle & Co., Chicago*

9 **ALKALOIDS**
Professor K. Wiesner, F.R.S., *University of New Brunswick*

10 **FREE RADICAL REACTIONS**
Professor W. A. Waters, F.R.S., *University of Oxford*

INDEX VOLUME

MTP International Review of Science

Physiology
Series One

Volume 8
Reproductive Physiology

Edited by **R. O. Greep**
Harvard Medical School, Boston

Butterworths · London
University Park Press · Baltimore

THE BUTTERWORTH GROUP

ENGLAND
Butterworth & Co (Publishers) Ltd
London: 88 Kingsway, WC2B 6AB

AUSTRALIA
Butterworths Pty Ltd
Sydney: 586 Pacific Highway 2067
Melbourne: 343 Little Collins Street, 3000
Brisbane: 240 Queen Street, 4000

NEW ZEALAND
Butterworths of New Zealand Ltd
Wellington: 26–28 Waring Taylor Street, 1

SOUTH AFRICA
Butterworth & Co (South Africa) (Pty) Ltd
Durban: 152–154 Gale Street

ISBN 0 408 70488 8

UNIVERSITY PARK PRESS

U.S.A. and CANADA
University Park Press
Chamber of Commerce Building
Baltimore, Maryland, 21202

Library of Congress Cataloging in Publication Data

Greep, Roy Orval, 1905–
Reproductive physiology.
(Physiology, series one, v. 8) (MTP international review of science)
1. Reproductive. I. Title. II. Series.
III. Series: MTP international review of science.
[DNLM: 1. Reproduction. W1PH951D v. 8 1974/
WQ205 R429 1974]
QP1.P62 vol. 8 [QP251] 599′.01′08s [599′.01′66]
ISBN 0–8391–1057–X 73–21883

First Published 1975

BUTTERWORTH & CO (PUBLISHERS) LTD

Typeset and printed in Great Britain by
REDWOOD BURN LIMITED
Trowbridge & Esher
and bound by R. J. Acford Ltd, Chichester, Sussex

Consultant Editors' Note

The International Review of Physiology, a review with a new format, is hopefully also new in concept. But before discussing the new concept, those of us who are joined in making this review a success must admit that we asked ourselves at the outset: Why should we promote a new review of physiology? Not that there is a paucity of reviews already, and not that the present reviews fail to fill important roles, because they do. Therefore, what could be the role of an additional review?

The International Review of Physiology has the same goals as all other reviews for accuracy, timeliness, and completeness, but it has new policies that we hope and believe will engender still other important qualities that are often elusive in reviews, the qualities of critical evaluation and instructiveness. The first decision toward achieving these goals was to design the new format, one that will allow publication of approximately 2500 pages per edition, divided into eight different sub-speciality volumes, each organised by experts in their respective fields. It is clear that this extensiveness of coverage will allow consideration of each subject in far greater depth than has been possible in the past. To make this review as timely as possible, a new edition of all eight volumes will be published every two years giving a cycle time that will keep the articles current. And in addition to the short cycle time, the publishers have arranged to produce each volume within only a few months after the articles themselves have been completed, thus further enhancing the immediate value of each author's contribution.

Yet, perhaps the greatest hope that this new review will achieve its goals of critical evaluation and instructiveness lies in its editorial policies. A simple but firm request has been made to each author that he utilise his expertise and his judgement to sift from the mass of biennial publications those new facts and concepts that are important to the progress of physiology; that he make a conscientious effort not to write a review consisting of annotated lists of references; and that the important material he does choose be presented in thoughtful and logical exposition, complete enough to convey full understanding and also woven into context with previously established physiological principles. Hopefully, these processes will bring to the reader each two years a treatise that he will use not merely as a reference in his own personal field but also as an exercise in refreshing and modernising his whole body of physiological knowledge.

Mississippi
Nairobi

A. C. Guyton
D. F. Horrobin

Preface

Advances in reproductive physiology are unpredictable and occur sporadically along an uneven front. A breakthrough in one area may lead to a great forward thrust while another area of equal importance suffers profound neglect. Such is the nature of research. In this volume we have tried not only to highlight those areas that are in a state of active ferment but also to call attention to problems that are in urgent need of further study. We have not sought complete coverage nor comprehensive treatment of selected areas. Instead, it was our aim to outline recent advances and point up newly emerging concepts and ideas.

The study of reproduction has been enlivened by many recent developments. The gonadotrophins, FSH, LH and HCG have all been isolated in pure form and their chemical structure determined. They are all glycoproteins composed of two non-identical subunits approximately equal in molecular weight and non-covalently bonded. Similarly, several of the pituitary regulating factors from the hypothalamus have been isolated, structurally characterised and synthesised. The introduction of extremely sensitive radioimmunoassay procedures have made possible for the first time the measurement in body fluids of all the hormones, whether peptide, protein or steroid, that are concerned with reproduction. Advances have been made also in our knowledge of the binding of hormones to protein receptors, and of the role of cyclic AMP, prostaglandins and various endocellular enzymes in the mechanism of hormone action. Knowledge of feed-back regulatory mechanisms has been clarified by use of specific hormone neutralising antibodies. Breakthroughs have also been made in our understanding of the hormonal control of the oviduct and its role in the transport, nourishment and early development of the fertilised ovum. The triggering by the foetus in initiating parturition is an interesting new twist. And last, but not least, important information relative to the genesis, maturation and capacitation of sperm has been gained.

The authors have been drawn from many different parts of the world in order to ensure balanced attention to research being done by an international cadre of scientists studying reproductive phenomena. Their ready willingness to participate is deeply appreciated.

In order that this review might attain its maximal potential impact in a fast moving field, reduced publication time has been a paramount consideration. The authors have cooperated exceedingly well in preparing their reviews within a reasonable time limit and holding to the deadline for submission of manuscripts. It is hoped that this, along with plans for rapid publication will provide an up-to-date, authoritative and critical survey of the field for research scientists and advanced students.

Massachusetts R. O. Greep

Contents

1
Hypothalamic Mediation of Neuroendocrine Regulation of Hypophysial Gonadotrophic Functions

B. FLERKÓ
University Medical School, Pécs, Hungary

1.1 INTRODUCTION

In contrast to the male, reproductive processes in the female mammals are characterised by cyclic alterations in the genital tract and in sexual receptivity. The recurrent reproductive periods of mature female mammals depend upon the rhythmical secretion of hormones by the anterior pituitary and ovary. The rhythm is, however, not intrinsic to these glands but imposed by brain mechanisms summarised in the following two sections.

In many species (e.g. cat, dog, ferret, sheep) these brain mechanisms are greatly influenced by afferentations from the external environment. These animals have sexual cycles that are dependent on external factors such as light, temperature, humidity, food, etc. In other mammals (e.g. human, monkey, guinea-pig, rat, hamster, mouse) neural mechanisms controlling the secretion of gonadotrophic hormones* operate fairly independently from external environmental influences. These species have sexual (menstrual or oestrous) cycles independent from the seasons. However, there are afferentations from the external environment that can influence the gonadotrophin-controlling brain mechanisms even in these species. Light, smell and tactile stimuli especially, play an important role in this respect.

On the other hand, many examples demonstrate the importance of afferent impulses from the 'milieu interieur' in the control of gonadotrophin secretion. Apart from neural signals originating from different parts of the reproductive tract, the hormonal feed-back appears to be of the greatest importance in the control of the cyclic release of GTH. Experimental data accumulated in the last two decades support the concept that the hormonal feed-back

* The term gonadotrophic hormones (GTH) used in this chapter covers two pituitary gonadotrophins, i.e. follicle-stimulating hormone (FSH) and luteinising hormone (LH). For discussion of their actions, see Ref. 54.

actions are also mediated mainly by gonadotrophin-controlling brain mechanisms. This is now termed indirect or neurohormonal feed-back, and will be discussed in Section 1.4.

The mechanism enabling the female anterior pituitary to release GTH in cyclic phases is apparently absent in the male, which does not display cyclic fluctuations in gonadotrophin secretion. A concept concerning the development of this status will be described in Section 1.5.

1.2 TONIC MECHANISM OF GONADOTROPHIN SECRETION

Current concepts invoke a dual brain mechanism controlling cyclic gonadotrophin secretion[4, 38, 56]. The first of these, termed a tonic mechanism by Barraclough and Gorski[4], stimulates the continuous ('tonic') discharge of GTH in sufficient quantity to maintain follicular growth and oestrogen secretion, but cannot initiate the ovulatory surge of gonadotrophins. Barraclough and Gorski[4] localised this mechanism in the ventromedial–arcuate nucleus.

1.2.1 Hypothalamic gonadotrophin-releasing factors

The hypothesis that hypothalamic control impulses are humorally mediated to the anterior pituitary was launched 3½ decades ago. The concept was clearly formulated later by Green and Harris[52] who proposed that the secretion of anterior pituitary hormones is controlled by specific substances of hypothalamic origin released into the capillary loops of the median eminence. The discovery of the various hypothalamic releasing and inhibiting factors furnished the best evidence for the Green–Harris theory.

The first evidence for the presence of LH-releasing activity in hypothalamic extracts was presented by McCann *et al.*[88]. With the aid of preparative electrophoresis, Schally and associates[119] separated the LRF activity of porcine hypothalamic extracts from the FSH-depleting activity demonstrated first by Igarashi and McCann[64]. In spite of these results, the point now open to question is the relationship between LRF and follicle-stimulating hormone-releasing factor (FRF), since the releasing decapeptide reported recently by Schally and co-workers[85] contains both activities. If these two activities do indeed reside within the same secreted molecule, then a tight coupling would exist at this point in the pituitary–gonadal axis. However, not all of the data on neural control suggest the simultaneous release of the two releasing activities, as has been pointed out by Schwartz and McCormack[120]. Also the fact that progesterone elevates FSH and not LH in rats with suprachiasmatic lesions is hard to reconcile with the concept of a single gonadotrophin-releasing factor[10]. Furthermore, experimental results, suggesting the existence of separate FRF- and LRF-producing neurones (Section 1.2.2) and of separate neural FSH- and LH-control mechanism (Section 1.3.1.1) do not appear to support the hypothesis of the exclusive existence of a single gonadotrophin-releasing factor.

1.2.2 Hypophysiotrophic area (HTA)

The hypophysiotrophic area expression is now generally used for the anatomical localisation of the tonic mechanism. The term hypophysiotrophic area (HTA) was coined by Halász *et al.*[58] for the half-moon shape region of the medial–basal hypothalamus (Figure 1.1) which maintains the normal structure and function of anterior pituitary tissue grafted into it. This suggests that the HTA contains nerve cells which synthesise the various hypothalamic-releasing and -inhibiting factors (RF and IF). These factors are transported

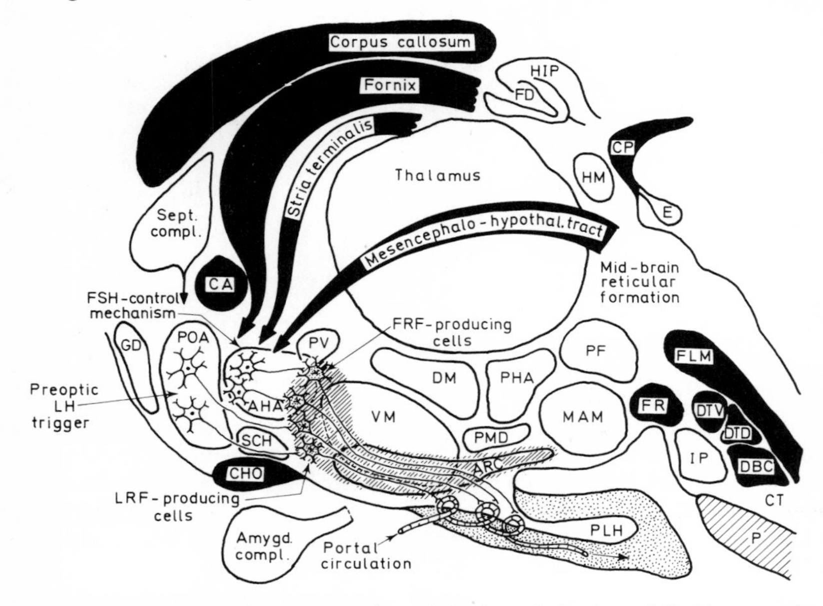

Figure 1.1 Schematic representation of the hypothalamic and limbic control mechanisms for FSH and LH secretion. Arrows show the direction of the blood flow in the hypothalamo–hypophysial portal system, the special capillary loops of which penetrate the median eminence. The FRF- and LRF-producing neurones of the 'tonic mechanism' are indicated by the neural units with dotted cell bodies. They are situated in the hypophysiotrophic area (hatched) and their nerve endings terminate on the capillary loops of the portal system. Perikarya of the neurones belonging to the hypothalamic 'cycle mechanism' are represented by clear cell bodies located in the preoptic and anterior hypothalamic area, respectively. Abbreviations used: AHA, anterior hypothalamic area; Amygd. compl., amygdaloid complex; ARC, arcuate nucleus; CA, anterior commissure; CHO, optic chiasma; CP, posterior commissure; CT, nucleus centralis tegmenti; DBC, decussatio brachiorum conjunctivorum; DM, dorsomedial nucleus; DTD, decussatio tegmenti dorsalis; DTV, decussatio tegmenti ventralis; E, pineal gland; FD, fascia dentata; FLM, fasciculus longitudinalis medialis; FR, fasciculus retroflexus; GD, gyrus diagonalis; HIP, hippocampus; HM, medial habenular nucleus; IP, interpeduncular nucleus; MAM, mamillary nucleus; P, pons; PF, nucleus parafascicularis thalami; PHA, posterior hypothalamic area; PLH, posterior lobe of the hypophysis; PMD, dorsal premamillary nucleus; POA, preoptic area; PV, paraventricular nucleus; SCH, suprachiasmatic nucleus; Sept. compl., septal complex; VM, ventromedial nucleus. (From Flerkó[41], by courtesy of Academic Press, New York.)

through the axons of the RF- and IF-producing nerve cells, and released at their axon terminals into the hypophysical portal circulation (Figure 1.1). In the rat with intrahypothalamic pituitary graft, the releasing and inhibiting factors might reach and stimulate the cells of the pituitary graft by diffusion.

This concept is in accord with morphological findings. The HTA corresponds cytoarchitectonically to the arcuate nucleus and to the suprachiasmatic region of the hypothalamus, i.e. to the distribution of nerve cells from which the fine-fibred tubero–infundibular tract or 'parvicellular neurosecretory system' to the median eminence arises. This discovery of Szentágothai[127] has recently been substantiated by electron-microscopical findings of Réthelyi and Halász[113] showing that, if the HTA was disconnected from the rest of the brain, degenerated nerve endings do not occur in the surface zone (*zona palisadica*) of the median eminence. These findings indicate that the axon terminals in the *zona palisadica* arise exclusively from neurones in the HTA, and that fibres of other origin do not terminate in this zone.

The finding of Halász *et al.*[58] that anterior pituitary grafts in the HTA not only release amounts of GTH sufficient to maintain the oestrous cycle, but also contain well-granulated, PAS-positive gonadotrophic basophiles, supports the view that FRF and LRF stimulate both synthesis and release of FSH and LH. The same might be suggested from the observation of Kobayashi *et al.*[76] that crude hypothalamic extracts promote granule accumulation in certain cells of dispersed pituitary cell cultures, as well as gonadotrophin release into the media. The experimental finding of Nikitovich-Winer *et al.*[101] points to the same direction. However, these experimental findings do not preclude the possibility of the existence of separate hypothalamic 'synthesising' factors[20].

The concept of Halász, claiming that the HTA is a functional unit of the hypothalamus, is consistent with the data showing that various releasing and inhibiting factors are localised almost exclusively in the medial–basal hypothalamus[94].

In order to localise the nerve cells synthesising the various releasing factors, Mess and Martini[94] lesioned different hypothalamic regions in rats and measured the content of the various releasing factors in the medial–basal hypothalamus. Their findings seem to suggest that neurones producing FRF are situated in the antero-dorsal, and those synthesising LRF mainly in the antero-ventral part of the HTA (Figure 1.1); whereas the whole HTA is involved in thyrotrophin-releasing factor production. No appreciable changes in corticotrophin-releasing factor (CRF) content were observed with any of the lesions. Several pieces of evidence support the view that CRF production is restricted to the median eminence region[56].

If it is true that FRF is produced in the antero-dorsal and LRF in the antero-ventral part of the HTA, an absence of FRF, but not of LRF, would be expected in the hypothalamic island made with the aid of the Halász knife. In consequence of this operation, the part of the HTA which contains FRF-synthesising nerve cells, i.e. the region just below the paraventricular nuclei (Figure 1.1), remains outside the island isolated completely from the rest of the brain. The results of the experiment with the above technique of Tima[129] showed that FRF stores were significantly reduced in the hypothalamic island 8 days after the operation, and even more reduced 15 days after isolation of the island which, however, contained more LRF 8 or 15 days after isolation than the HTA of the control, unoperated rats. This latter finding might be explained by the absence of neural afferentation from the rest of the brain which can stimulate LRF release in the intact animal.

The ependyma of the third ventricle at the level of the arcuate nucleus differs from ependyma lining found in other parts of the ventricle system. These special ependymal cells, called ependymal tanycytes, have long processes which often end on capillaries within the hypothalamus or on the capillary loops penetrating the median eminence. On the basis of this, and of their peculiar cytology and histochemistry, it has been suggested that the tanycytes might be secretory cells involved in the hypothalamic control of the anterior pituitary. However, as pointed out by Brawer[13], the evidence implicating the arcuate tanycytes in the hypothalamo–hypophysial neuroendocrine control is inadequate, and the structural peculiarities of these cells cannot yet be interpreted functionally due to the lack of direct experimental evidence.

1.2.3 Functional capacity of the hypophysiotrophic area

Neural isolation of the HTA—with the aid of the Halász knife—from the rest of the brain has proved to be a useful approach for studying the functional capacity of the HTA in the control of the anterior pituitary. Interruption of all neural connections of the HTA is not followed by atrophy of the target organs, and the operation induces only slight changes in the function of the testis, thyroid and adrenal cortex. Spermiogenesis is maintained, basal TSH secretion appears to be slightly impaired and propylthyouracil treatment results in moderate thyroid hypertrophy. Corticosterone levels in the peripheral plasma were found to be normal or slightly elevated, but never decreased, a few weeks after neural isolation of the HTA. Adrenal compensatory hypertrophy develops following unilateral adrenalectomy, and there is pituitary ACTH response to ether stress. LH content of the anterior lobe increases, castration cells are formed in the pituitary after gonadectomy and only a slight retardation in body growth occurs. However, the animals do not ovulate, ovarian compensatory hypertrophy fails to develop after hemispaying and the diurnal ACTH rhythm as well as the cyclic release of gonadotrophic hormones are absent[56, 127].

These observations indicate that the HTA by itself is able to maintain, even to regulate to some extent, the basal (tonic) secretion of the trophic hormones. In order to secure the rhythmic output of ACTH and GTH, however, the HTA needs afferent impulses which can modulate, i.e. enhance or inhibit, the activity of the RF- and IF-producing neurones. Neural structures mediating such afferentations to the FRF- and LRF-producing neurones are concentrated in the preoptic–anterior hypothalamic area, HTA and limbic system. The assembly of these hypothalamic and limbic neurones may be termed the 'cycle mechanism' of brain control for gonadotrophin secretion, since it is indispensable in the maintenance of the cyclic release of GTH.

1.3 CYCLE MECHANISM OF GONADOTROPHIN SECRETION

1.3.1 Hypothalamic parts of the cycle mechanism

1.3.1.1 Preoptic–anterior hypothalamic area (PAHA)

Electrolytic lesions in the anterior hypothalamus, interrupting neural connections between preoptic and hypophysiotrophic areas, produced in polycyclic animals a state characterised by inability to mobilise GTH sufficient for ovulation and formation of corpora lutea. Since it was a general belief previously that ovulation and formation of corpora lutea was induced by a sudden increase of LH output, many authors assumed that the lesions destroyed an LH-release apparatus in the preoptic–anterior hypothalamic area (PAHA)[38].

A 24 h rhythm in the LH-release apparatus has been demonstrated by Everett and Sawyer[34]. Their finding, that progesterone-induced release of GTH occurred at a definite time of the day, indicated that the principal factor inducing gonadotrophin release was not the steroid itself but some neural process which was elicited or facilitated by progesterone[33]. The ovulatory activation of the hypophysis could be prevented by administering barbiturates[34], which suggested that the neural mechanism was located somewhere in the hypothalamus. Indeed, electrical stimulation of the preoptic region in persistent oestrous rats induced ovulation[14]. Later, it was shown by Everett[32] that iron deposited from the electrode tip and carried radially electrophoretically from the anode is the effective agent in stimulation involving electrolysis with electrodes of ferrous alloys. The anodic stimulus through a steel electrode is thus irritative, amounting to a 'slow burn', its action continuing long after brief passage of the electric current. Such irritative foci, when centred unilaterally almost anywhere in the medial preoptic area or in the anterior hypothalamic area, seemed equally effective in inducing ovulation if the minimum diameters of the foci were approximately 0.5 mm. A quantitative relationship between the size of an electrochemical lesion of the medial preoptic area and the amount of gonadotrophin released is indicated by several pieces of indirect evidence. It seems that the larger the electrochemical lesion, i.e. the greater the amount of the preoptic-tuberal system involved, the more gonadotrophin is released per unit time. Direct evidence for this assumption has been furnished by Everett *et al.*[35] and Turgeon and Barraclough[131].

Until 1970, it was uncertain whether the PAHA itself is a centre of the release apparatus responsible for the output of the ovulatory quota of GTH, or whether the pathway simply passes through the region as from the septum and other components of the limbic system[32]. Köves and Halász[82] investigated this question in recent studies. The anterior, lateral and superior connections of the preoptic area were interrupted bilaterally in adult female rats so that the PAHA itself remained in contact with the HTA, and occurrence of ovulation was tested. Tubal ova were found in the majority of the animals, and all ovaries contained fresh corpora lutea. However, the number of ova seen in the oviducts was less than in the controls, and the animals with preoptic isolation exhibited somewhat irregular vaginal cycles. Bilateral transection in the caudal suprachiasmatic plane, separating completely the preoptic–suprachiasmatic region from the HTA, blocked ovulation in 100% of the rats, and led to the development of the persistent vaginal oestrus syndrome. Unilateral arched cut in the retrochiasmatic area did not interfere with ovulation and with cyclic gonadotrophin release. This has been confirmed in its essentials by Kaasjager *et al.*[67].

From the above, it can be concluded that it is the PAHA in the normal female rat which responds under proper environmental and hormonal circumstances by an activation of the more terminal tonic mechanism to elicit ovulatory discharge of GTH. In the absence of this mechanism, the tonic mechanism still functions normally, and LH as well as FSH are secreted at a basal rate which maintains follicular growth and oestrogen secretion without ovulation and corpus luteum formation. One half of the preoptic–tuberal pathway, conveying the impulses from the preoptic area to the tonic mechanism, appears to be sufficient for inducing ovulation. The findings, however, that ovulation was sub-normal in rats with preoptic isolation, and that these animals had somewhat irregular vaginal cycles[82], indicate that neural structures outside of the PAHA are also involved in the control of the completely normal and well-timed cyclic ovulatory output of GTH.

(a) *Preoptic–LH trigger*—The experiments of Terasawa and Sawyer[128] suggested that increased neuronal activity in the HTA occurring after preoptic stimulation was correlated with activation of the cyclic release of GTH with an emphasis on LH[74]. Kalra *et al.*[68] found no FSH release after the stimulation of the medial preoptic area, whereas Clemens *et al.*[17] did in some cases. Both groups of authors emphasise that whenever an FSH increase occurred it was always accompanied by increased LH, but LH was not consistently accompanied by FSH. Cramer and Barraclough[21] have found that electrical activation of the medial preoptic area can induce a release of LH that does not differ markedly from the patterns of LH release observed in the normal cycling animal. These findings suggest that the centre for the neural mechanism that triggers the burst of ovulatory LH release is located in the preoptic area, and may be termed the preoptic LH-trigger or LH-control mechanism (Figure 1.1).

It has been shown by Kalra *et al.*[68] that elevated levels of LH were first detected 30 min after stimulation in the HTA but only after stimulating for 60 min at more rostral sites. An identical delay in release of LH was noted in reports of other investigators[17, 21]. Similarly, Everett[32] found that 45–60 min was needed following electrochemical stimulation to release sufficient GTH to bring about full ovulation. Since elevation in multi-unit activity in HTA was observed immediately after similar stimulation of the preoptic area[128], it appears likely that the neurones in the preoptic–tuberal system are active for a period of time prior to release of sufficient LRF to evoke an elevation of plasma LH.

(b) *Anterior hypothalamic FSH-control mechanism*—In 1956, Flerkó[36, 37] suggested the presence of a neural mechanism in the anterior hypothalamus through which a slight (physiological) elevation of the oestrogen level in the blood inhibits FSH secretion. On the other hand, absence of compensatory ovarian hypertrophy following anterior hypothalamic lesions in rats[38] suggested that the anterior hypothalamic neural mechanism might also play a role in that process by which reduction of the oestrogen level in the blood enhances FSH release.

The onset of the breeding season in ferrets was accelerated by anterior hypothalamic lesion[25]. It has been assumed that the lesions induce oestrus by the destruction of a neural mechanism which, during sexual quiescence, restrains FSH secretion. If this hypothesis is correct, the onset of puberty might be explained on the basis of removal of an FSH-inhibitory mechanism situated in the anterior hypothalamus. Indeed, lesions in the anterior hypothalamic area of infantile rats produced precocious puberty in the experiments of many investigators[38].

On the basis of the above findings, an FSH-control mechanism has been assumed in the anterior hypothalamic area (Figure 1.1). This hypothesis has recently been supported by Kalra *et al.*[68] who have shown that FSH release was stimulated only when the stimulating electrodes were located in the anterior hypothalamic area. Stimulation in the preoptic area and septal complex, which was very effective in evoking LH release, failed to alter FSH levels. Studies with the deafferentation technique of Halász and Gorski[57] as well as Köves and Halász[81] reported that no compensatory ovarian hypertrophy occurred in rats bearing a frontal cut that separated the anterior hypothalamic area from the HTA. Also Kalra *et al.*[70] observed that separation of the anterior hypothalamic area from the HTA inhibited post-castration rise of FSH in immature parabiotic rats.

1.3.1.2 Other hypothalamic areas

(a) *HTA–tubero–infundibular dopaminergic system*—It has long been suspected that the catecholamines may have a role in the control of release of the gonadotrophins. Sawyer *et al.*[118] found that the intravenous injection of certain adrenergic blocking agents into the rabbit shortly after coitus can prevent the release of an ovulatory dose of gonadotrophins. More recent work has shown that drugs such as reserpine, which deplete monoamines, can also block ovulation[6] and lead to the development of pseudo-pregnancy in animals. If the depletion of monoamines is prevented by the administration of monoamine oxidase inhibitors such as pargyline or of precursors of catecholamines, the effects of reserpine are reversed which indicates that they are due to the depletion of amines[86].

It has been shown a few years ago that, besides the RF- and IF-producing neurones, the tubero–infundibular (TI) tract contains a number of dopamine (DA) containing neurones, the cell bodies of which are localised mainly in the arcuate nucleus and send their axons to the *zona palisadica* of the median eminence. In view of the extremely high density of DA nerve endings in the neighbourhood of the capillary loops penetrating the median eminence, it might be imagined that DA is released into the portal vessels and exerts its effect directly on the anterior pituitary cells. However, many experimental findings (see, Hökfelt and Fuxe[62]) rather favour an action of DA at the median eminence level. The TIDA nerve endings are often seen in close contact with non-monoamine terminals characterised by their content of agranular vesicles (probably terminals of RF- and IF-producing neurones) offering a morphological basis for axo–axonic transmitter influences[62].

The TIDA neurones display marked changes in rate of disappearance of DA from their nerve terminals in various endocrine states[62]. These changes which occur physiologically (e.g, during pregnancy, lactation and the ovarian cycle) as well as in states induced by various experimental procedures (hypophysectomy, castration, etc.) or by drugs or hormones (especially hormonal steroids) suggest that the TIDA system can influence the discharge of the gonadotrophin-inhibiting and -releasing factors in the hypophysial portal circulation.

A controversy exists among authors concerning the direction of this influence. Fuxe and Hökfelt[49] raised the hypothesis that the TIDA neurones inhibit the release of FRF and LRF, probably by an axo–axonic influence at the median eminence level. On the basis of their recent findings, Hökfelt and Fuxe[62] assumed that the TIDA system also inhibits prolactin secretion, in this case by a stimulating effect on prolactin-inhibiting factor (PIF) secretion. On the other hand, Kordon and Glowinski[79], Porter and associates[109] and McCann and co-workers[87] concluded from their studies that TIDA neurones stimulate secretion not only of PIF but also of FRF and LRF. However, some doubt has been raised by Ahren and co-workers[1] regarding the acceptance of DA as the neurotransmitter for LRF release. DA turnover in the median eminence was lower during the early afternoon of pro-oestrus than during metoestrus or dioestrus. Moreover, epinephrine induced ovulation, but DA did not, when injected into the third ventricle of pentobarbital-blocked pro-oestrous rats[115]. Since a marked activation of the TIDA system was observed after treatment with oestrogen or combined oestrogen–progesterone or testosterone in castrated rats[62], which will result in a reduction of FSH and LH secretion, it seems reasonable to assume that the TIDA system is involved—probably as a hormone-receptor system of the HTA—in the negative feed-back action of these hormones (see Section 1.4.2.2).

In addition to the short TIDA system, localised entirely in the HTA, it became evident quite early that there are a number of longer noradrenaline- and serotonin-containing neurones in the hypothalamus which terminate in various hypothalamic nuclei[62]. Also Björklund and co-workers[11] have shown a large group of noradrenaline axons in the hypothalamus which originate in areas outside of the HTA. The noradrenergic neurones are clearly sensitive to ovarian steroids and FSH, and seem to show changes in activity in relation to the ovarian cycle[62]. It appears, therefore, highly probable that these neurones are involved in the limbic and preoptic-tuberal pathways mediating neuroendocrine, especially ovulatory, influences on the FRF- and LRF-producing neurones of the HTA. Changes in multiple unit activity in the HTA have been shown to be correlated with the stimulation of an ovulatory surge of LH[128], and the injection of adrenaline or noradrenaline caused a highly repeatable biphasic change in neural activity of the median eminence[136]. Furthermore, the destruction of noradrenergic neurones following deafferentation of the HTA can be correlated with depression in the tonic secretion of LH[136], suggesting the involvement of noradrenaline in the transmission of stimulatory effects on the release of FSH and LH. Also the most recent experimental results of Kalra and McCann[69] suggest that transmission across a noradrenaline synapse may be necessary for preoptic stimulation to release LRF. Serotonin-containing neurones have been implicated in the inhibition of LH secretion[80], and Ladosky and Gaziri[82a] have raised the involvement of serotonin in the metabolism of the hypothalamus during the period of sexual differentiation of the region.

(b) *Lateral hypothalamus* — Observations from various laboratories have suggested that certain cells in the lateral hypothalamic area are sensitive to sex steroids and contribute to the cycle mechanism. Cross[22] assumed that the cells of the arcuate nuclei, a part of the HTA, would be maintained in a state of normal FSH- and LH-releasing activity by tonic impulses from the lateral hypothalamic area. This assumption is not supported by the findings of Halász and Gorski[57] whose rats bearing HTA isolated from the lateral hypothalamic area ovulated normally, whereas the separation of the PAHA from the HTA abolished ovulation and compensatory ovarian hypertrophy. Furthermore, while a significant increase was recorded by Cross and Dyer[23] in firing frequency of anterior hypothalamic units in pro-oestrus over that in oestrus, metoestrus or dioestrus in diencephalic island preparations, no significant change was seen in units from the lateral hypothalamic area in any day of the 4-day cycle.

(c) *Posterior hypothalamus* — Kordon[78] has also localised FSH-inhibiting structures in the anterior hypothalamus. Simultaneously, however, on the basis of his findings that in rats with persistent oestrus from anterior hypothalamic lesions formation of corpora lutea could be induced by bilateral premamillary lesions, he postulated a neural mechanism in the premamillary region which would continuously stimulate FRF-producing neurones.

This postulate has not been supported by the results of Illei-Donhoffer and co-workers[65] because (i) not only bilateral but also unilateral premamillary lesions have been followed by formation of corpora lutea from ruptured follicles. The specificity of premamillary lesions is further diminished by the fact that formation of true corpora lutea was also induced by extensive hypothalamic lesions which destroyed all hypothalamic structures indispensable to the normal ovulatory release of gonadotrophic hormones. This luteinisation can be explained by the phenomenon observed by Bishop *et al.*[9], i.e. that lesions in the median eminence resulted in an outpouring of GTH, especially in that of LH. There were more than adequate amounts of LH to produce ovulation on the basis of plasma levels observed following electrochemical stimulations that induced full ovulation[68]. (ii) Formation of corpora lutea from ovulated follicles was found only one or two days after the second hypothalamic lesion. All corpora lutea present in the ovaries beyond this period of time were either persistent or regressive or, when newly forming, not true but atretic. This, and the fact that normal, or at least approximately regular, vaginal cycling did not occur in any of the rats after the second hypothalamic lesion[65], speaks strongly against the assumption that premamillary lesions would lead to repeated ovulation, i.e. to restitution of more or less normal cycles.

1.3.2 Extrahypothalamic parts of the cycle mechanism

In current concepts[38,56], the higher level of neural control for the anterior pituitary includes all the brain structures which can modulate the activity of the tubero–infundibular neurones producing the releasing and inhibiting factors. With respect to the FRF- and LRF-producing neurones, brain structures of this type are located not only in the HTA and PAHA but also in the limbic system—including the mid-brain reticular formation, often called the 'caudal limbic system'—and in the epithalamo–epiphysial complex. Some parts of the brain belonging to the limbic system and epithalamo–epiphysial complex constitute the so-called extrahypothalamic cycle mechanism because also these structures are indispensable in the maintenance of the regular cyclic release of GTH and, hence, in that of the well-timed ovulation.

1.3.2.1 Limbic system

(a) *Amygdaloid nuclear complex* — Koikegami and associates[77] were the first to report that stimulation of medial amygdalar nuclei induced ovulation, a finding confirmed by many investigators. However, the amygdalar afferents do not seem to be indispensable to the simple ovulatory release of GTH. This is indicated by the finding of Köves and Halász[82] who showed that preoptic deafferentation did not block spontaneous ovulation in the rat. Furthermore, rats with long-term transection of either the stria terminalis, the cortico-hypothalamic tract, or both, had regular oestrous cycles and ovulated normally[133].

The hypothesis that the influence of the amygdala would be facilitatory on gonadotrophin release is diametrically opposed to the views of Elwers and Critchlow[30, 31] and Eleftheriou and Zolovick[29] who believe that the amygdala inhibits GTH secretion. Elwers and Critchlow[30] reported that effective amygdalar lesion associated with increased uterine weight shared a common area of destruction involving parts of the medial nuclei and an area between them containing the convergent fibres of the stria terminalis. Furthermore, they showed that the above lesion advanced the date of vaginal opening and onset of oestrous cycles as effectively as the anterior hypothalamic lesions, and they were able to enhance the amygdalar inhibitory influence and delayed puberty by electrical stimulation of the cortico–medial amygdala[31]. The significant inhibition of the compensatory ovarian hypertrophy in rats with electrolytic lesions in the cortical nucleus of the amygdala[122] suggests also an inhibitory amygdalar action over hypothalamo–hypophysial gonadotrophic function.

(b) *Hippocampal formation* — According to Kawakami *et al.*[73], stimulation of the hippocampus facilitates the release of GTH, yielding an increased secretion of progesterone. In contrast to this, Velasco and Taleisnik[132] have concluded that stimuli arising from the hippocampus are capable of suppressing gonadotrophin release. In their experiments, electrochemical stimulation of the ventral or dorsal hippocampus during pro-oestrus reduced the incidence of spontaneous ovulation in rats. However, when the medial cortico-hypothalamic tract—probably carrying afferents from the hippocampus to the HTA—was interrupted a few weeks earlier, hippocampal stimulation was no longer able to inhibit spontaneous ovulation. Stimulation of the hippocampus also prevented the ovulation normally observed after stimulation of the arcuate nucleus. Also more recent results of the same authors[133] seem to support the assumption that the influence on gonadotrophin secretion of the hippocampus is inhibitory.

(c) *Mid-brain reticular formation* — Experiments performed to investigate the mechanism of the ovulation-blocking effect of morphine, chlorpromazine and reserpine pointed to the possibility of an involvement of the mid-brain reticular formation in the gonadotrophin-releasing mechanism[5,6].

In the experiment of Appeltauer and co-workers[2], animals bearing lesions in the periaqueductal grey matter had a potentiation of the inhibiting effects of oestrogen on gonadotrophin secretion. This resulted in a marked decrease in the weights of the testicles and of the ventral prostate, with more marked regressive changes in the Leydig cells than in the seminiferous epithelium. Since the administration of reserpine to rats bearing lesions in the periaqueductal grey has been found to bring about an increase in the weight of the testicles, it may be inferred that the caudal pole of the limbic system may regulate either activating or inhibiting influences that control gonadotrophin secretion. A dual role for mesencephalic structures in the control of gonadotrophin secretion in rats has also been postulated by Carrer and Taleisnik[16]. Whereas electrochemical stimulation during pro-oestrus of the ventral tegmentum or medial raphe nucleus significantly decreased the incidence of spontaneous ovulation, stimulation of the dorsal tegmentum induced ovulation in rats which had been made persistent oestrous by exposure to constant light.

(d) *Septal complex* — Ovulation by stimulation of the septum pellucidum was elicited in persistent oestrous rats induced by continuous illumination[14]. Everett[32] later found that in the septal complex (including the medial and lateral septal nuclei, the nucleus accumbens and the caudal portion of the medial paraolfactory area) it was necessary to use somewhat larger currents (i.e. forming a larger stimulative focus than in the preoptic area of the atropine-blocked pro-oestrous rat), but precise location was not critical, in order to induce ovulation. On the basis of these findings, a diffuse neuronal system has been postulated, highly dispersed throughout the septal complex, converging somewhat as it passes through the PAHA and sharply converging in the medial basal tuberal region. That the system is multi-synaptic is clearly indicated by the fact that rats can maintain ovulatory cycles after disconnection of the PAHA rostrally, dorsally and laterally[67,82].

It is emerging also from the work of Hagino and Yamaoka[55] that the septal complex is one of the most important mechanisms for integrating information from the limbic and reticulo–thalamo–cortical system and for mediating them to the hypothalamus.

In their study using urethane anaesthesia, electrical stimulation of terminating area in mesencephalon of the medial forebrain bundle caused a predominantly inhibitory effect on the spontaneous septal neurone activity. This stimulus travels along with the medial forebrain bundle. However, another pathway such as the non-specific diffuse thalamic projection system might be involved in the spontaneous septal neurone activity, since stimulation of the anterior ventral nucleus of the thalamus also resulted predominantly in an inhibitory effect on the spontaneous septal neurone activity without interference with the sleep stage. Parenteral injection of 50 μg of dexamethasone sodium phosphate tends to reduce the spontaneous septal neurone activity, and alters the threshold upon dorsal hippocampal stimulation. However, local injection of 4 μg in 1 μl of dexamethasone into the septum on the day of dioestrus or pro-oestrous blocks ovulation and alters the plasma level of corticosterone on the following day.

To rule out the possibility that the failure of ovulation or alteration of plasma corticosterone is due to the action of dexamethasone after transport through the general circulation, dexamethasone was injected into the dorso–medial nucleus of the thalamus or cortex. This regimen had no such effect on ovulation or plasma corticosterone on the following day. This evidence suggests that dexamethasone acts upon the septum to alter the neural mechanism, and subsequently makes it possible to alter the release of CRF, FRF and LRF. On the other hand, the anterior amygdaloid stimulus caused a predominantly inhibitory effect on spontaneous septal neurone activity, while a dorsal hippocampal stimulus resulted in 50–50 excitation or inhibition, suggesting the existence of an inhibitory interneurone in the septum. Irrespective of the method employed, this hippocampal stimulus has an antagonistic effect on the septal neurone with the anterior amygdaloid stimulation.

From these findings, it can be inferred that sensory input from the mid-brain limbic area travels along with the medial forebrain bundle and/or non-diffuse thalamic projection system to reach the septal complex. At this level, this input is combined with the information from the dorsal hippocampus and/or anterior amygdala, and subsequently the forebrain signal mediated by the preoptic–tuberal tract makes it possible to maintain the rhythmic release of corticotrophin- and gonadotrophin-releasing factors. If so, it is tempting to speculate that the circadian ACTH rhythm might also play a role in timing ovulation and in the maintenance of cyclic gonadotrophin release, as seems to be the case with the cyclic mating behaviour of the female rat[100].

1.3.2.2 Epithalamo–epiphysial complex

Another extrahypothalamic part of the brain that participates in the control of gonadotrophin secretion is the epithalamo–epiphysial complex. Moszkowszka[97] assumed that the pineal gland normally inhibits the synthesis and release of pituitary GTH in three different ways. The so-called F_3 fraction of pineal extract inhibits FSH release by direct action on the pituitary; serotonin acts through the inhibition of the hypothalamic-releasing factors, and melatonin inhibits LH release 'in a still unclarified way'. In an attempt to elucidate the mechanism through which melatonin exerts its influence on the anterior pituitary, Mess *et al.*[92] stereotaxically implanted fragments of pineal tissue or crystals of melatonin and of 5-hydroxytryptophol in the HTA, in the mid-brain reticular formation and in the pituitary gland of castrated male rats. Placement of pineal fragments or of indole compounds in these two brain areas was followed by a significant reduction of pituitary LH stores, but melatonin was unable to reduce pituitary LH when directly implanted in the pituitary gland. It is suggested by these results that these indol compounds of pineal origin may play a role in the control of LH secretion, possibly by acting on specific receptors localised in the HTA and in the mid-brain reticular formation. This assumption has recently been supported by the results of Mess *et al.*[93]. Pinealectomy resulted in appearance of irregular vaginal cycles and ovulation, confirmed by the presence of tubal ova, in persistent oestrous rats from anterior hypothalamic lesions. In the most recent experiments of Mess *et al.* (personal communication), a few of the previously anovulatory rats ovulated and gave birth repeatedly 1–7 weeks after pinealectomy, which indicates that some mechanism inhibiting gonadotrophin synthesis and/or release has been removed by pinealectomy.

1.4 MEDIATION OF NEUROENDOCRINE SIGNALS

1.4.1 Neural signals

In the polycyclic animals, the cycle mechanism operates fairly independently from external environmental influences. However, there are also factors from the external environment that influence the cycle mechanism in these species, and in this way alter the phasing of the oestrous cycles within the period of reproductive activity. Light is a dominant factor in this respect[39].

Light afferentations can reach the anterior pituitary in various ways, and influence its gonadotrophic activity in various directions.

Since the fundamental influence of light upon the gonadotrophic function of the pituitary became known, the old hypothesis of direct optic fibres terminating in the hypothalamus is frequently mentioned. Following unilateral retinal destruction or a unilateral enucleation, analysis of the Fink–Heimer stained sections of the experimental rats suggests that fibres leaving the optic pathway pass to the PAHA, suprachiasmatic, ventromedial, arcuate and ventral premamillary nucleus[96, 125]. Also radioactivity in the hypothalamus, especially in the region of the medial forebrain bundle, in the arcuate and premamillary nuclei after intraocular administration of the serotonin precursor[103], and over both suprachiasmatic nuclei after intraocular injection of tritiated leucine[96], appears to support the view that there would be a direct projection from the retina to the hypothalamus in the rat.

In material based on six cats with extirpation of one or both eyes, however, Szentágothai[127] was not able to detect a single degenerated fibre in the whole hypothalamus, although degeneration of optic fibres in the lateral geniculate body, in the superior colliculi and in some parts of the mesencephalic reticular formation was clear enough in the same material. This is in accordance with the findings of many neuroanatomists including Nauta and van Straaten[99] and Campos-Ortega and Clüver[15] who also did not find degenerating fibres in the hypothalamus of different species after enucleation.

There are, of course, ample opportunities for optic fibres to establish indirect, yet rather short, connections with the hypothalamus. Perhaps the most powerful afferent pathway of the hypothalamus comes from the mid-brain. Thus, an optic influence reaching the mid-brain reticular formation directly, or conveyed by the anterior colliculi, can easily be relayed to the hypothalamus. Another possibility is the secondary pathway arising from the lateral geniculate body. Furthermore, it has been supposed earlier that the most ventral part of the supraoptic commissure of Gudden connects the two lateral geniculate bodies. In cats with fairly well isolated lesions of the lateral geniculate bodies, Szentágothai could indeed trace very fine-calibred degenerated fibres to the lateral hypothalamic area immediately dorsally from the optic tract, but he could follow only a few of them into the ventral supraoptic commissure[127].

The gonadotrophin-stimulating action of light is more clearcut in birds than in other classes of vertebrates. Benoit[8] was able to show in ducks that a significant portion of visible radiations of higher wavelength (essentially the red) can penetrate across the tissues and reach the hypothalamus. A sharp stimulation of the testes still occurs when immature male ducks are given a light stimulus to the lateral side of the head, even after the optic nerves have been severed. A strong testicular growth is the consequence of direct light stimulation of the anterior hypothalamus with the aid of a quartz rod introduced across the eyesocket. It has also been demonstrated with the aid of intracerebral photoelectrodes in rats, rabbits, dogs and sheep that light can penetrate the skull and reach the hypothalamus[50]. Stimulated by both retinal afferent impulses and penetrating radiations, the hypothalamus is apparently able to enhance the gonadotrophic activity of the pituitary gland.

Data exist suggesting that light can also influence gonadotrophin secretion through the pineal gland. Thus pinealectomy at the appropriate time prevented the deficiencies in gonadal function which develop in response to light deprivation or blinding in male[112] or female[124] hamsters. Blinded female rats showed a decreased capacity for ovarian compensatory hypertrophy; this deficiency was prevented by pinealectomy[123]. Taking into account the inhibitory action of melatonin, mentioned in Section 1.3.2.2, these findings can well be explained on the basis of the observation of Wurtman *et al.*[138], i.e. that continuous darkness for 6 days produced a fivefold increase in the melatonin-synthesising enzyme, hydroxyindole-*O*-methyl transferase (HIOMT), as compared with animals kept in continuous light. However, what is the mechanism whereby constant illumination suppresses gonadotrophin output and induces the state of anovulatory persistent vaginal oestrus in rats needs to be elucidated. It follows immediately from the above experimental findings that this effect of constant light on the pituitary cannot be brought about by pineal melatonin mediation. This is indicated also by the findings that the pineal glands of rats maintained in constant light showed an approximate 50% decrease of HIOMT activity as compared to animals kept under normal diurnal lighting conditions[138].

The limbic system receives afferent impulses primarily through olfactory and somaesthetic pathways. It is therefore easy to understand that olfactory and various tactile stimuli play a considerable part in the regulation of gonadotrophin secretion.

In oestrous rabbits, Sawyer[116] found that impulses from the olfactory bulbs were essential for a type of pharmacologically induced ovulation. In female mice with intact olfactory bulbs, the oestrous cycle is synchronised by the addition of males or the odour of males[137]. These effects fail to occur in females deprived of their sense of smell. Moreover, this so-called Whitten effect can be reproduced by placing the females in a box recently vacated by males. It seems certain, therefore, that olfactory signals are involved in both cases, as they are in the so-called pregnancy-block phenomenon. Parkes and Bruce[104] have shown that the odour of an alien male mouse other than the successful mate has the capacity to cause abortion and re-start oestrous cycles after conception. The olfactory nature of the pregnancy-blocking effect exerted by the alien males was evidenced by its non-appearance in females if the sense of smell had been destroyed.

The chemical nature of the substances involved is unknown, but in the case of the oestrus acceleration effect[137] it is known that the active substance is in the urine of the male mouse but not in the urine of the castrated mouse. This has been substantiated by the experiments of Scott and Pfaff[121]. In behavioural tests, female mice spent more time sniffing tubes containing urine from normal males than those with urine from castrated males. In electrophysiological experiments, neurones in the olfactory bulb, preoptic area and lateral hypothalamus responded to these biologically significant odours. However, no neurones were observed that responded exclusively to urine odours and only a few neurones showed consistently different responses to the odours from normal and castrated males.

Apparently, odour responses are mediated by the olfactory bulb and fed through limbic pathways into the hypothalamus. In the latter respect, the mid-brain reticular formation especially seems to play an essential role. Responses to odours were recorded from single units in the mesencephalic reticular formation and in the EEG. Changes in firing rates of units in the mesencephalic reticular formation were more closely associated with changes in the EEG than firing rates of units in the preoptic area or olfactory bulb[108].

Tactile stimuli from the external as well as from the 'internal' environment can influence the hypothalamic cycle mechanism and alter phasing of the oestrous cycles. In birds, the number of eggs laid in a clutch seems to be regulated through either visual or tactile stimuli from the ventral body surface Sterile mating as well as several other procedures involving neural stimulation will cause rats to stop cycling and to become pseudo pregnant. Inserting glass rods into the cervix uteri induces development of corpora lutea in rats and this effect follows even if the ovaries have been transplanted into another part of the body. The result, therefore, is not the consequence of any direct nervous excitation of the ovary but the interoceptive neural signal transmitted by spinal, mid-brain reticular and hypothalamic pathways to the anterior pituitary. Other interoceptive neural signals originating from different parts of the reproductive tract include afferent impulses produced by stimuli affecting the oviduct or by the mechanical distension of the uterine horn. However, among factors from the internal environment which regulate the secretion of GTH, the actual sex hormone level in the blood appears to be of the greatest importance.

1.4.2 Hormonal signals

Although Hohlweg and Junkmann[61] raised the possibility more than four decades ago that the gonadal hormones might influence gonadotrophic functions through a hypothetical 'sexual centre' located somewhere in the brain, it was generally believed until the early fifties that the peripheral hormonal milieu influences synthesis and release of GTH exclusively by direct hormonal actions on the anterior pituitary. In the light of experimental data accumulated in the last two decades, it is now well established that the feed-back action of sexual steroids is mediated, at least in part, through multiple neuronal elements situated in the nerual mechanisms controlling gonadotrophin secretion. This is now termed indirect or neurohormonal feed-back. Evidence for the existence of an indirect or neurohormonal feed-back has been summarised and discussed by Flerkó[38, 39] and Szentágothai *et al.*[127].

There may well be three levels at which the feed-back effects of sexual steroids act to modify gonadotrophin secretion. The lowest of these is the anterior pituitary, while the other two are situated in the brain. Of these neural mechanisms, the lower seems to be the tonic mechanism in the HTA and the higher the cycle mechanism including all the parts of the brain which are capable of modulating the activity of the HTA neurones producing gonadotrophin-inhibiting and -releasing factors.

1.4.2.1 Direct steroid feed-back

The observation that sexual steroids can be bound to the adenohypophysis[28, 45, 46, 71] simply indicates that they may act directly on certain anterior pituitary cells. Rose and Nelson[114] showed that oestrogen perfused into the hypophysial fossa inhibited the castration reaction. Bogdanove[12] also studied the castration reaction and concluded that there could be a direct inhibitory effect of oestrogen on the pituitary. Certainly this is the case if a larger amount of oestrogen is acting on the anterior pituitary. The inhibitory action of oestrogen on gonadotrophin secretion in doses of 25–30 μg/day cannot be eliminated by hypothalamic lesions[127], but it is possible to block the inhibitory effect of small (physiological) doses of oestrogen by destroying the anterior hypothalamus[37]. On the other hand, Flerkó and Bárdos[42] have shown in rats with anterior hypothalamic lesions that pituitary hypertrophy and characteristic cytological changes are the results of the direct action of even small amounts of oestrogen on the adenohypophysis, whereas simultaneous inhibition of gonadotrophin secretion is an indirect effect mediated by anterior hypothalamic neurones.

The other newly recognised aspect of direct oestrogen action on the pituitary gonadotrophic cells is that the steroid acts to lower the pituitary threshold to LRF and probably also to FRF[27, 135]. Testosterone and progesterone appear to have an opposite effect. Measurements of the amount of LH released into the medium by LRF addition to pituitaries removed from donor male rats, pre-treated *in vivo* by castration or testosterone, have shown that LH release was greater from pituitaries taken from castrated males, and less from pituitaries removed from testosterone-treated males, than that from intact males[134]. Pre-treatment of rabbits with progesterone can, at least partially, block the effects of *in vivo* injection of LRF in eliciting LH release[60]. Thus, one of the main physiological importances of the direct action of gonadal steroids on the adenohypophysis seems to be that they can change the sensitivity to the gonadotrophin-releasing factors of the pituitary gonadotrophic cells.

1.4.2.2 External or long-loop feed-back

If the mode of coupling between the different units of the neurohormonal feed-back loop is considered, it is apparent that two types of coupling frequently used in control engineering exist in the mammalian organism. These are 'external' and 'internal' feed-back.

The 'external' or 'long-loop' feed-back is the basic coupling through which information is supplied from one characteristic parameter of the functional state of the target glands back into the control unit, i.e. into the limbic–hypothalamo–pituitary system. In general, the information supplied is the change of sex steroid level in the blood. However, information need not arise from the target gland, i.e. from the ovary itself, but may originate in any other organ or function activated by the ovarian hormones. The existence of such a feed-back in the brain–pituitary–ovarian system can be inferred from the observations mentioned at the end of Section 1.4.1. In this case, of course, information reaches the limbic and hypothalamic mechanisms controlling gonadotrophin secretion from the genital tract through purely nervous pathways (neural feed-back). Since the functional state of the genitalia depends on the secretion of ovarian hormones, the existence of such secondary neural feed-back loops in the control of gonadotrophin secretion is apparent.

(a) *Positive or stimulating feed-back* — In the case of a 'positive' or 'stimulating' feed-back, the gonadotrophin release is simultaneously increasing with the increase of the hormone output from the ovary. Theoretical considerations alone would support the idea that positive feed-back occurs in the control of cyclic gonadotrophic functions. The very existence of a function constantly oscillating between extremes, as is the case with the female gonad, points to the existence of positive feed-back in the mechanism controlling it.

Indeed, it was demonstrated by Sawyer *et al.*[117] and by Everett and Sawyer[33] that the effects of oestrogen and progesterone in advancing spontaneous ovulation could be eliminated by neural blocking agents. This clearly indicated that there was a neural link in the positive feed-back action of ovarian steroids on gonadotrophin secretion. Barraclough and Gorski[4], Barraclough, Yrarrazaval and Hatton[7], Döcke and Dörner[26] and Bishop *et al.*[10] have shown later that the medial preoptic and suprachiasmatic area in the rat is the specific site at which ovarian steroids act to facilitate ovulation.

(b) *Negative or inhibiting feed-back* — The external feed-back in control systems of biological importance is in most cases 'negative' or 'inhibiting'. In the case of the brain–pituitary–ovarian system, the hormone output of the system is decreased when the control input from the ovary is increased and vice versa, so that any deviation of the function induces an opposite deviation correcting the initial deviation of the function.

The first evidence for the existence of an external negative neurohormonal feed-back through anterior hypothalamic neurones in the control of gonadotrophin secretion was given by Flerkó[36, 37] who demonstrated in parabiosis experiments that the inhibitory effect of oestradiol (1 μg/day) on the castration-induced rise in gonadotrophin, especially in FSH, output could be significantly diminished by electrolytic lesions placed in the anterior hypothalamus of juvenile rats. The results of his experiments[38] suggested, furthermore the presence of oestrogen-sensitive nerve cells in the PAHA. The postulate that this part of the brain would contain oestrogen-sensitive neurones[48] was later supported by the finding that individual nerve cells in this part of the brain accumulate oestrogen[3,95,107,126], and that the anterior hypothalamus, including the preoptic area, takes up and retains oestradiol in the same way as the peripheral oestrogen-responsive organs[28,46,71].

Depression of the activity of the FRF- and LRF-producing neurones in the HTA by continuous oestrogen action appears to be an important factor in inducing the anovulatory condition following anterior hypothalamic lesion[40]. When the continuous negative oestrogen feed-back is greatly reduced or eliminated, FRF- and LRF-producing neurones again stimulate the anterior pituitary to release GTH sufficient for luteinisation. In this way, Flerkó and Bárdos[43] succeeded in inducing formation of corpora lutea in the formerly polyfollicular ovaries of rats made anovulatory by anterior hypothalamic lesions. It seemed, therefore, reasonable to assume that either the FRF- and LRF-producing neurones, or some other neural element in the HTA, that can influence the activity of the FRF- and LRF-producing neurones, are also sensitive to the feed-back action of gonadal hormones. This assumption was supported by the experimental results of Lisk[83,84] and many other investigators[39,127]. Further support for this assumption came from findings that a considerable number of neurones in the HTA, especially in the arcuate nucleus, accumulate oestradiol[107,126], and the medial basal part of the hypothalamus shows a pattern of uptake and retention of tritiated oestradiol which is similar to the pattern found in the peripheral oestrogen-responsive organs[46]. A functional role for oestradiol in the cells which concentrate it has been strongly suggested by the anatomical correspondence between the sites of oestradiol-binding cells and the sites of oestradiol-implant effects. Furthermore, as follows from the foregoing sections, lesion, electrical stimulation and electrophysiological activity studies have also implicated PAHA and HTA, which have peak numbers of oestradiol-binding cells, in the neurohormonal control of pituitary gonadotrophic function.

It can be assumed with considerable certainty that the sensitivity or responsiveness to sexual steroids of the neurones in the PAHA is not identical with that of the neurones in the HTA. Ovarian compensatory hypertrophy was blocked in rats with a frontal cut behind the anterior hypothalamic area, but this deafferentation did not interfere with the pituitary response to castration[81]. Similar observations have been reported in animals with complete deafferentation of the HTA. Furthermore, ovarian compensatory hypertrophy did not occur in these rats, but at the same time pituitary LH content increased and castration cells developed in the anterior pituitary after bilateral ovariectomy[57].

These findings suggest that the neurones of the preoptic–anterior hypothalamic cycle mechanism are more sensitive to changes in the blood sex steroid level than are the hormone-sensitive neurones in the HTA. This may account for the fact that HTA by itself is unable to maintain the cyclic release of gonadotrophic hormones. To do this, the tonic mechanism, i.e. the FRF- and LRF-producing neurones in the HTA, needs afferent impulses from the more sensitive steroid-receptor neurones of the preoptic–anterior hypothalamic cycle mechanism. According to the varying ovarian steroid level in the blood, the highly sensitive steroid-receptor neurones modulate (i.e. enhance or inhibit) the tonic activity of the FRF- and LRF-producing neurones and, in this way, secure the cyclic release of FSH and LH. If the sensitivity or responsiveness of the neurones of the preoptic–anterior hypothalamic cycle mechanism is greatly reduced—as in the male[105] or in androgen sterilised female rats[106]—the tonic mechanism by itself can maintain only a male-type, i.e. a non-cyclic, release of GTH.

Experimental findings suggest that limbic structures containing a number of steroid-binding neurones[89, 126] are also deeply influenced by gonadal steroid feed-back[73, 122, 133].

1.4.2.3 Internal or short-loop feed-back

'Internal' or 'short-loop' feed-back is a coupling in which the output of the control unit is tapped and one branch is brought back into the control input of the unit. This principle is often applied in combination with external feed-back to secure an appropriate sensitivity and stability of the whole control system.

In the brain–pituitary–target gland control mechanism, the principle of internal neurohormonal feed-back would be realised if information of some parameter of anterior lobe function were fed back directly into the HTA, and were to exert a direct influence on anterior pituitary function. Anatomical evidence[127] supports this idea in principle, and Halász and Szentágothai[127] were the first to show that such a mechanism of internal neurohormonal feed-back really exists in the case of adrenocorticotrophic functions. The existence of a negative internal feed-back in gonadotrophic functions has been suggested by Kawakami and Sawyer[72] and later supported by many experimental results. The findings of David *et al.*[24], Corbin[19] and other investigators imply that FSH or LH implanted into the HTA may have induced a partial inhibition of pituitary FSH and LH, respectively, through a negative feed-back effect on FRF- and LRF-synthesis and/or release. The possibility of the existence of positive internal LH and FSH feed-back has been raised by Ramirez and Sawyer[111] and by Ojeda and Ramirez[102], respectively.

1.4.2.4 Ultrashort feed-back

Subcutaneous administration of crude rat hypothalamic extracts containing FRF and devoid of any FSH and sex steroid contamination was found by Hyyppä *et al.*[63] to give rise to a considerable reduction of FRF stores in the hypothalamus of castrated-hypophysectomised rats. It is suggested that the brain contains receptors sensitive to circulating levels of FRF, and that FRF might directly regulate its own production, through what might be called an 'ultrashort' feed-back effect. More recent observations of Martini and associates (personal communication) indicate that the thyrotrophin-releasing factor (TRF) also exerts an 'ultrashort' feed-back effect on hypothalamic stores of TRF when administered to hypophysectomised–thyroidectomised animals. These data, if confirmed, will be particularly conclusive since a synthetic preparation of TRF was used, and this obviously rules out the possibility of non-specific effects due to the presence of contaminants in the crude hypothalamic extracts.

1.5 SEXUAL DIFFERENTIATION OF BRAIN MECHANISMS CONTROLLING GONADOTROPHIN SECRETION

1.5.1 Perinatal androgen action inducing the male pattern of gonadotrophin secretion

On the basis of very suggestive experiments, Yazaki[139] first came to the conclusion that the non-cyclicity of the hypothalamo–hypophysial function in the adult male rat is only established at about the third day after birth, and the presence of testes in the first few days of postnatal life determines the male specificity of the hypothalamus. The same conclusion was also reached by Harris and Levine[59] and Gorski and Wagner[51]. The experimental results of these authors have also suggested that the hypothalamus of the newborn rat of either sex has the inherent ability to maintain a cyclic release of gonadotrophins. It is only in the first few postnatal days that normal male animals under the influence of testicular androgen, or female rats that are given testosterone, lose the ability to release GTH in a cyclic manner and thereby ensure ovulation. It seems probable that the same conditioning by androgen also occurs in the human, but the brain structures, instrumental in the maintenance of the cyclic gonadotrophin release and ovulation, may be organised at an earlier stage, so that the male or female pattern of gonadotrophin secretion might be established earlier in the foetal life.

1.5.2 Female rats without cycle mechanism

The neural mechanism enabling the female anterior pituitary to release gonadotrophic hormones in cyclic phases is apparently absent in the male which does not display cyclic fluctuations in gonadotrophin secretion. In this respect, the male rat resembles the female in which electrolytic lesions or a frontal cut separated the limbic and preoptic–anterior hypothalamic cycle mechanism from the tonic mechanism, or the female which received early postnatal androgen treatment. Under both experimental conditions, female rats show persistent vaginal oestrus associated with polyfollicular ovaries. Furthermore, pituitaries of both lesioned and androgenised rats contain gonadotrophins sufficient for ovulation[130], but they are unable to release an ovulatory quota of them. The identity of the two experimental syndromes has suggested that identical brain structures might suffer damage in both lesioned and androgen-sterilised rats.

Indeed, Barraclough and Gorksi[4] have shown that the preoptic–LH trigger was deranged in female rats which received early postnatal androgen treatment. On the basis of the concept concerning the preoptic–anterior hypothalamic cycle mechanism, described in Section 1.3.1.1, it would be expected that not only the preoptic–LH trigger but also simultaneously the anterior hypothalamic FSH control mechanism might be affected by such treatment. This assumption has been supported by the results of experiments[105, 106] showing that the sensitivity or responsiveness to oestradiol of the anterior hypothalamic neurones contributing to the hypothalamic FSH control mechanism was considerably reduced in adult males as well as in androgen-sterilised female rats. Simultaneously, many authors (see, Flerkó[41]) reported that responsiveness to oestradiol of the peripheral oestrogen-responsive tissues (uterus, vagina, anterior pituitary) was reduced in rats injected with testosterone in the first few postnatal days. These findings indicate that neural and non-neural target tissues of oestrogen are similarly affected by the early postnatal androgen action which reduces sensitivity or responsiveness to oestradiol of these tissues. The brain may also be relatively insensitive to progesterone after androgenisation[18].

1.5.3 Possible mechanism of perinatal androgen action

Jensen *et al.*[66] and King[75] suggested that uptake and retention of oestradiol occurs at receptor sites that are specific for oestrogens. The specificity of action of oestradiol on the oestrogen-responsive tissues has been explained by the ability of these tissues to take up and retain oestradiol in an unconverted form. The inhibition of oestradiol-induced uterine growth by various doses of antiuterotrophic compounds, like nafoxidine, was found to parallel reduction of oestradiol uptake and provided evidence that the oestrogen-binding phenomenon is closely involved in the biological action of the hormone. With the exception of limited glial cell proliferation in adult animals, brain cells do not divide. Therefore, the action of oestradiol on neurones is essentially different from the action of oestradiol on the uterus, where cell proliferation and growth are very prominent.

As pointed out by McEwen and associates[90], since the nuclear binding of oestradiol and corticosterone has been seen by autoradiography to be identified with neurones and to a lesser extent with glial cells, it is appropriate that the major emphasis on hormonal effects should be on neuronal metabolism. Among the most important neuronal enzyme systems which might be influenced by hormones are the enzymes responsible for biosynthesis or metabolism of neurotransmitters like dopamine, noradrenaline, etc. (for details, see McEwen *et al.*[90]). The receptor which enables the postsynaptic cell to respond to the neurotransmitters could be considered a part of this system. Hormone action in the nucleus results in formation of proteins which may be concerned with neurotransmitter biosynthesis or destruction or with receptors for neurotransmitters in the postsynaptic region. Alterations either in the metabolism of a neurotransmitter or the receptor which responds to it could have profound consequences for neuronal processes underlying endocrine regulation[90].

A preliminary experiment[45] revealed that the oestradiol-binding capacity of pituitaries and uteri of androgen-sterilised rats was significantly reduced as compared to intact controls. This raised the possibility that early postnatal androgen action might interfere with the development of the oestrogen-binding or receptor proteins and, in this way, interfere with oestradiol uptake and/or retention by the oestrogen-responsive tissues. Since the findings mentioned in Section 1.4.2.2 indicate that the middle and anterior hypothalamus, including the preoptic area, are direct target tissues for oestradiol, and the sensitivity to oestradiol of these brain areas has been found to be reduced in androgenised rats[105,106], one might postulate that the early postnatal androgen action might damage the specific trapping mechanism in the oestrogen-sensitive hypothalamic neurones. If so, the oestradiol-binding capacity of the hypothalamic areas containing oestrogen-responsive neurones should be reduced in androgen-sterilised rats. This has been confirmed by the findings of Flerkó *et al.*[46] who showed that the oestradiol-binding capacity of the anterior and middle hypothalamus, as well as that of the non-neural target tissues (uterus and anterior pituitary) was significantly reduced in rats which received 1.25 mg testosterone phenylpropionate two days after birth. No change occurred in the posterior hypothalamus and parietal cortex of the same animals.

Similar findings have been reported by many investigators (see Flerkó[41]) some time later. On the other hand, using male and female rats of equal group body weight but of different age, Green and co-workers[53] did not find any difference between sexes in the binding of oestradiol by any tissue, and concluded that the effect observed by many authors might reflect weight differences rather than a blockage of oestrogen-receptors sites by neonatal androgenisation, as has been suggested by Flerkó *et al.*[46] and by others[41]. In a subsequent experiment of Flerkó and associates[44], however, the oestradiol-binding capacity of the neural (anterior and middle hypothalamic) and of the non-neural (uterine) target tissues was significantly reduced in androgen sterilised rats as compared to controls of approximately equal body weight.

The finding that androgen-sterilised rats had significantly lower anterior and middle hypothalamic radioactivity levels than controls of either equal age or of equal body weight supports the hypothesis that early postnatal androgen might interfere with the normal development of oestrogen-receptor proteins and, hence, with the normal uptake and/or retention of oestradiol by the hypothalamic oestrogen-responsive neurones. In this way, these neurones might become desensitised and functionally inactive in mediating positive and negative oestrogen feed-back on FRF- and LRF-producing neurones, and through these neurones on pituitary cells producing gonadotrophic hormones. It appears, therefore, highly probable that the perinatal androgen action, by reducing the oestradiol-binding capacity of the neurones contributing to the hypothalamic cycle mechanism, 'switches off' the cycle mechanism leaving the tonic mechanism intact, which by itself, however, is able to maintain only a 'male-type' i.e. a non-cyclic gonadotrophin secretion.

Naftolin *et al.*[98] have reported recently the aromatisation of androstenedione to oestrone by the anterior hypothalamus of human male foetuses and of adult male and female rats, and have suggested that oestrogens would mediate brain differentiation.

This assumption does not disagree with the above hypothesis, since it has been shown by McGuire and Lisk[91] that early postnatal oestrogen action reduces the oestradiol-binding capacity of the anterior hypothalamus in the same way as early postnatal androgen. It is well known that oestrogens, in contrast to androgens in the male, are not secreted, at least in measurable amounts, in the female organism before the end of the critical period. This would explain why functional elimination of the cycle mechanism in the male, but not in the female, occurs during the perinatal period. Furthermore, oestradiol benzoate given to 5 day-old rats on weight basis is more potent than testosterone propionate in inducing non-cyclic gonadotrophin release when the animals mature.

However, 0.1 μg oestradiol administered daily over 10 days to newborn male rats did not enhance the 'masculinisation' of the brain; on the contrary, this amount of oestradiol alone or with 0.5 or 2.5 μg progesterone prevented the 'masculinising' effect of testicular androgen[47], a finding which does not seem to support the assumption of Naftolin and associates. Taking into account the fact that a number of steroids, if given in sufficiently high doses for prolonged periods immediately after birth, are capable of eliciting the anovulatory persistent vaginal oestrus syndrome, it does not appear likely that the oestrogens were exclusively responsible for inducing this syndrome or for 'imprinting' the male pattern of gonadotrophin secretion on the brain of the newborn male rats.

An ovulatory surge of gonadotrophins can be elicited by stimulation of the preoptic area in the male rat or in the female rat treated with androgens neonatally[4]; neither possesses a spontaneous trigger for the induction of ovulation. This therefore raises the possibility that the functional differences between the male and the female preoptic areas are partly due to some difference in the neural connections of the preoptic area, a conclusion which is in agreement with anatomical demonstration by Raisman and Field[110] of sexual dimorphism in the mode of termination of the afferent fibres to the preoptic area.

1.6 SUMMARY AND CONCLUSIONS

Current concepts invoke a dual mechanism for the neural control of the cyclic secretion of gonadotrophic hormones (GTH).

The first, termed the tonic mechanism, maintains a tonic basal discharge of FSH and LH. The tonic mechanism consists of tubero–infundibular neurones in the hypophysiotrophic area (HTA) of the hypothalamus which synthesise and release into the hypophysial portal circulation the FSH- and LH-releasing factors (FRF and LRF).

The second mechanism modulates the activity of the FRF- and LRF-producing neurones according to afferentations from the external and internal environment. This mechanism involves brain structures in the HTA (tubero–infundibular dopaminergic neurones) in the preoptic–anterior hypothalamic area, limbic system and epithalamo–epiphysial complex, and may be termed a cycle mechanism, being indispensable in the maintenance of the cyclic release of FRF–FSH and LRF–LH, respectively.

Among signals from the 'milieu interieur' ovarian steroid feed-back from the external environment, light and smell influences seem to be especially important in the control of cyclic gonadotrophin secretion and ovulation. At least a part of the neurones contributing to the cycle mechanism are hormone-sensitive, i.e. contain hormone-binding proteins. Thus these neurones can mediate both hormonal and neural signals to the same extent on the FRF- and LRF-producing neurones of the tonic mechanism, which is the final common pathway of neuroendocrine influences towards the gonadotrophin-secreting cells of the anterior pituitary.

Direct action of gonadal steroids on the anterior pituitary seems to change the sensitivity to LRF, and probably also to FRF, of the gonadotrophin-producing pituitary cells. Negative, indirect or neurohormonal steroid feed-back, continuously correcting the initial deviation of the FRF–FSH and LRF–LH output, is one of the main driving forces in the maintenance of the cyclic gonadotrophin release. This negative external feed-back combined with negative internal feed-back of FSH and LH makes the reaction of the brain–pituitary–ovarian system sluggish, whereas the well-established positive neurohormonal steroid feed-back facilitates the neural trigger function of the cycle mechanism rendering possible the quick release of an ovulatory quota of GTH.

The cycle mechanism, enabling the female pituitary to release gonadotrophic hormones in cyclic phases, does not function in the male. This appears to be a hormonally induced rather than a genetically determined status of the male brain.

It has been shown that the non-cyclicity of the gonadotrophic function in the adult male rat is established only around birth, and the presence of testes in this period of life determines the male specificity of the brain mechanism controlling gonadotrophin secretion. It is only in the first few postnatal days that intact male rats under the influence of testicular androgen, or female rats that are given testosterone, lose the ability to release GTH in a cyclic manner. Reduction of the oestradiol-binding capacity by perinatal steroid action on the neurones contributing to the cycle mechanism, and the consequent loss of neurohormonal steroid feed-back, instrumental in the maintenance of the cyclic release of GTH, appears to play a decisive role in this respect.

References

1. Ahrén, K., Fuxe, K., Hamberger, L. and Hökfelt, T. (1971). Turnover changes in the tubero–infundibular dopamine neurons during the ovarian cycle of the rat, *Endocrinology*, **88,** 1415
2. Appeltauer, L. C., Reissenweber, N. J., Dominguez, R., Griño, E., Sas, J. and Benedetti, W. L. (1966). Effects of estrogen on the pituitary–gonadal axis of rats bearing lesions in the periaqueductal gray matter, *Acta Neuroveg.*, **29,** 75
3. Attramadal, A. (1964). Distribution and site of action of oestradiol in the brain and pituitary gland of the rat following intramuscular administration, *Int. Cong. Ser. No.* 83, 612 (Amsterdam: Excerpta Med. Found.)
4. Barraclough, C. A. and Gorski, R. A. (1961). Evidence that the hypothalamus is responsible for androgen-induced sterility in the female rat, *Endocrinology*, **68,** 68
5. Barraclough, C. A. and Sawyer, C. H. (1955). Inhibition of the release of pituitary ovulatory hormone in the rat by morphine, *Endocrinology*, **57,** 329
6. Barraclough, C. A. and Sawyer, C. H. (1957). Blockade of the release of pituitary ovulatory hormone in the rat by chlorpromazine and reserpine: possible mechanism of action, *Endocrinology*, **61,** 341

7. Barraclough, C. A., Yrarrazaval, S. and Hatton, R. (1964). A possible hypothalamus site of action of progesterone in the facilitation of ovulation in the rat, *Endocrinology*, **75,** 838
8. Benoit, J. (1964). The role of the eye and of the hypothalamus in the photostimulation of gonads in the duck, *Ann. N. Y. Acad. Sci.* ,**117,** 204
9. Bishop, W., Fawcett, C. P., Krulich, L. and McCann, S. M. (1972). Acute and chronic effects of hypothalamic lesions on the release of FSH, LH and prolactin in intact and castrated rats, *Endocrinology*, **91,** 673
10. Bishop, W., Kalra, P. S., Fawcett, C. P., Krulich, L. and McCann, S. M. (1972). The effects of hypothalamic lesions on the release of gonadotrophins and prolactin in response to oestrogen and progesterone treatment in female rats, *Endocrinology*, **91,** 1704
11. Björklund, A., Falck, B., Hromek, F., Owman, C. and West, K. A. (1970). Identification and terminal distribution of the tubero–hypophyseal monoamine fibre systems in the rat by means of stereotaxic and microspectrofluorimetric techniques, *Brain Res.*, **17,** 1
12. Bogdanove, E. M. (1964). The role of the brain in the regulation of pituitary gonadotrophin secretion, *Vitam. Horm.*, **22,** 205
13. Brawer, J. R. (1972). The fine structure of the ependymal tanycytes at the level of the arcuate nucleus, *J. Comp. Neurol.*, **145,** 25
14. Bunn, J. P. and Everett, J. W. (1957). Ovulation in persistent-oestrous rats after electrical stimulation of the brain, *Proc, Soc. Exp. Biol. Med.*, **96,** 369
15. Campos-Ortega, J. A. and Clüver, P. F. de V. (1968). The distribution of retinal fibres in *Galago crassicaudatus*, *Brain Res.*, **7,** 487
16. Carrer, H. F. and Taleisnik, S. (1970). Effect of mesencephalic stimulation on the release of gonadotrophins. *J. Endocrinol.*, **48,** 527
17. Clemens, J. A., Sharr, C. J., Kleber, J. W. and Tandy, W. A. (1971). Areas of the brain stimulatory to LH and FSH secretion, *Endocrinology*, **88,** 180
18. Clemens, L. G., Shyrne, J. and Gorski, R. A. (1970). Androgen and development of progesterone responsiveness in male and female rats, *Physiol. Behav.*, **5,** 673
19. Corbin, A. (1966). Pituitary and plasma LH of ovariectomized rats with median eminence implants, *Endocrinology*, **78,** 893
20. Corbin, A., Milmore, J. E. and Daniels, E. L. (1970). Further evidence for the existence of a hypothalamic follicle stimulating hormone synthesising factor, *Experienta*, **26,** 1010
21. Cramer, O. M. and Barraclough, C. A. (1971). Effect of electrical stimulation of the preoptic area on plasma LH concentrations in proestrous rats, *Endocrinology*, **88,** 1175
22. Cross, B. A. (1964). Electrical recording techniques in the study of hypothalamic control of gonadotrophin secretion, *Int. Cong. Ser. No. 83*, 513 (Amsterdam: Excerpta Med. Found.)
23. Cross, B. A. and Dyer, R. G. (1972). Cyclic changes in neurons of the anterior hypothalamus during the rat oestrous cycle and the effect of anesthesia. *Steroid Hormones and Brain Function*, 95 (C. H. Sawyer and R. A. Gorski editors) (Los Angeles: Univ. Calif. Press)
24. David, M. A., Fraschini, F. and Martini, L. (1965). Parallélisme entre le contenu hypophysaire en FSH et le contenu hypothalamique en FSH-RF (FSH-releasing factor), *C. R. Acad. Sci. (Paris)*, **261,** 2249
25. Donovan, B. T. and van der Werff ten Bosch, J. J. (1956). Oestrus in winter following hypothalamic lesions in the ferret, *J. Physiol. (London)*, **132,** 57
26. Döcke, F. and Dörner, G. (1965). The mechanism of induction of ovulation by oestrogens, *J. Endocrinol.*, **33,** 491
27. Dörner, G. and Döcke, F. (1966). The influence of intrahypothalamic and intrahypophysial implantation of oestrogen or progestogen on gonadotrophin release, *Int. Cong. Ser. No. 111*, 194 (Amsterdam: Excerpta Med. Found.)
28. Eisenfeld, A. J. and Axelrod, J. (1965). Selectivity of oestrogen distribution in tissues, *J. Pharmacol. Exp. Theor.*, **150,** 469
29. Eleftheriou, B. E. and Zolovick, A. J. (1967). Effect of amygdaloid lesions on plasma and pituitary levels of luteinizing hormone, *J. Reprod. Fert.*, **14,** 33
30. Elwers, M. and Critchlow, V. (1960). Precocious ovarian stimulation following hypothalamic and amygdaloid lesions in rats, *Amer. J. Physiol.*, **198,** 381

31. Elwers, M. and Critchlow, V. (1966). Delayed puberty following electrical stimulation of amygdala in female rats, *Amer. J. Physiol.*, **211**, 1103
32. Everett, J. W. (1972). Brain, pituitary gland and the ovarian cycle, *Biol. Reprod.*, **6**, 3
33. Everett, J. W. and Sawyer, C. H. (1949). A neural timing factor in the mechanism by which progesterone advances ovulation in the cyclic rat, *Endocrinology*, **45**, 581
34. Everett, J. W. and Sawyer, C. H. (1950). A 24 h periodicity in the 'LH-release apparatus' of female rats, disclosed by barbiturate sedation, *Endocrinology*, **47**, 198
35. Everett, J. W., Tyrey, L. and Krey, L. C. (1972). Quantitative relation between preoptic electrochemical stimulation and LH release, *Int. Cong. Ser. No. 256*, 203 (Amsterdam: Excerpta Med. Found.)
36. Flerkó, B. (1956). Die Rolle hypothalamischer Strukturen bei der Hemmungswirkuing des erhöhten Östrogenblutspiegels auf die Gonadotrophinsekretion, *Acta Physiol. Acad. Sci. Hung.*, **9**, Suppl. 17
37. Flerkó, B. (1957). Le rôle des structures hypothalamiques dans l'action inhibitrice de la folliculine sur la sécrétion de l'hormone folliculo-stimulante, *Arch. Anat. Micr. Morph. exp.*, **46**, 159
38. Flerkó, B. (1962). Hypothalamic control of hypophyseal gonadotrophic function, *Hypothalamic Control of the Anterior Pituitary*, 1st ed., 192 (J. Szentágothai, B. Flerkó, B. Mess and B. Halász, editors) (Bupadest: Akadémiai Kiadó)
39. Flerkó, B. (1966). Control of gonadotrophin secretion in the female, *Neuroendocrinology*, 613 (L. Martini and W. F. Ganong, editors) (New York: Academic Press)
40. Flerkó, B. (1968). The experimental polyfollicular ovary, *Endocrinology and Human Behaviour*, 119 (R. P. Michael, editor) (London: Oxford University Press)
41. Flerkó, B. (1971). Steroid hormones and the differentiation of the central nervous system, *Current Topics Experimental Endocrinology*, Vol. 1, 59 (L. Martini and V. H. T. James, editors) (New York: Academic Press)
42. Flerkó, B. and Bárdos, V. (1960). Pituitary hypertrophy after anterior hypothalamic lesions, *Acta Endocrinol.*, **35**, 375
43. Flerkó, B. and Bárdos, V. (1961). Luteinization induced in 'constant oestrus rats' by lowering oestrogen production, *Acta Endocrinol.*, **37**, 418
44. Flerkó, B., Illei-Donhoffer, A. and Mess, B. (1971). Oestradiol-binding capacity in neural and non-neural target tissues of neonatally androgenized female rats, *Acta Biol. Acad. Sci. Hung.*, **22**, 125
45. Flerkó, B. and Mess, B. (1968). Reduced oestradiol-binding capacity of androgen-sterilized rats, *Acta Physiol. Acad. Sci. Hung.*, **33**, 111
46. Flerkó, B., Mess, B. and Illei-Donhoffer, A. (1969). On the mechanisms of androgen-sterilization, *Neuroendocrinology*, **4**, 164
47. Flerkó, B., Petrusz, P. and Tima, L. (1967). On the mechanism of sexual differentiation of the hypothalamus. Factors influencing the 'critical period' of the rat, *Acta Biol. Acad. Sci. Hung.*, **18**, 27
48. Flerkó, B. and Szentágothai, J. (1957). Oestrogen sensitive nervous structures in the hypothalamus, *Acta Endocrinol.*, **26**, 121
49. Fuxe, K. and Hökfelt, T. (1969). Catecholamines in the hypothalamus and the pituitary gland, *Frontiers in Neuroendocrinology, 1969*, Vol. 1, 47 (W. F. Ganong and L. Martini, editors) (London: Oxford University Press)
50. Ganong, W. F., Shepherd, M. D., Wall, J. R., von Brunt, E. E. and Clegg, M. T. (1963). Penetration of light into the brain of mammals, *Endocrinology*, **72**, 962
51. Gorski, R. A. and Wagner, J. W. (1965). Gonadal activity and sexual differentiation of the hypothalamus, *Endocrinology*, **76**, 226
52. Green, J. D. and Harris, G. W. (1947). The neurovascular link between the neurohypophysis and adenohypophysis, *J. Endocrinol.*, **5**, 136
53. Green, R., Luttge, W. G. and Whalen, R. E. (1969). Uptake and retention of tritiated estradiol in brain and peripheral tissues of male, female and neonatally androgenized female rats, *Endocrinology*, **85**, 373
54. Greep, R. O. (1961). Physiology of the anterior hypophysis in relation to reproduction. *Sex and Internal Secretions*, 3rd ed., Vol. 1, 240 (W. C. Young, editor) (Baltimore: William and Wilkins)
55. Hagino, N. and Yamaoka, S. (1972). Neural interaction of ACTH–adrenal rhythm and ovulatory cycles in the rat, *Int. Cong. Ser. No. 256*, 2 (Amsterdam: Excerpta Med. Found.)

56. Halász, B. (1969). The endocrine effects of isolation of the hypothalamus from the rest of the brain, *Frontiers in Neuroendocrinology, 1969*, 307 (W. F. Ganong and L. Martini, editors) (London: Oxford University Press)
57. Halász, B. and Gorski, R. A. (1967). Gonadotrophic hormone secretion in female rats after partial or total interruption of neural afferents to the medial basal hypothalamus, *Endocrinology*, **80,** 608
58. Halász, B., Pupp, L. and Uhlarik, S. (1962). Hypophysiotrophic area in the hypothalamus, *J. Endocrinol.*, **25,** 147
59. Harris, G. W. and Levine, S. (1965). Sexual differentiation of the brain and its experimental control, *J. Physiol.* (*London*), **181,** 379
60. Hilliard, J., Schally, A. V. and Sawyer, C. H. (1971). Progesterone blockade of the ovulatory response to intrapituitary infusion of LH–RH in rabbits, *Endocrinology*, **88,** 730
61. Hohlweg, W. and Junkmann, K. (1932). Die hormonalnervöse Regulierung der Funktion des Hypophysenvorderlappens, *Klin. Wochensch.*, **11,** 321
62. Hökfelt, T. and Fuxe, K. (1972). On the morphology and the neuroendocrine role of the hypothalamic catecholamine neurons, *Brain–Endocrine Interaction. Median Eminence: Structure and Function. Int. Symp. Munich 1971*, 181 (Basel: Karger)
63. Hyyppä, M., Motta, M. and Martini, L. (1971). 'Ultrashort' feedback control of follicle-stimulating hormone-releasing factor secretion, *Neuroendocrinology*, **7,** 227
64. Igarashi, M. and McCann, S. M. (1964). A hypothalamic follicle-stimulating hormone-releasing factor, *Endocrinology*, **74,** 446
65. Illei-Donhoffer, A., Tima, L. and Flerkó, B. (1970). Ovulation in persistent oestrous rats, *Acta Biol. Acad. Sci. Hung.*, **21,** 197
66. Jensen, E. V., Desombres, E. R., Hurst, D. J., Kawashima, T. and Jungblut, P. W. (1967). Oestrogen-receptor interactions in target tissues, *Arch. Anat. Microsc. Morph. Exp.*, **56 Suppl. No. 3–4,** 547
67. Kaasjager, W. A., Woodbury, D. M., van Dieten, J. A. M. J. and van Rees, G. P. (1971). The role played by the preoptic region and the hypothalamus in spontaneous ovulation and ovulation induced by progesterone, *Neuroendocrinology*, **7,** 54
68. Kalra, S. P., Ajika, K., Krulich, L., Fawcett, C. P., Quijada, M. and McCann, S. M. (1971). Effects of hypothalamic and preoptic electrochemical stimulation on gonadotrophin and prolactin release in pro-oestrous rats, *Endocrinology*, **88,** 1150
69. Kalra, S. P. and McCann, S. M. (1972). Modification of brain catecholamine levels and LH release by preoptic stimulation, *Int. Cong. Ser. No. 256*, 202 (Amsterdam: Excerpta Med. Found.)
70. Kalra, S. P., Velasco, M. E. and Sawyer, C. H. (1970). Influences of hypothalamic deafferentation on pituitary FSH release and oestrogen feedback in immature parabiotic rats, *Neuroendocrinology*, **6,** 228
71. Kato, J. and Villee, C. A. (1967). Preferential uptake of oestradiol by the anterior hypothalamus of the rat, *Endocrinology*, **80,** 567
72. Kawakami, M. and Sawyer, C. H. (1959). Induction of behavioural and electroencephalographic changes in the rabbit by hormone administration or brain stimulation, *Endocrinology*, **65,** 631
73. Kawakami, M., Seto, K. and Yoshida, K. (1966). Influence of the limbic system on ovulation and on progesterone and oestrogen formation in rabbit's ovary, *Jap. J. Physiol.*, **16,** 254
74. Kawakami, M., Terasawa, E., Seto, K. and Wakabayashi, K. (1971). Effect of electrical stimulation of the medial preoptic area on hypothalamic multiple unit activity in relation to LH release, *Endocrinol. Jap.* **18,** 13
75. King, R. J. B. (1967). Fixation of steroids to receptors, *Arch. Anat. Microsc. Morph. Exp.*, **56 Suppl. No. 3–4,** 570
76. Kobayashi, T., Kobayashi, T., Kigawa, T., Mizuno, M. and Amenomori, Y. (1963). Influence of rat hypothalamic extract on gonadotrophic activity of cultivated anterior pituitary cells, *Endocrinol. Jap.*, **10,** 16
77. Koikegami, H., Yamada, T. and Usei, K. (1954). Stimulation of the amygdaloid nuclei and periamygdaloid cortex with special reference to its effects on uterine movements and ovulation, *Folia Psychiat. Neurol. Jap.*, **8,** 7
78. Kordon, C. (1967). Contrôle nerveux du cycle ovarien, *Arch. Anat. Microsc. Morph. Exp.*, **56 Suppl. 3–4,** 458

79. Kordon, C. and Glowinski, J. (1969). Selective inhibition of superovulation by blockade of dopamine synthesis during the 'critical period' in the immature rat, *Endocrinology*, **85,** 924
80. Kordon, C., Javoy, F., Vassent, G. and Glowinski, J. (1968). Blockade of superovulation in the immature rat by increased brain serotonin, *Europ. J. Pharmacol.*, **4,** 169
81. Köves, K. and Halász, B. (1969). Data on the location of the neural structures indispensable for the occurrence of ovarian compensatory hypertrophy, *Neuroendocrinology* **4,** 1
82. Köves, K. and Halász, B. (1970). Location of the neural structures triggering ovulation in the rat, *Neuroendocrinology*, **6,** 180
82a. Ladosky, W. and Gaziri, L. C. J. (1970). Brain serotonin and sexual differentiation of the nervous system, *Neuroendocrinology*, **6,** 168
83. Lisk, R. D. (1960). Oestrogen-sensitive centres in the hypothalamus of the rat, *J. Exp. Zool.*, **145,** 197
84. Lisk, R. D. (1962). Testosterone-sensitive centres in the hypothalamus of the rat, *Acta Endocrinol.*, **41,** 195
85. Matsuo, H., Baba, Y., Nair, R. M. G., Arimura, A. and Schally, A. V. (1971). Structure of the porcine LH- and FSH-releasing hormone. I. The proposed amino-acid sequence, *Biochem. Biophys. Res. Commun.*, **43,** 1334
86. Mayerson, B. J. and Sawyer, C. H. (1968). Monoamines and ovulation in the rat, *Endocrinology*, **83,** 170
87. McCann, S. M., Kalra, P. S., Donoso, A. O., Bishop, W., Schneider, H. P. G., Fawcett, C. P. and Krulich, L. (1972). The role of monoamines in the control of gonadotropin and prolactin secretion, *Brain–Endocrine Interaction. Median Eminence: Structure and Function. Int. Symp. Munich 1971*, 224 (Basel: Karger)
88. McCann, S. M., Taleisnik, S. and Friedman, H. M. (1960). LH-releasing activity in hypothalamic extracts, *Proc. Soc. Exp. Biol. Med.*, **104,** 432
89. McEwen, B. S. and Pfaff, D. W. (1970). Factors influencing sex hormone uptake by rat brain region. I. Effects of neonatal treatment, hypophysectomy and competing steroid on oestradiol uptake, *Brain Res.*, **21,** 1
90. McEwen, B. S., Zigmond, R. E. and Gerlach, J. L. (1972). Sites of steroid binding and action in the brain, *Structure and Function of Nervous Tissue*, Vol. 5, 205 (New York: Academic Press)
91. McGuire, J. L. and Lisk, R. D. (1969). Oestrogen receptors in androgen- or oestrogen-sterilized female rats, *Nature (London)*, **221,** 1068
92. Mess, B., Fraschini, F., Piva, F. and Martini, L. (1966). The pineal body and the control of LH secretion, *Int. Cong. Ser. No. 111*, 361 (E. B. Romanoff and L. Martini, editors) (Amsterdam: Excerpta Med. Found.)
93. Mess, B., Heizer, A., Tóth, A. and Tima, L. (1971). Luteinization induced by pinealectomy in the polyfollicular ovaries of rats bearing anterior hypothalamic lesions, *Ciba Foundation Symposium on the Pineal Gland*, 229 (G. E. W. Wolstenholme and J. Knight, editors) (London: Churchill)
94. Mess. B. and Martini, L. (1968). The central nervous system and the secretion of anterior pituitary trophic hormones, *Recent Advances in Endocrinology*, 8th ed., 1 (V. H. T. James, editor) (London: Churchill)
95. Michael, R. P. (1962). Oestrogen-sensitive neurons and sexual behaviour in female cats, *Science*, **136,** 322
96. Moore, R. Y., Karapas, F. and Lenn, N. J. (1971). A retinohypothalamic projection
97. Moszkowszka, A. (1966). New data concerning pineal-pituitary relationships and
97. Moszkowszka, A. (1966). Mew data concerning pineal–pituitary relationships and pituitary gonadotrophic function, *Int. Cong. Ser. No. 111*, 361, (E. B. Romanoff and L. Martini, editors) (Amsterdam: Excerpta Med. Found.)
98. Naftolin, F., Ryan, K. J. and Petro, Z. (1972). Aromatisation of androstenedione by the anterior hypothalamus of adult male and female rats, *Endocrinology*, **90,** 295
99. Nauta, W. J. H. and van Straaten, J. J. (1947). The primary optic centres of the rat. An experimental study by the 'bouton' method, *J. Anat.*, **81,** 127
100. Nequin, L. G. and Schwartz, N. B. (1971). Adrenal participation in the timing of mating and LH release in the cyclic rat, *Endocrinology*, **88,** 325
101. Nikitovich-Winer, M. B., Evans, J. E. and Kiracofe, G. H. (1966). Hypophysiotrophic effect of hypothalamic humoral factors, *Int. Cong. Ser. No. 111*, 95 (Amsterdam: Excerpta Med. Found.)

102. Ojeda, R. and Ramirez, V. D. (1968). Automatic control of LH and FSH by short-circuits in the immature rats, *Int. Cong. Ser. No. 157*, 113 (Amsterdam: Excerpta Med. Found.)
103. O'Steen, W. K. and Vaughan, G. M. (1968). Radioactivity in the optic pathway and hypothalamus of the rat after intraocular injection of tritiated 5-hydroxytryptophan, *Brain Res.*, **8,** 209
104. Parkes, A. S. and Bruce, H. M. (1961). Olfactory stimuli in mammalian reproduction, *Science*, **134,** 1049
105. Petrusz, P. and Flerkó, B. (1965). On the mechanism of sexual differentiation of the hypothalamus, *Acta Biol. Acad. Sci. Hung.*, **16,** 169
106. Petrusz, P. and Nagy, É. (1967). On the mechanism of sexual differentiation of the hypothalamus; decreased hypothalamic oestrogen sensitivity in androgen-sterilized female rats, *Acta Biol. Acad. Sci. Hung.*, **18,** 21
107. Pfaff, D. W. (1968). Autoradiographic localization of radioactivity in rat brain after injection of tritiated sex hormones, *Science*, **161,** 1355
108. Pfaff, D. W. and Pfaffmann, C. (1969). Olfactory and hormonal influences on the basal forebrain of male rat, *Brain Res.*, **15,** 137
109. Porter, J. C., Kamberi, J. A. and Mical, R. S. (1971). The neurovascular link of the hypothalamic–hypophysial system and the role of monoamines in the control of gonadotrophin release, *Int. Cong. Ser. No. 238*, 331 (Amsterdam: Excerpta Med. Found.)
110. Raisman, G. and Field, P. M. (1971). Sexual dimorphism in the preoptic area of the rat, *Science*, **173,** 731
111. Ramirez, V. D. and Sawyer, C. H. (1965). Fluctuations in hypothalamic LH–RF (luteinizing hormone-releasing factor) during the rat oestrous cycle, *Endocrinology*, **76,** 82
112. Reiter, R. J., Sorrentino, S., Jr. and Hoffman, R. A. (1970). Early photoperiodic conditions and pineal antigonadal function in male hamsters, *Int. J. Fert.*, **15,** 163
113. Réthelyi, M. and Halász, B. (1970). Origin of the nerve endings in the surface zone of the median eminence of the rat hypothalamus, *Exp. Brain. Res.*, **11,** 145
114. Rose, S. and Nelson, J. F. (1957). The direct effect of oestradiol on the pars distalis, *Austr. J. Exp. Biol. Med. Sci.*, **35,** 605
115. Rubinstein, L. and Sawyer, C. H. (1970). Role of catecholamines in stimulating the release of pituitary ovulating hormones in rats, *Endocrinology*, **86,** 988
116. Sawyer, C. H. (1955). Rhinencephalic involvement in pituitary activation by intraventricular histamine in the rabbit under Nembutal anesthesia, *Amer. J. Physiol.*, **180,** 37
117. Sawyer, C. H., Everett, J. W. and Markee, J. E. (1949). A neural factor in the mechanism by which oestrogen induced the release of luteinizing hormone, *Endocrinology*, **44,** 218
118. Sawyer, C. H., Markee, J. E. and Hollinshead, W. H. (1947). Inhibition of ovulation in the rabbit by the adrenergic-blocking agent dibenamine, *Endocrinology*, **41,** 395
119. Schally, A. V., Bowers, C. Y., White, W. F. and Cohen, A. I. (1967). Purification and *in vivo* and *in vitro* studies with porcine luteinizing hormone-releasing factor (LRF), *Endocrinology*, **81,** 77
120. Schwartz, N. B. and McCormack, C. E. (1972). Reproduction: gonadal function and its regulation, *An. Rev. Physiol.*, **34,** 425
121. Scott, J. W. and Pfaff, D. W. (1970). Behavioural and electrophysiological responses of female mice to male urine odours, *Physiol. Behav.*, **5,** 407
122. Smith, S. W. and Lawton, I. E. (1972). Involvement of the amygdala in the ovarian compensatory hypertrophy response, *Neuroendocrinology*, **9,** 228
123. Sorrentino, S., Jr. and Benson, B. (1970). Effect of blinding and pinealectomy on the reproductive organs of adult male and female rats, *Gen. Comp. Endocrinol.*, **15,** 242
124. Sorrentino, S., Jr. and Reiter, R. J. (1970). Pineal-induced alteration of estrous cycles in blinded hamsters, *Gen. Comp. Endocrinol.*, **15,** 39
125. Sousa-Pinto, A. and Castro-Correia, J. (1970). Light microscopic observations on the possible retinohypothalamic projection in the rat, *Exp. Brain. Res.*, **11,** 515
126. Stumpf, W. E. (1970). Estrogen-neurons and estrogen-neuron systems in the periventricular brain, *Amer. J. Anat.*, **129,** 207

127. Szentágothai, J., Flerkó, B., Mess, B. and Halász, B. (1968). *Hypothalamic Control of the Anterior Pituitary*, 3rd (revised, enlarged) edition (Budapest: Akadémiai Kiadó)
128. Terasawa, E. and Sawyer, C. H. (1969). Changes in electrical activity in the rat hypothalamus related to electrochemical stimulation of adenohypophyseal function, *Endocrinology*, **85,** 143
129. Tima, L. (1971). On the site of production of releasing factors, *Mem. Soc. Endocr., No. 19*, 895 (H. Heller and K. Lederis, editors) (London: Cambridge University Press)
130. Tima, L. and Flerkó, B. (1968). Ovulation induced by autologous pituitary extracts in androgen- and light-sterilized rats, *Arch. Anat. Hist. Embryol. Norm. Exp.*, **51,** 669
131. Turgeon, J. L. and Barraclough, C. A. (1972). Plasma LH after graded POA electrochemical stimulation, *Int. Cong. Ser. No. 256*, 203 (Amsterdam: Excerpta Med. Found.)
132. Velasco, M. E. and Taleisnik, S. (1969). Effect of hippocampal stimulation on the release of gonadotrophin, *Endocrinology*, **85,** 1154
133. Velasco, M. E. and Taleisnik, S. (1971). Effects of interruption of amygdaloid and hippocampal afferents to the medial hypothalamus on gonadotrophin release, *J. Endocrinol.*, **51,** 41
134. Wakabayashi, K. and McCann, S. M. (1970). *In vitro* responses of anterior pituitary glands from normal, castrated and androgen-treated male rats to LH-releasing factor (LRF) and high potassium medium, *Endocrinology*, **87,** 771
135. Weick, R. F., Smith, E. R., Dominguez, R., Dhariwal, A. P. S. and Davidson, J. M. (1971). Mechanism of stimulatory feedback effect of estradiol benzoate on the pituitary, *Endocrinology*, **88,** 293
136. Weiner, R. I., Gorski, R. A. and Sawyer, C. H. (1972). Hypothalamic catecholamines and pituitary gonadotropic function. *Brain–Endocrine Interaction. Median Eminence: Structure and Function. Int. Symp. Munich 1971*, 236 (Basel: Karger)
137. Whitten, W. K. (1956). The effect of removal of the olfactory bulbs on the gonads of mice, *J. Endocrinol.*, **14,** 160
138. Wurtman, R. J., Axelrod, J. and Phillips, L. L. (1963). Melatonin synthesis in the pineal gland: control by light, *Science*, **142,** 1071
139. Yazaki, J. (1960). Further studies on endocrine activity of subcutaneous ovarian grafts in male rats by daily examination of smears from vaginal grafts, *Annot. Zool. Jap.*, **33,** 217

2
Hormone Antibodies— An Appraisal of their Use in Reproductive Endocrinology

N. R. MOUDGAL
Indian Institute of Science, Bangalore

2.1 INTRODUCTION

The use of antagonists and inhibitors in studies on the biochemical action of hormones, though of recent origin, has provided us with information of considerable significance. Improvement in the methods for the production and characterisation of antisera permits us to add, to this growing list, the antibodies to individual hormones. It is to be hoped that a wider application of this novel tool will provide us with a clearer understanding of the mechanism of action of hormones.

Gonadotrophin and steroid hormone antibodies of differing specificity are being used lately in studying the physiological role of individual hormones. Reproductive physiology and endocrinology abound in several instances where the precise role of a hormone in a specific event like follicle maturation, ovulation and corpus luteum function, implantation, spermatogenesis etc. is ambiguous and as such needs clarification. It is the purpose of this review to analyse the recent attempts made by several investigators, using immunological methods, towards a better understanding of the physiological role of the hormones involved in regulating mammalian reproduction. The earlier studies have not been covered thoroughly, except wherever necessary, and the reader is referred, for more information on these to reviews by Li *et al.*[1], Hayashida[2] and Midgley[3]. This presentation precludes the consideration of radioimmunoassay of gonadotropic[4] and steroid hormones as well as the use of sperm and reproductive tissue-specific antibodies in studies on reproductive physiology since these have been the subject of periodic reviews[5]. One-third of the present review is concerned with the methodology used in the preparation and analysis of antisera for specificity, the remainder discusses in detail the advances made in reproductive endocrinology using the immunological approach.

2.2 PRODUCTION OF ANTIBODIES TO HORMONES

2.2.1 Protein hormones as antigens

Gonadotropic hormones by virtue of their size and carbohydrate content have proved to be good antigens. Though the antigenicity of these hormone proteins was realised as early as 1936 by Parkes and Rowlands[6], definitive work had to await the availability of purified preparations and the development of better immunisation techniques. Methods used to raise antibodies to these hormones have varied from group to group but generally it is now an established practice to administer the hormones to rabbits, over a period of several weeks, as emulsions of saline with Freunds complete adjuvant[7]. Test bleeds taken after several immunisation injections are checked for antibody (Ab) titre by the qualitative precipitin reaction as described by Kabat and Meyer[7]. A satisfactory level of immunisation is considered to have been achieved if the qualitative precipitin test is positive using an antiserum dilution of 1:32. A booster injection of hormone in saline is then given, followed by ear-vein bleeding 7–10 days later[8]. In our opinion it is practical to keep those rabbits which exhibit a very high antibody titre as active antibody producers for a period of 12–18 months by exposing them periodically to booster injections of antigen.

It is needless to stress here that it is ideal to use well characterised, homogeneous hormone proteins as antigens as these will give rise to (unlike the impure hormone preparations) a minimum of antibodies to contaminants[1]. Normally to raise a sufficient quantity of antiserum of good titre in rabbits, 5–15 mg of protein/rabbit is required[1]. Vaitukaitis *et al.*[9], however, have recently described a method to produce antisera of good titre in rabbits by giving as little as 20–100 μg of the immunogen. This method has been shown to give specific antisera useful for radioimmunoassay. It is, however, not known whether it produces sufficiently high antibody titre to be of use in physiological neutralisation experiments.

2.2.2 Antigenicity of steroid–protein conjugates

Unequivocal proof for the antigenic nature of steroid molecules, when used as steroid–protein conjugates, was furnished only in the late 1950s by Leberman and co-workers[10] in the United States and by Goodfriend and Sehon[11] in Canada. These attempts showed that the antibody obtained was of low titre and with the exception of oestrogen, the antibodies to other steroids were to a great extent, non-specific. While the Canadain group confined itself to the production of oestrogen and testosterone antibodies, the Columbia group developed methodology for coupling a broad spectrum of steroids to proteins. Except in a few cases where position 2 or 3 was used, most investigators have preferred to use position 17 for coupling[11]. The steroid–protein conjugate was prepared using classical methods like the mixed anhydride, carbamide and Schotten–Bauman reaction, where a reactive functional group of steroid is linked to the ε-amino group of the lysine residues in the protein. The high degree of specificity exhibited by the oestrogen antibodies is apparently due to the aromatic A ring of this C_{18} steroid. The reason for observing a much less specificity among antibodies raised to C_{19} and C_{21} steroids is probably, as suggested by Gross[12], due to immunological specificity being more dependent on the particular carbonyl or phenol group occupied rather than on the whole steroid molecule. Gross[12] has proposed a method of preparing an oestradiol protein conjugate, wherein the polar functional groups at positions 3 and 17 of the steroid are kept intact. The method consists in preparaing 2 or 4 oestrogen-azobenzoic acid and coupling it to the protein via the carbodi-imide reaction. The antibody raised to such a molecule cross-reacts with all the three oestrogens equally well. This simple method has been tested in our laboratory and found to give satisfactory results.

Recently progesterone derivatives have also been prepared using the OH group at position C_{11} instead of C_{17} as the reactive functional group[13,14]. By this method, the entire steroid nucleus and the functional groups, being left intact, are expected to confer a greater degree of specificity on the antibody formed. Any antiserum to a steroid, prior to its use in physiological experiments, will have to be subjected to intensive characterisation, particularly in regard to its ability to inhibit the action of other closely related steroids. Hitherto oestrogen antiserum was the only one that had been used in physiological neutralisation experiments without fear of obtaining equivocal results[15,16]. From the characterisation studies[13,14], it appears that the antiserum to the 11 α-hemisuccinyl progesterone could also be used to specifically neutralise endogenous progesterone activity.

2.3 CHARACTERISATION OF HORMONE ANTISERA FOR SPECIFICITY

2.3.1 Absorption of antibodies to non-specific protein contaminants

It is generally observed that notwithstanding the use of 'pure' hormone protein as antigens, the antisera tend to have in addition to specific antibodies to the hormone, antibodies to contaminants like homologous serum and tissue proteins[1]. It appears to be difficult to *completely* rid hormone preparations of contaminating homologous tissue proteins, the physico-chemical methods not being able to detect them if these are present at a concentration of 0.1% or less[1]. Even at this level of contamination, however, the non-specific proteins seem to elicit an antibody response. Detection of these non-specific antibodies can mostly be done by the Ouchterlony agar–gel double-diffusion test[8]. They can be removed, by repeated addition in small quantities, of either purified tissue and serum proteins or crude tissue and serum protein extracts. Completion of absorption can be ascertained by the absence of precipitin reaction in the Ouchterlony test. The identification and removal by absorption of antibodies to contaminants is best illustrated by quoting the example of antisera to NIH–FSH–S series. In Figure 2.1 is shown how a crude antiserum can be cleaned by the step-wise removal of non-specific antibodies[17]. The absorption technique described above usually ensures the removal of precipitating, agglutinating and complement-fixing antibodies but not those of the soluble type.

2.3.2 Detection of antibodies to specific hormone contaminants and their removal

Since the pituitary is the source of several protein and polypeptide hormones, it is essential to make sure that whenever these hormones are used as antigens, the corresponding antisera do not contain antibodies to other pituitary hormone contaminants. As pointed out earlier, the presence of contaminating

Figure 2.1 Agar-gel double-diffusion pattern showing the step-wise purification of antiserum (A/S) to ovine NIH–FSH–S4. Wells 3, 6, 9 and 12 contained antiserum 'A', 'B', 'C' and 'D' respectively. Wells 1, 7 and 13 contained NIH–FSH–S4. Wells 4 and 10 contained 3.5 M ammonium sulphate fraction of 0.3 M KCl extract of sheep placenta. Wells 2, 8 and 14 contained ovine LH. Wells 5 and 11 contained normal sheep serum. All antigens used were of 2 mg/ml concentration. The photograph was taken 72 h after development.

A/S 'A'–unabsorbed FSH antiserum
A/S 'B' NSS absorbed FSH antiserum
A/S 'C'–NSS + organ extract absorbed FSH antiserum
A/S 'D'–NSS + organ extract + LH absorbed FSH antiserum

Note that the A/S devoid of antibodies to contaminants did not give any precipitin line against FSH. The antibody titre towards FSH, however, did not change due to the serial absorption technique followed here. (After Jagannadha Rao and Moudgal, see Ref. 17.)

hormones in concentrations of 0.1% or less is sufficient to cause an antibody response. An example of this is the presence of LH antibodies in antisera to NIH–FSH–S series[17]. If the antibody were to be of the precipitating type, it could easily be detected by the Ouchterlony test and removed by the simple serial absorption technique (Figure 2.1). In this method, in order to ensure total removal of all contaminating antibodies, a slight excess of the contaminating hormone protein is generally added. This can be avoided if a preliminary quantitative precipitin test[7] is carried out with the contaminant. For instance, from Figure 2.2 it can be seen that addition of LH at equivalence (35 μg

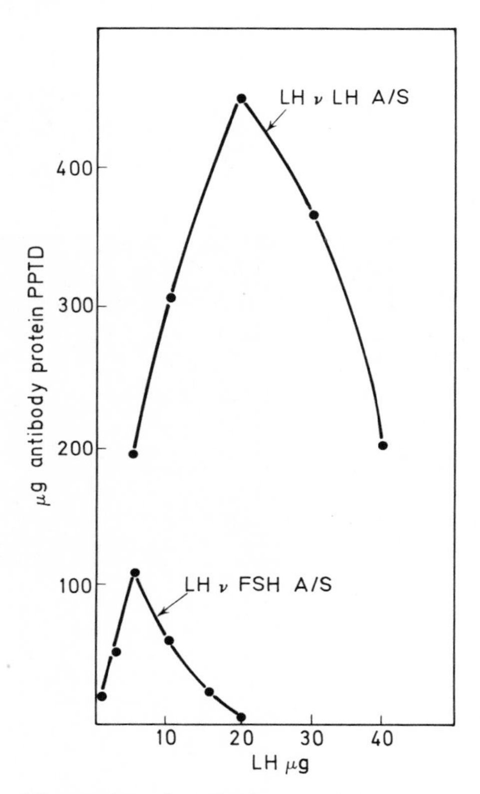

Figure 2.2 Quantitative precipitin reaction of ovine LH with rabbit antiserum (A/S) to ovine LH and ovine FSH.
Different amounts of LH were incubated with 50 μl of LH A/S or 200 μl of FSH A/S, and the antibody precipitate processed according to the methods described earlier[8]. (After Jagannadha Rao, Madhwa Raj and Moudgal, see Ref. 19.)

of LH per ml of FSH antiserum) to FSH antiserum should remove all precipitating type of antibodies to this hormone.

In addition to the above, the presence of soluble antibodies to contaminants can be ascertained by using highly purified ^{125}I-labelled samples of protein hormones. Incubation for 24 h at 37 °C of 10 μl or more of an undiluted rabbit antiserum with ^{125}I-labelled hormone protein, addition of sufficient quantity of antiserum to rabbit γ-globulin followed by incubation, centrifugation and gamma spectrometry will furnish us with information on the presence of soluble antibodies to the suspected contaminants[18]. Generally it has been our experience (Table 2.1) that while LH antiserum prepared

Table 2.1 Detection of antibodies to hormone contaminants in the antiserum preparations

10 µl of undiluted antiserum is incubated for 24 h at 37 °C with [^{125}I] labelled purified hormone antigen–antibody complex precipitated by the addition of goat antiserum to rabbit γ-globulin and the radioactivity in the precipitate quantitated by gamma spectrometry. The labelled hormones used were all of ovine origin.

Group	*Antiserum to*	*[^{125}I] Labelled hormone*	*% Net binding*
I	Ovine LH	LH	83.0
II	Ovine LH	FSH	2.0
III	Ovine LH	Prolactin	2.0
IV	Ovine FSH (unabsorbed)	FSH	51.0
V	Ovine FSH (absorbed free of LH antibodies)	FSH	42.0
VI	Ovine FSH (unabsorbed)	LH	58.0
VII	Ovine FSH (absorbed free of LH antibodies)	LH	6.0

Based on data from Madhwa Raj and Moudgal[18] and Jagannadha Rao *et al.*[48].

in this laboratory is free of antibodies to highly purified FSH, the FSH antiserum is not necessarily free of LH antibodies. It is possible to remove soluble antibodies to contaminating hormones by using an insoluble form of the contaminant. This has been achieved by Jagannadha Rao, Muralidhar and Moudgal (unpublished observations) by using either insoluble hormone polymers or hormone-coated tanned sheep red blood cells. The antisera to steroid–protein conjugates can be absorbed free of antibodies to carrier protein by resorting either to the serial absorption technique or to the use of insolubilised carrier protein preparations.

2.3.3 Establishing specificity of antisera by biological means

In addition to the above immunochemical tests, it is necessary to cross-check the specificity of the antiserum in biological systems. For example, the absence of antibodies to FSH and prolactin in the LH antiserum used, could be ascertained by including in the test system a group(s) which receives in addition to the antiserum a large excess of the suspected hormone contaminant, non-reversal of the antiserum effect demonstrating the lack of contaminating antibodies. Examples of such tests for specificity are generally included in experimental investigations carried out in this laboratory[18, 19].

2.3.4 Determination of antibody titre

Different groups of workers have used a variety of methods to estimate the antibody titre of the antiserum. These could be the haemagglutination test[7], the quantitative precipitin test[7], and the radio-active labelled hormone binding test[18]. We prefer to use the latter two since they make use of undiluted antiserum (used later for hormone neutralisation) and as such furnish an objective assessment of the antibody content. It is our experience that to obtain effective biological neutralisation of a hormonal activity, depending upon the expected hormone concentration in the blood, 10–20% more than the optimal amount of antibody should be used.

2.3.5 Analysis of cross-reactivity of antisera

It is now well established that antisera to gonadotrophins are essentially species non-specific[8, 17]. The degree of cross-reactivity between an antiserum and the test hormone of other species, however, could be vastly different, and this could also vary from batch to batch of antiserum (Table 2.2.) This information is valuable, since, based on this, the amount of antiserum to be given to neutralise a specific hormone in a test species can be computed. Cross-reactivity of antisera is of great advantage as it permits use of rigorously characterised antisera to a highly purified hormone preparation of one species (like an antiserum to sheep LH or FSH) for neutralisation of specific hormone activity in other species like rats, mice, hamsters, guinea-pigs, rabbits etc.

Table 2.2 Cross-reactivity of ovine LH antiserum with heterologous LH preparations

Source of pituitary extract	*Antiserum batch*	*% Cross-reactivity**
1 Rat	2–I	53.0
2 Rat	23–I	27.0
3 Rat	23–IV	37.0
4 Hamster	23–XI	40.0
5 Hamster	23–XIV	33.0

* % Cross-reactivity was assessed by carrying out quantitative precipitin test according to Kabat and Mayer[7]. 0.01 M phosphate buffered saline (pH 7.0) extract of pituitaries was used as source of heterologous LH preparations (based on data from Madhwa Raj and Moudgal[18] and Jagannadha Rao *et al.*[19])

2.4 USE OF ANTIBODIES AS INHIBITORS OF HORMONE ACTION

2.4.1 Analysis of active *v.* passive immunisation as a method of inhibiting hormone activity

Bourdel and Li[20], and Hayashida[21] showed that the gonadal function of the female and male rat could be effectively inhibited by injecting them daily with relatively large volumes of antisera to purified LH. Since antibodies produced to LH and FSH are species non-specific, the animals used for immunisation can be expected to exhibit, as a consequence of neutralisation of endogenous hormonal activity, an impairment in the development of gonads. Using this technique Laurence and co-workers[22-24] have tried to evaluate the relative roles of LH and FSH in gonadal development in both male and female rats and rabbits. Following immunisation with LH or FSH a variety of reproductive functions like oestrous cycle, implantation in the female[22,23] and spermatogenesis in the male[24] have been shown to be affected. Generally immunisation with LH appeared to produce much greater damage than that with FSH.

As was pointed out earlier, depending upon the homogeneity or purity of a protein hormone, immunisation of an animal with it leads to the production of less antibodies against contaminants. If, however, the contaminant happens to be a hormone of good antigenicity (e.g. LH in FSH preparations) it will be difficult to arrive, from active immunisation studies, at unequivocal conclusions as to the specific role of a particular gonadotrophin in a physiological process. Further, it may not be possible to exclude that the overall effect observed in the immunised animal (e.g. impairment in gonadal development) is not influenced by other non-specific immune reactions taking place due to antibodies to the contaminant present in the original antigen.

In contrast to the above, passive immunisation, as a method of inhibiting a specific hormone action, is vastly superior. In this, by using antisera thoroughly characterised for their specificity, neutralisation of a single hormone at a time can be achieved. The specificity of the antiserum effect can be further enhanced by using antisera raised to heterologous instead of a homologous hormone. Thus the use of ovine LH antisera for neutralising endogenous rat or monkey LH has the advantage that the chances of antibodies to non-specific tissue proteins contributing to the overall effect are minimal. Further, the use of minimal effective doses of the antisera permits the study of effects of short-term deprivation of hormone support. Characterised antisera in very small amounts have also been used in biochemical experiments such as elucidation of the biosynthesis of gonadotrophins[25], location of tissue bound to gonadotrophins by tagged or labelled antibody technique[26,27] and in studies on the mechanism of LH and FSH stimulation of steroidogenesis using tissue slices[28] and single cell suspensions[29,30].

2.4.2 Duration of antibody effect

In the passive immunisation experiments, the duration for which the antibody activity lasts in the body depends upon both the antibody titre of the serum and on the half-life of the γ-globulin. According to Kabat and Meyer[7], the half-life of γ-globulin varies from 2 days in mouse to 11 days in man. Bindon[31] reported that PMS neutralising activity in mice could be detected weeks after an initial injection of anti-PMS serum. Further, it has been observed by us that when large amounts (0.4 ml or more) of a high titre LH antiserum is given to rats, the antibody activity in the serum at the end of 24 h remains virtually unaltered[32]. We have observed at the same time that when minimal effective doses of antisera are given, the animals recover from the initial antibody treatment very soon and are able to once again respond to the endogenous gonadotrophin stimulus[33]. Schwartz and co-workers[34] have made similar observations while studying the effect of LH antiserum on the oestrus cycle of rats. It thus appears that by using high titre antibody and regulating the amount of antiserum injected, it is possible to obtain short-term hormone neutralisation.

2.5 USE OF GONADOTROPHIN ANTIBODIES IN DELINEATING THE RELATIVE ROLE OF LH AND FSH IN OVULATION

The active immunisation studies of Laurence and Ichikawa[22] and Talaat and Laurence[23] showed that rabbits and rats immunised with either LH or FSH exhibited anoestrus cycles and ovulation blockade. As pointed out earlier, active immunisation studies do not permit us to draw unequivocal conclusions as to the direct involvement of one hormone or the other in a specific physiological event.

LH antisera subjected to varying degrees of characterisation have been used by several investigators in studying the ovulation process in rabbits[35], mice[36, 37], rats[20, 38-40], hamsters[41] and pigs[42]. Schwartz and Gold[40] observed that administration of a single injection of ovine LH antiserum at 1 p.m. of pro-oestrus resulted, in addition to the blockade of ovulation normally seen the next day, in the blockade of ovulation for the next 4–7 days. This was explained as due to the ability of LH antiserum to neutralise both surge and tonic levels of LH, the latter being necessary for the maintenance of oestrogen secretion by the follicle. Ovulation blockade continued until all the antibody was catabolised, the half-life of rabbit γ-globulin in rat being 10 days. In contrast to this when hypothalamic suppressor drugs like pentobarbital were used, since they affect only the release of pituitary LH in surge form, the follicular oestrogen, dependent upon tonic LH, was able to bring about a fresh release of an ovulating level of pituitary LH within 24 h of the initial blockade. The involvement of an oestrogen trigger, in the release of 'LH surge' has been clearly demonstrated by Ferin *et al.*[15] using a specific antiserum to oestrogen.

It is well known that just prior to ovulation the pituitary of several mammals, including man, releases a surge of LH and FSH[43]. While it has been shown by several independent investigators that neutralisation of LH with a specific LH antibody results in blockade of ovulation, the role of FSH in the induction of ovulation is still dubious. The ability of FSH alone to induce ovulation has been tested using both hypophysectomised, PMS-primed, immature animals[44, 45] and adult cycling animals blocked on pro-oestrus with pentobarbital[46, 47]. The dosage of FSH used (mostly NIH ovine preparations) has been varied, minimal doses of FSH (whose LH contamination has been considered too small to cause ovulation on its own) also resulting in marginal ovulation[46]. Further, Harrington and Elton[47] using FSH treated with 6 M urea or chymotrypsin to destroy its contaminating LH activity, still observe an ovulation response. In contrast to this is the observation that FSH treated with minimal amount of LH antibody to 'cleanse' it of its LH

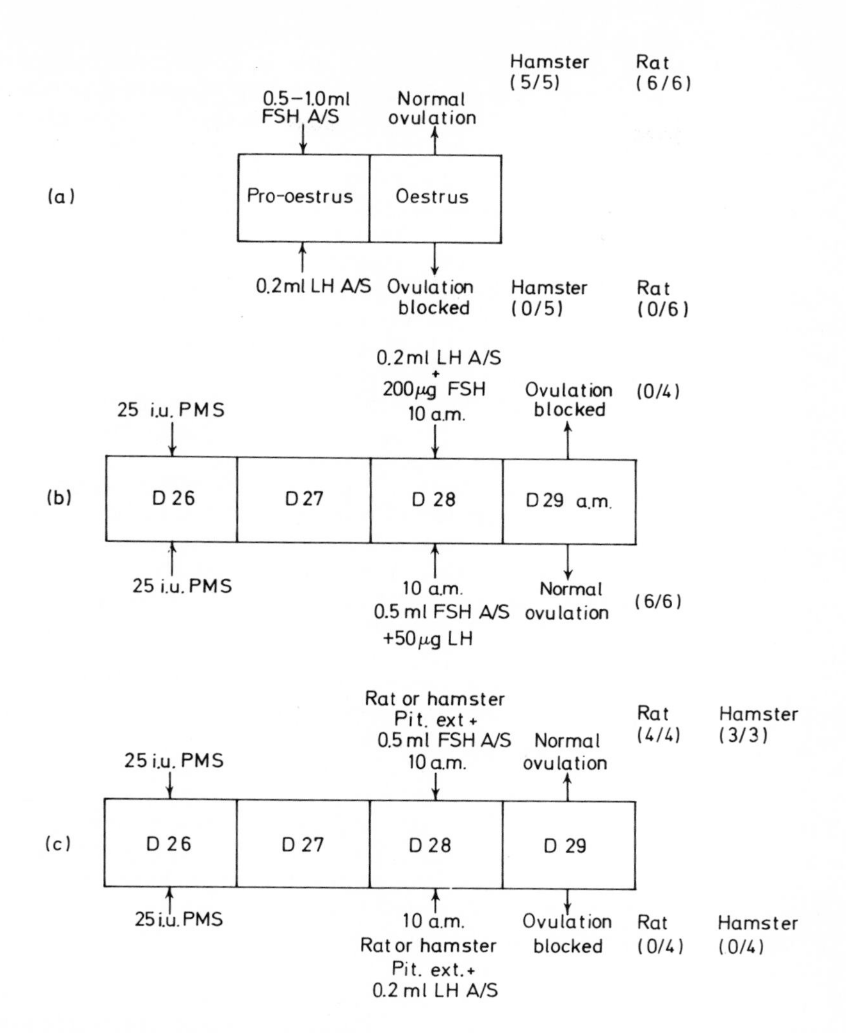

Figure 2.3 Studies on the effect of ovine FSH and LH antiserum (A/S) on ovulation in the intact rat and hamster.
(a) Effect of FSH and LH A/S on ovulation in cycling rats and hamsters.
(b) Induction of ovulation in PMS-primed, immature hamsters using as ovulating hormone, ovine LH or FSH treated with FSH or LH A/S respectively.
(c) Induction of ovulation in PMS-primed, immature rats using as source of ovulating hormone, pituitary extracts of rats and hamsters treated with FSH A/S or LH A/S.

Numbers in parenthesis indicate $\frac{\text{Number of animals showing ovulation}}{\text{Number of animals used in the expt}}$

(After Jagannadha Rao *et al.* see Ref. 48.)

contamination is unable to cause significant ovulation in an 'LH free' system[45]. NIH ovine LH treated with a specific FSH antiserum to neutralise its contaminating FSH activity, on the contrary is still able to evoke an ovulatory response in both rats and hamsters[41, 48]. Similarly in a PMS primed immature rat, hamster or rat pituitary extract failed, in the presence of LH antiserum, to exhibit ovulation inducing activity, the same in the presence of FSH antiserum bringing about normal ovulation[48]. Finally, on administration of minimal doses of well characterised and specific FSH or LH antiserum on the pro-oestrus day of cycling rats and hamsters, only the latter was able to block ovulation[41, 48]. Some of these results are graphically represented in Figure 2.3. The reason for the discrepancy between the antiserum studies and other techniques perhaps lies in the fact that in none of the earlier experiments was the ovulating ability of FSH tested in an 'LH free' system. The residuary LH activity of the PMS used for priming the follicle may act additively with the test dose of FSH, resulting in normal ovulation. Similarly the use of hypothalamic depressor drugs is known to result mostly in the blockade of only 'surge' and not 'tonic' LH release.

The work of Goldman and Mahesh[49] in hamsters using an antiserum raised to LH, but having equipotency to neutralise LH and FSH, shows that the ability of the antiserum to block ovulation is reduced if it is absorbed prior to use with NIH–FSH–S1. This observation is sited as a support to the general contention that FSH independent of its LH contamination has ovulation inducing ability. The experiments of Goldman and Mahesh, however, will have to stand the following criticism. Firstly, considering that the antigen used to absorb their gonadotrophin antiserum (NIH–FSH–S1), is known to be definitely contaminated with LH, it is surprising they did not observe a reduction in LH antibody titre following absorption with it; in addition to not testing the inhibitory effect of the LH antiserum with minimal effective doses, it is known that the bioassay used to detect any reduction in LH antibody titre is not necessarily specific for this hormone. More sensitive immunochemical techniques like haemagglutination or quantitative precipitin test, should have been used to demonstrate an absence of change in the titre of antibody to LH. Secondly, no mention is made of their attempt, if any, to obtain a 'specific FSH antibody' by absorping their all-purpose gonadotrophin antiserum free of all LH antibody with a purified LH preparation.

In the light of these objections, the experiments using specific and well characterised FSH and LH antisera, seem to provide the more plausible answer, namely, that in both rat and hamster, LH is the primary ovulation inducing stimulus. The above generalisation that LH is the prime ovulation inducing stimulus in rodents can perhaps be extended to the primate also. Moudgal, Macdonald and Greep[50] have shown, using an antiserum to HCG, which was earlier established to be cross-reactive with monkey LH, that ovulation and corpus luteum formation can be inhibited when the antiserum is administered to female monkeys (*Macaca fascicularis*) from days 10–13 of the cycle. Periodic examination by laparotomy for ovulation point, change in menstrual cycle length and reduction in the plasma progesterone level were used as parameters to determine the efficacy of the antiserum in blocking ovulation (Figure 2.4(a)).

Sasamoto[51] and Madhwa Raj and Moudgal[52] using HCG and LH antiserum respectively have independently shown that the gonadotrophin stimulus to induce ovulation in PMS primed immature mice[51] and rats[52] need be present only for the first 60–120 min, the ovulation process becoming autonomous thereafter. A study of the sequence of events occurring during the first 60 min of LH action would perhaps help us establish the nature of the physiological trigger behind ovulation.

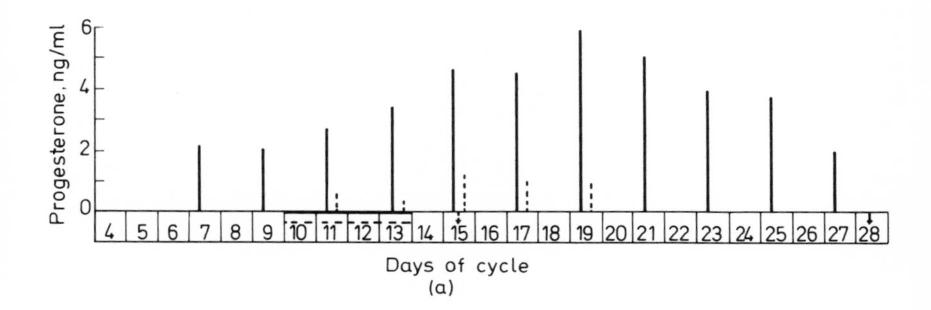

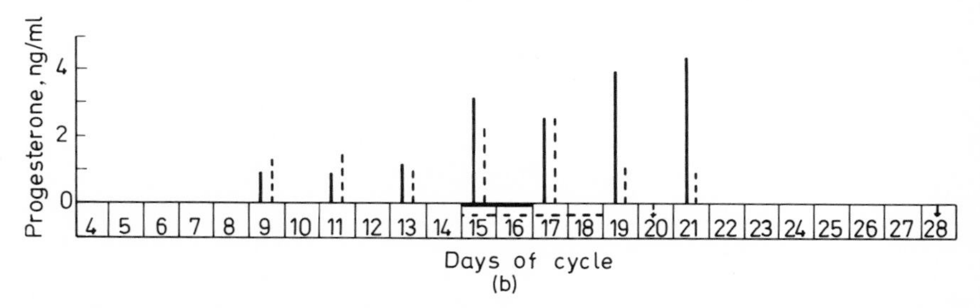

Figure 2.4 Experiment demonstrating the ability of HCG antiserum (A/S) to block ovulation and disrupt luteal function in cycling monkeys *Macaca fascicularis.* The first day of vaginal bleeding was taken as the termination of the cycle indicated by an ↓. Blood was withdrawn every alternate day of the experimental period for progesterone estimation. (a) Studies on ovulation: Normal rabbit serum (NRS) or HCG A/S (2 ml per day) was given from days 10–13 of the cycle.
Solid bars refer to NRS and broken bars refer to HCG A/S group. Ovulation as seen by periodic laparotomy was blocked in HCG A/S group. (After Moudgal *et al.*, see Ref. 50)
(b) Effect of HCG A/S on corpus luteum function: HCG A/S (2 ml per day) was given for 2–4 days beginning from day 15 of the cycle.
Solid bars refer to a batch of HCG A/S of low titre and low cross-reactivity given for 2 days only; broken bars refer to a batch of HCG A/S of high titre and high cross-reactivity given for 4 days. (After Moudgal *et al.*, See Ref. 72.)

Schwartz[34] based on her experience administering LH and FSH antiserum on various days of cycle to rats, is of the opinion that the FSH release that occurs concomitantly with LH just prior to ovulation, has perhaps a significant role in promoting maturation of the follicle in time for the next ovulation to occur. Our experience using specific FSH antiserum in hamsters seems to support such a possibility. Treatment with FSH antiserum was shown to bring about inhibition of follicular maturation in experimental hamsters[19].

(a) Blockade of implantation in the rat with LH A/S:

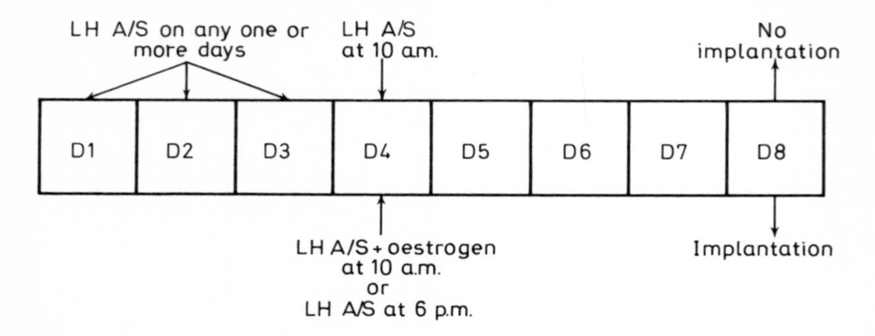

(b) Blockade of implantation in the hamster with LH A/S:

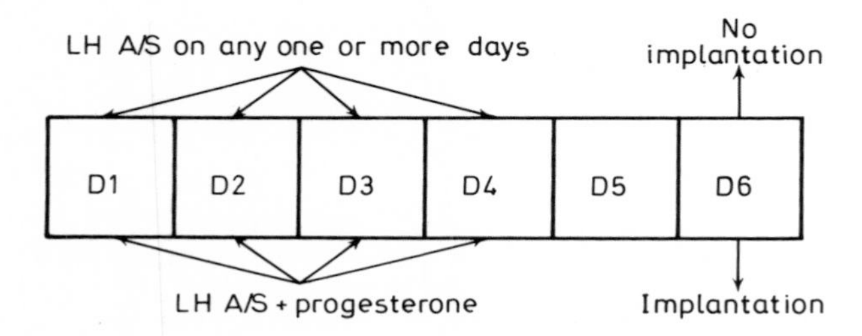

(c) Demonstration of delayed implantation in the rat with a minimum effective dose (MED) of LH A/S:

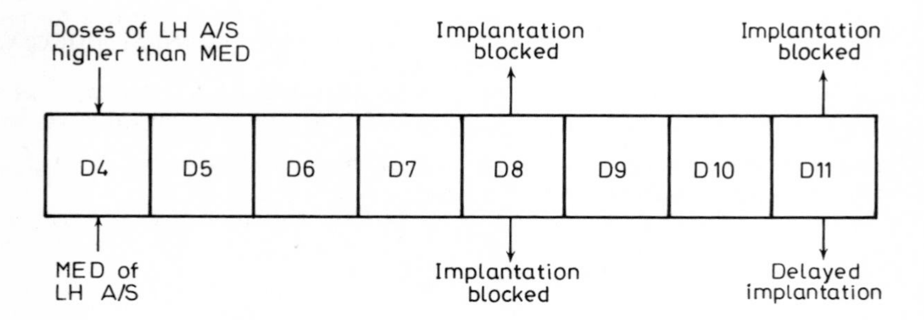

Figure 2.5 Studies on implantation in rat and hamster using antiserum (A/S) to ovine LH (After Madhwa Raj *et al.*, See Ref. 33 and Jagannadha Rao *et al.*, See Ref. 19.)

2.6 STUDIES ON IMPLANTATION

Implantation of the fertilised blastocyst occurs between the evening of day 5 and morning of day 6 of pregnancy in most rodents and is dependent upon a functional ovary and steroid secretion from it. Investigations carried out by Moudgal and collaborators in pregnant rats[53, 33] and hamsters[54, 19] with well characterised and specific LH and FSH antibodies, show clearly the relatively important role LH must be having in the implantation process. Administration of a single or multiple injection of LH antiserum to pregnant rats[33], anytime before 6 p.m. of day 4, results in an inhibition of implantation observed at laparotomy on day 8. The observation that LH antiserum injections at 10 a.m. of day 4 was still effective in inhibiting implantation, considered with the fact that oestrogen in minimal amounts was able to reverse this inhibition, suggested to us that LHs role in stimulating oestrogen secretion between 10 a.m. and 6 p.m. of day 4 in the pregnant rat is a physiological event of some significance[53].

Using an entirely different model system of pregnant rats with autotransplanted pituitaries, Macdonald, Armstrong and Greep[55] showed that exogenously administered LH in a delay vehicle would cause implantation. In confirmation of our finding that LH has an important function on day 4 of pregnancy in the rat, Yoshinaga *et al.*[56] measuring plasma oestrogen levels on various days of pregnancy, observed that a short burst of oestrogen secretion occurs on day 4. The involvement of oestrogen in the implantation process of the rat has also been shown by the administration of specific oestrogen antibodies on days 3 and 4 of pregnancy in the rat[57].

Natural implantation in the rat can be delayed by creating artificial conditions of oestrogen lack, e.g. overectomising rats on day 1 of pregnancy and maintaining blastocysts in a viable state with minimal amounts of progesterone[58]. The use of LH antiserum in minimal effective doses has made it possible to study natural delay in implantation in intact pregnant rats[33]. Injection of antiserum in large amounts blocks implantation completely, probably due to its prolonged and total inhibitory effect on steroidogenesis. The delayed implantation study in addition to showing that the antiserum effect is dose and time dependent, underscores its lack of blastotoxic effects. Even though the level of plasma LH during days 1–4 of pregnancy is quite low[59] there thus appears to be within this, a threshold for the amount of LH needed to cause optimal oestrogen secretion. This threshold level remains to be determined.

Injection of minimal amounts of LH antiserum on any day between days 1–5 of pregnancy[54] also causes blockade of implantation in the hamster. Progesterone, the steroid responsible for implantation in this species, can overcome the antiserum inhibition. Further supplementation with oestrogen, seems to have a facilitatory influence on the implantation process. It has however, not been possible to pinpoint, in the hamster study, the day and time that LH acts on the ovary to produce threshold levels of steroid necessary for implantation. The results on the implantation studies in the rat and hamster are summarised in Figure 2.5.

LH antiserum has been shown to block implantation in pregnant mouse also[37, 60]. However, there appears to be some confusion as to the ability of oestrogen, the steroid capable of inducing implantation in this species, to overcome LH antiserum blockade. It is likely that in the mouse study, a dose of antiserum in excess of the minimal amount has been used with the result that its inhibitory effect on both oestrogen and progesterone synthesis is total as well as prolonged. Consequently, even if a single injection of oestrogen is able to reverse the antibody inhibition of implantation, the implanted blastocysts would not have survived till autopsy due to lack of progestational support.

The injection of large doses of specific and cross-reacting FSH antiserum during pre-implantation period had no effect on the implantation process of the rat[33] and hamster[54]. This would essentially suggest that FSH has no effect on active ovarian steroidogenesis in these two species. In addition, this negative effect can be taken as proof of the absence of antibodies to LH in the FSH antiserum used and also as evidence to the lack of effect of non-specific serum proteins and antibodies of unknown specificity (likely to be present in FSH antiserum) on implantation. These results obtained using highly characterised FSH antiserum, are quite in contrast to those observed in the active immunisation studies using as antigens FSH preparations obviously contaminated with LH[23].

2.7 USE OF GONADOTROPHIN ANTIBODIES TO DELINEATE THE ROLE OF LH, FSH AND PROLACTIN IN THE MAINTENANCE OF PREGNANCY, PSEUDO-PREGNANCY AND LACTATION

2.7.1 Studies on early pregnancy in the rat, hamster and mouse

Classical studies in the rat and hamster, using techniques of hypophysectomy and hormone replacement therapy, have indicated that all the three pituitary gonadotrophins, namely, prolactin, FSH and LH are needed for the continuation of gestation. Whereas prolactin was ascribed the relatively important function of being the primary luteotrophin (in maintaining optimal progesterone production from the corpus luteum), FSH and LH were considered necessary adjuncts responsible for the secretion of oestrogen[61, 62]. Studies on the effect of LH antiserum on implantation, however, suggested that deprival of LH would result in an inhibition of both oestrogen and progesterone secretion.

Several investigators have observed that administration of LH antiserum during the post-implantation period to pregnant rats[18, 63], hamsters[19], mice[60], rabbits[64] and guinea-pigs (Jagannadha Rao, and Moudgal—unpublished results) results in a termination of gestation. The efficacy of the antiserum to disrupt pregnancy seems to be restricted for the first 11 days in the case of the rat, hamster and mouse, 20–25 days in guinea-pigs and 18 days in rabbits (Figure 2.6). In other words, during the first one-third to one-half of the gestation period, neutralisation of endogenous LH results in termination of pregnancy. Depending upon the antibody titre, a single injection of minimal amounts of antiserum is sufficient to cause termination of gestation. During this period the LH support seems so very essential for continuation of gestation that deprival of it even for 2–4 h, as demonstrated in rats[18] and hamsters[19], is detrimental to the progress of pregnancy. The earliest time interval, at which visible signs of foetal damage could be seen, was 12 h following antiserum injection, the implantation sites appearing blue in colour and the uterine lumen becoming engorged with blood. The specificity of the antiserum effect was checked in several ways. An important check

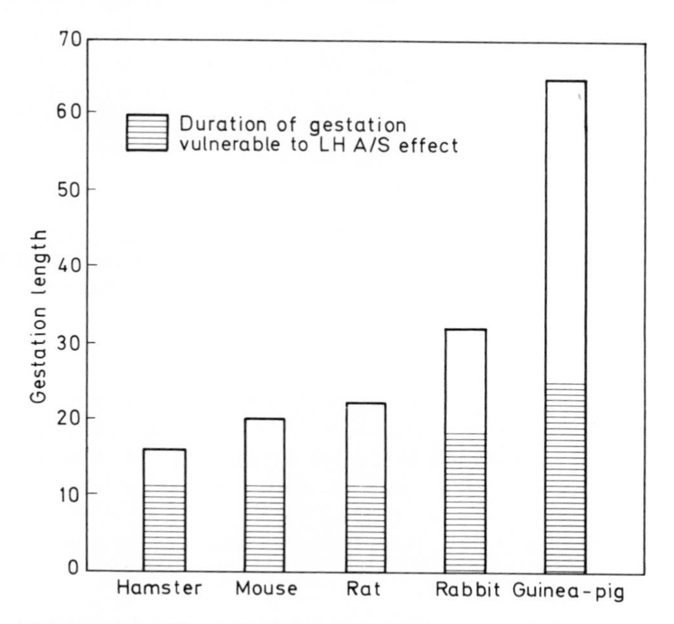

Figure 2.6 Figure depicting the duration of gestation vulnerable to LH antiserum effect in different species of animals tested

used by Moudgal and collaborators[18, 19] was to see whether a simultaneous injection of excess amounts of prolactin, FSH or LH could override the LH antiserum inhibition. Only LH could reverse the antiserum blockade.

Steroid supplementation experiments have revealed that, with the exception of rabbit[64], treatment with oestrogen alone does not overcome the antiserum inhibition. The effect of LH on the rabbit corpora lutea thus appears to be indirect, mediated via its ability to promote oestrogen secretion. On the other hand in rats[18, 64], hamsters[19], and mice[60], progesterone alone can override the effects of the LH antibody. In these animals additional supplementation with oestrogen has been shown to have a facilitatory effect.

Histological examination of the ovary of animals 12–72 h following LH antiserum treatment revealed the onset of degenerative changes in the corpora lutea, and the Graafian follicles. The hamster once again leads by showing degenerative changes at very early time intervals following LH deprival. Degenerative changes in the ovary are apparent within 12–24 h of LH antiserum treatment, the corpora lutea and Graafian follicles virtually disappearing by 72 h (Jagannadha Rao and Moudgal,—unpublished observations). Loewit *et al.*[63] have shown, using histochemical techniques, that 20 α-steroid dehydrogenase activity appears greatly enhanced in rats whose gestation is terminated due to LH antiserum treatment. The elevation in the activity of this enzyme, which controls the conversion of progesterone to its 20 α-dihydro derivative, has been suggested by some workers to indicate the onset of active luteolysis[65, 63]. The period of LH deprival which would lead to non-reversible luteolysis is yet to be determined, but preliminary observations reveal that this is subject to species variation.

Relatively large doses of ovine FSH antiserum administered singly or consecutively for 3 days in early or late pregnancy did not affect pregnancy either in rat[18] or hamster[19]. However, the ovary of the hamster, showed histological changes with a significant reduction in primary and secondary follicles; the luteal tissue and Graffian follicles remaining undamaged. Assuming that for gonadotropin antisera to terminate gestation, it should be capable of significantly reducing ovarian progesterone secretion, it is highly probable that FSH antiserum does not effect progesterone synthesis. The FSH antiserum used in the above studies was free of antibodies to LH and was earlier shown to neutralise the FSH activity of rat and hamster pituitary extracts[17]. Even though sufficiently large volumes of ovine FSH antiserum was administered in these studies, to pregnant rats and hamsters, the possibility that it was not enough to neutralise the FSH activity of heterologous species cannot be excluded. Since the ovine FSH antigen-antibody system, as characterised by Jagannadha Rao and Moudgal[17] forms a soluble complex it has been difficult to quantitate its cross-reactivity with rat or hamster pituitary FSH. Antisera to sheep prolactin has been observed to be species specific[2] and as such cannot be used to neutralise the endogenous prolactin activity of other species of animals. Well characterised and high titre antiserum to rat prolactin has been administered in sufficiently high doses to both pregnant rats and hamsters and found ineffective in terminating gestation (Madhwa Raj—personal communication). The effect of this antiserum on progesterone secretion from the ovary, however, is not known. Antiserum to human placental lactogen has been shown to be effective in terminating gestation of rats when administered from days 1–5 or 16–21 of pregnancy[66]. This antiserum, however, has been shown to be contaminated with antibody to LH.

2.7.2 Studies on pseudopregnancy in the rat

Pseudopregnancy and decidualisation are well-studied physiological events shown to be dependent upon a proper functioning of the corpus luteum. Using the decidual cell reaction as a marker, a variety of experimental models have been set up to prove the luteotropic effect of prolactin[67]. These experimental models are dependent upon creating a condition where, either by producing excessive and continuous endogenous prolactin (autotransplanted pituitaries) or by administering to hypophysectomised rats repeatedly, a relatively large dose of homologous or heterologous prolactin, the corpora lutea are continuously kept under the tropic influence of prolactin.

In order to assess the role of LH in the above process, Kiracofe *et al.*[68], Moudgal *et al.*[69], Maneckjee *et al.*[70] and Chang *et al.*[71] independently studied the effect of LH antiserum on decidual cell reaction by injecting it into intact pseudopregnant rats from either days 1–4 or 5–8 of pseudopregnancy. While there was a general agreement that neutralisation of LH inhibited the decidual cell response, Kiracofe *et al.*[68] observed that only oestrogen could reverse the LH antiserum effect. In contrast, the other groups showed that progesterone by itself was able to reverse the antiserum effect, additional supplementation with oestrogen having a beneficial effect on the degree of decidualisation[69-71]. The differences in these observations could perhaps be ascribed to the use of antisera of inadequate characterisation and poor antibody titre by Kiracofe *et al.* The model system used by Moudgal *et al.*[70] (pregnant animals wherein, in one uterine horn natural implantation was allowed to occur while in the other ligated horn, decidualisation was induced by the injection of histamine hydrochloride) permitted one to conclude that both decidualisation and implantation are directly comparable in being dependent upon LH. The production of both steroids—oestrogen needed for sensitising the uterus prior to traumatisation, and progesterone needed for maximal decidualisation and maintenance of implanted blastocysts, thus seem dependent on optimal LH support.

2.7.3 Studies on late pregnancy, parturition and lactation

It was earlier mentioned that LH antiserum was effective in terminating gestation if given before day 11 of pregnancy in the rat, hamster and mouse. The daily injection of LH antiserum from days 12 to 16 in the hamster[19] or days 12 to 22 in the rat[18] had no deleterious effects on the course of gestation or parturition. The ovaries of the hamster, however, continued to be influenced by LH antiserum in a manner similar to the pre-day 11 period by showing a reduction in ovarian weight and degenerative changes in the luteal and follicular apparatus[19]. The overall observations in the rat and hamster suggested several possibilities—(a) the functioning of the different ovarian compartments—in particular the corpus luteum—ceases, by mid-gestation, to be influenced by pituitary LH, but is probably under the control of placental tropic factors (b) the corpus luteum becomes autonomous by mid-gestation and starts a deceleration in its activity, a process essential for the production of low levels of progesterone close to parturition and (c) the functioning of the corpus luteum during the second half of gestation continues to be influenced by LH though to a relatively lesser extent, the consequent reduction in luteal activity not having a deleterious effect on the course of gestation (pregnancy by this time not being dependent upon high progesterone levels). Madhwa Raj and Moudgal[18] have critically discussed the likelihood of anyone of the above possibilities being correct in the case of the rat. Evidences for the partial autonomy of the rat corpus luteum at mid-gestation and the relative lack of effect of LH deprival on the functioning of corpora lutea of late pregnancy are forthcoming and will be discussed in greater detail below (see Section 2.7.4).

Administration of LH antiserum, for two consecutive days, to lactating rats with 2 or 6 pups did not appear to influence lactation *per se* as judged by the daily increase in pup weights[72]. Using the same qualitative criteria for lactation, a highly characterised rat prolactin antiserum administered after parturition does not appear to influence lactation (Madhwa Raj—personal communication). A further critical study with antiprolactin is required before it can be definitely concluded that this antiserum does not affect the synthesis and secretion of milk proteins. In a recent study in our laboratory an interesting correlation between prolactin and LH in inducing decidualisation in lactating rats has been observed. It is known that the larger the number of suckling pups, the lesser the degree of decidualisation observed[73]. This inhibition in decidual cell response has been correlated to an increased secretion of prolactin and a consequent suppression of pituitary LH release[73]. The use of LH antiserum in this experimental set-up has shown that LH neutralisation in both the small and large suckling groups results in a 50–25% inhibition of decidual cell response, The increase observed in decidualisation following withdrawal of suckling stimulus on day 5 of lactation, can apparently be ascribed to an increased release of LH as the augmentation in response can be abolished by a simultaneous administration of LH antiserum (Maneckjee and Moudgal—unpublished observations).

2.7.4 Analysis of corpus luteum function in the above physiological states and the effect of LH deprival on it

The use of specific LH antibody has permitted us to critically evaluate for the first time the role of this hormone in regulating the functioning of the corpus luteum. The observation that the circulating level of LH, except at the time of pre-ovulatory surge, is extremely low had raised doubts about the contribution of this hormone towards the maintenance of corpus luteum[43,59]. Recent experiments of Moudgal, Macdonald and Greep[74] on the ability of HCG antiserum, when administered from days 15–18 of the menstrual cycle of the subhuman primate *Macaca fascicularis*, to significantly reduce progesterone output and thereby foreshorten the cycle length, show very clearly the essential part tonic levels of LH must be playing in maintaining luteal function (Figure 2.4(b)). Similar observations have been made in the heifer by Snook *et al.*[75].

Experimental proof for the hypothesis of Madhwa Raj and Moudgal[18] that the rat ovaries of early and late pregnancy are affected to different degrees by LH deprival is forthcoming. Moudgal, Behrman and Greep[32] using two different parameters of luteal functionality, namely cholesterol turnover (Figure 2.8) and progesterone output (Figure 2.7) showed that the ovary of the day 8 pregnant rat is affected by lack of LH to a greater extent than that of day 15. The psuedopregnant rat ovary appears to be equally sensitive to lack of LH and under the influence of LH antiserum shows a significant reduction in progesterone output (Figure 2.7). These studies further revealed that the 150–200% increase in ovarian cholesterol ester stores seen in the ovaries of the day 8 pregnant rat following LH deprival could be correlated with change in the activity of the enzymes cholesterol esterase (−90%) and cholesterol esterol synthetase[28] (+150%) (Figure 2.8.)

Behrman and Armstrong[76] and Behrman *et al.*[77] using hypophysectomised rats treated with LH have similarly observed that while LH activates the enzyme cholesterol esterase it appears to inhibit the activity of cholesterol-ester synthetase.

The ovaries of lactating rats under intense suckling stimulus (each nursing 6–8 pups) are under the influence of high levels of prolactin, the LH influence on the luteal function at this stage being considered insignificant. Despite this, neutralisation of LH by antiserum treatment resulted in a marked reduction in progestin secretion[72], emphasising once again that LH need not necessarily be in large amounts to have a salutory effect on luteal function (Figure 2.7).

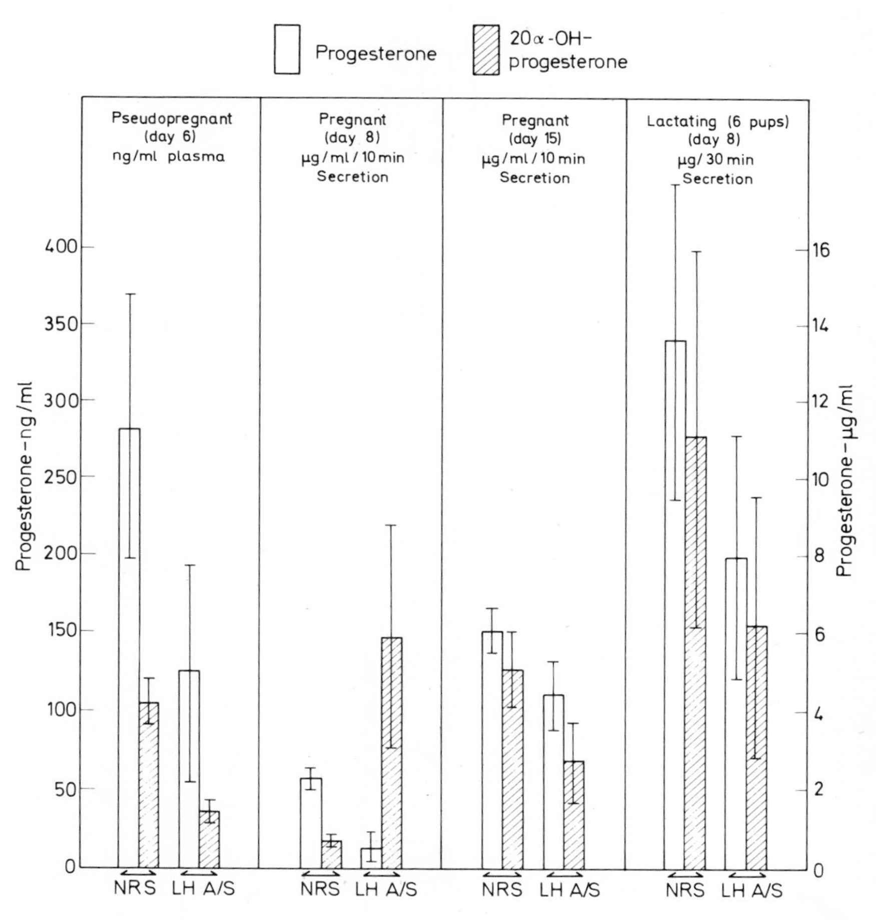

Figure 2.7 Histogram depicting the effect of LH antiserum (A/S) on progestin levels in pseudopregnant, early and late pregnant and lactating rats. 0.3 to 0.4 ml of LH A/S given on the day noted. Lactating rats received 2 injections of LH A/S on days 6 and 7. Procedural details of blood collection of analyses as described by Moudgal *et al.*, (see Ref. 32); Behrman *et al.*, (see Ref. 28); and Yoshinaga *et al.*, (see Ref. 72)

Rothchild[78] recently observed that the ovaries of rats hypophysectomised and hysterectomised on day 12 of pregnancy, in confirmation of our suggestion of luteal autonomy[18], continue to secrete substantial amounts of progesterone (enough to override an oestrogen challenge and maintain vaginal di-oestrus) for a minimum period of 5 days. No LH antiserum was given to these rats to remove any tissue bound LH. Yoshinaga[79] has observed that ovaries of day 15 pregnant rats, while they show a significant reduction in their ability to secrete progesterone 24 h following LH antiserum treatment, regain their original capacity as soon as the antiserum effect wears out (in approximately 48 h). Moudgal *et al.*[74] in their studies on the primate corpus luteum similarly observed that while injection of a low titre HCG antiserum on day 15 and 16 of cycle resulted in an immediate but short-

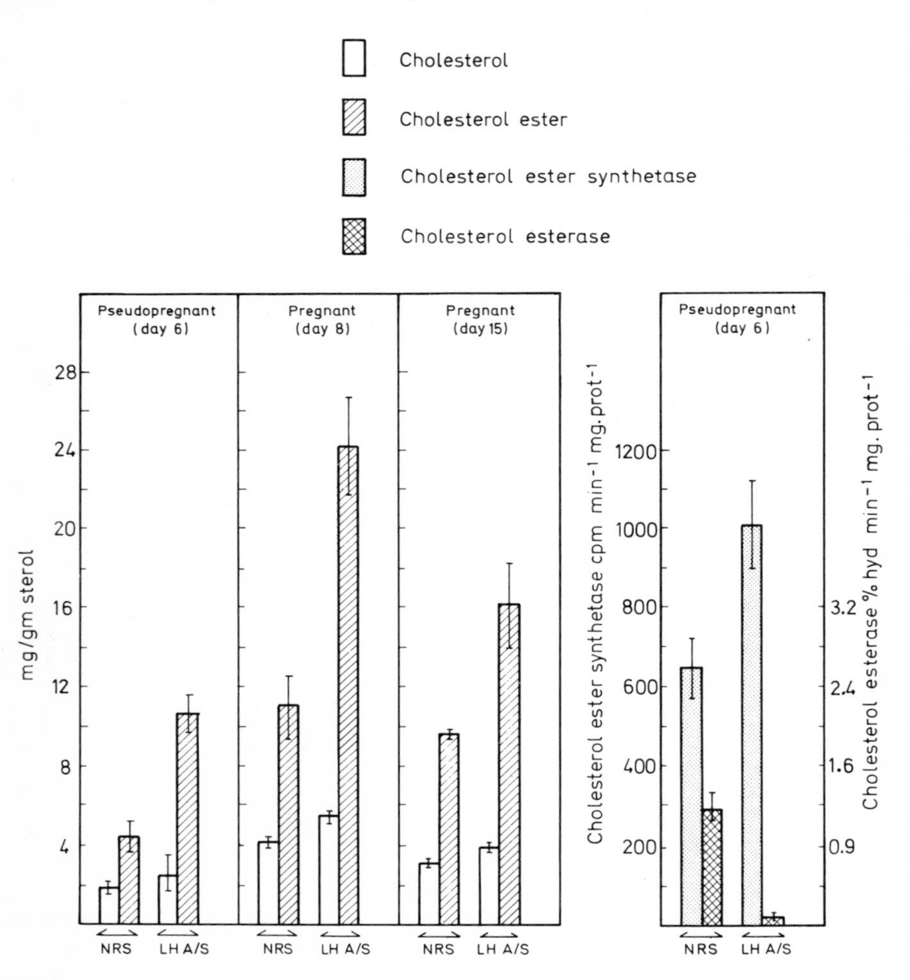

Figure 2.8. Effect of LH deprival on cholesterol turnover in luteinised rat ovaries. All rats were autopsied 24 h after the injection of 0.4 ml of NRS or LH A/S (After Moudgal *et al.*, see Ref. 32 and Behrman *et al.*, see Ref. 28.)

lived lowering in plasma progesterone level, injection of a high titre antiserum for four consecutive days (day 15 through 18) led to continued low plasma progesterone levels and termination of cycle suggesting that irreversible luteolysis had started (Figure 2.4(b)). Short-term deprivation of endogenous LH support does not appear to affect ovarian responsiveness to LH in subsequent *in vitro* incubation experiments[28]. The *in vitro* responsiveness to LH of pseudopregnant rat ovaries treated 90 min prior to sacrifice with LH antiserum is at times greater than the saline controls, perhaps due to the accumulated precursor stores (cholesterol and cholesterol esters) brought about by the lack of endogenous LH[28].

Despite the innumerable evidences available in support of the ability of LH to increase cyclic AMP production[80] and progesterone synthesis in both *in vivo* and *in vitro* systems of various species[81], LH even today is dubbed as a luteolytic agent. Deprival of endogenous LH, however low its circulating level, has been shown here to lead to a general reduction in both progesterone and 20 α-dihydroprogesterone secretion, the level of the latter increasing markedly only in situations where gestation is terminated and luteolysis has apparently set in (Figure 2.7). It is interesting that in contrast to the earlier observations[82, 83], which led to the general assumption that LH is the prime luteolytic agent in the rat and the hamster, the immunological approach has clearly demonstrated that LH *deprival* leads to luteolysis. The cause for this apparent discrepancy perhaps lies in the experimental model used by earlier investigators[79, 80]. This consisted of administering daily fairly large doses of LH in saline, luteolysis occurring after several days of continuous treatment. The immunological studies thus allow us to conclude that the total deprival of LH has a variable effect on the ability of corpora lutea of different ages and species to survive. Their responsiveness to fresh LH stimulus after varying periods of abstinence may give us an idea as to the vulnerability or fragility of the luteal tissue. A systematic study of the ultra structural and biochemical changes in the corpus luteum following LH and/or prolactin withdrawal could provide us with basic information on luteolysis.

2.8 USE OF GONADOTROPHIN ANTIBODIES IN STUDIES ON THE MALE

Both active and passive immunisation studies have been carried out in the rat, rabbit and mouse to pin-point the locus of action of LH and FSH. It is evident from these studies that neutralisation of LH will initially lead to cessation of interstitial cell function, the consequent inhibition of testosterone production causing arrest of spermatogenesis and loss of libido[2, 21, 35, 37]. Histochemical evidence for interference in the production of androgen following active or passive immunisation with LH has been drawn from the rapid decline in Δ^5-3β-hydroxysteroid dehydrogenase activity of the interstitial tissue[24]. According to Talaat and Laurence[24] passive immunisation with LH antiserum affects the recipient's reproductive capacity much more than the active immunisation procedure. Administration of LH antiserum to 30 day male guinea-pigs and 50 day male hamsters for 6 and 11 days respectively, essentially brings about changes in the testes and accessory gland function similar to that noticed for rats and rabbits[84]. This inhibition of gonadal function, as observed in the hamster, is apparently not permanent and is fully restored if the animals are rested for 30 days after the last LH antiserum injection[84].

Histological examination of the testes[24,84] removed at the termination of the experiment with LH antibody revealed uniform interruption of spermatogenesis. Focal interruption of the spermatogenic cycle was observed in the seminiferous tubules, sertoli cells of the affected tubules showing vacuolation with spermatids appearing multinucleated. The affected tubules appeared much smaller in diameter than the control tubules. The interstitial cells appear to be extensively fibroblastic in nature.

The experience with FSH antibodies in the male have been very disappointing[24,84]. Hitherto no one has been able to observe, following active immunisation with FSH or after injection of FSH antiserum, a clear-cut diminution in testicular and accessory gland function. Earlier, classical work of Greep *et al.*[85] using hypophysectomised animals treated with FSH suggested that this is primarily involved in the stimulation of spermiogenic cells and growth of the seminiferous tubules. Recent work of Means and Vaitukaitis[86] and Dorrington *et al.*[87] using sophisticated methods, clearly points to seminiferous tubules as the target tissue for FSH. While FSH can increase the cyclic 3′5′ AMP level and protein kinase activity of the seminiferous tubule preparation, the action of LHs appears to be restricted to Leydig cells[87]. Even though the levels of FSH used in some of these systems are low, it is difficult to conclude from this that the test system and FSH preparation used were free of contaminating LH. Immunologically 'LH free systems' and 'LH free FSH' preparations may have to be used before assigning seminiferous tubules and sertoli cells as specific target tissues in the male for FSH action. Theoretically it should be possible to show, using a potent and specific FSH antiserum, damage to the seminiferous tubules and consequently spermatogenesis. Perhaps the lack of success hitherto with FSH antibodies[24,84] has been due to use of FSH antisera of low antibody titre. It is our experience that most of the antibodies seen following immunisation with FSH are directed to contaminants (these including LH). Strict measures taken to remove these contaminating antibodies have provided us with antisera specific to FSH but of very low neutralising capacity.

Talaat and Laurence[24] in their active immunisation studies with LH do not rule out the possibility of generating (due to FSH contamination of LH preparation) by this method, antibodies to FSH. Further, active immunisation studies carried out by Laurence and his group with ovine FSH in the female[23] show clear interference with several functions like ovulation, implantation etc.—physiological events which have been shown both by classical and immunological methods to be LH dependent. While this would suggest that effects they have seen in the female immunised with FSH could all be due to LH antibody contamination, it is at the same time surprising that they have not observed following immunisation of the male with FSH the effect due to contaminating LH antibody. This would only suggest that renewed efforts will have to be made to obtain high titre and specific FSH antibody to be ultimately used in localising FSH action in the male.

2.9 USE OF ANTIBODIES IN STUDIES ON GONADAL DEVELOPMENT AND DIFFERENTIATION

Passive transfer of antibodies to gonadotrophins or steroids to prenates via the pregnant mother and to neonates during the first few days of life have been tried to pin-point those events in gonadal development which are under the regulatory control of gonadotrophins. Lunenfeld and Eshkol's[88] study in female mice using a rat gonadotrophin antiserum and reversal with either 'LH free' FSH or HMG (representing FSH + LH), shows that while antiserum injection for the first 7 days of life arrests development of follicles (by not allowing them to develop beyond the 40 cell stage), 'LH free' FSH alone can reverse this condition. Presence of additional LH as in HMG does not seem to better FSHs ability to overcome the antiserum blockade. If the antigonadotrophin treatment was, however, continued till day 14 of infancy, arrest in follicular development appeared much more marked, and FSH while reversing the effect was able to further the development of the follicle. The reversal with HMG appeared to be superior to FSH since it permitted, probably due to the presence of LH in it, the formation of antral space.

While Goldman and Mahesh[89] did not find any change in the oestrous cycle of rats treated neonatally from days 1–5 with a generalised gonadotrophin antiserum, Jagannadha Rao and Moudgal[90] observed that the fertility of the rats (ability to become pregnant) could be affected by injecting them during days 1–11 of life with 100 μl LH antiserum per day. FSH antiserum injection during the same period was, however, without effect[90].

Goldman and Mahesh[89], injecting an antiserum raised to sheep LH (but having according to them equipotency in neutralising both LH and FSH activities) on days 1–5 of life into male neonates, observed that this resulted in infertility of rats as judged by mating experiments carried out at 80–170 days of age. While antiserum injection on days 7–11 did not affect fertility, injection of 10 i.u. HCG or 0.2 mg testosterone on day 1 conferred protection against the injurious effect of the antiserum injection on days 1–5. Thus the sterilising effect of the antiserum is apparently related to its ability to suppress testosterone production in the neonate testis during the first few days of life. Essentially similar results have been obtained by injecting neonate male rats from days 1–11 of their life with highly specific LH antiserum[90]. FSH antiserum tried at the dose levels of 100 μl per day had no effect on the fertility of the male neonates[90]. Attempts to study masculine differentiation in male rat foetuses by injecting testosterone antibodies into the mother from the 13th to 20th day of gestation have been made[91] and the results seem to suggest that the antiserum is capable of preventing testicular growth and masculine differentiation of the anogenital area. Whether this would influence the future fertility of the male neonates was unfortunately not studied. Jagannadha Rao and Moudgal[90], however, observe that injection into mothers of 0.5 ml of LH antiserum per day, from days 12–22 of gestation, had no effect on the gonadal and accessory gland weight in both the male and female litter mates. This raises the question of how effective a transmission of antibodies across the placental membrane can be obtained by injecting antisera to mothers.

2.10 CONCLUDING REMARKS

It may not be far fetched to state at this stage that the immunological approach to the study of reproductive biology as reviewed here, shows that it has yielded results of essential and fundamental importance to the understanding of the various physiological processes. Provided one has a high titre and specific antibody, injection of it is one of the best ways to selectively 'de-hormonise' a system. Techniques in the production, characterisation and use of antisera are continuously being improved. As such their use in sophisticated physiological and biochemical researches like studies on localisation of gonadotrophin synthesising cells in the pituitary and placenta[26,27], gonadotrophin biosynthesis[25], their mechanism of action at the cell level[28-30], demonstration of the presence of a short feed-back system between hypothalamus/pituitary[92], isolation of LH and FSH free of each others contamination by using suitable immunosorbents[93] and other related interesting projects is gaining ground. The discovery of a high degree of homology between the α-subunits of LH, TSH and FSH[94,95], has caused concern with regard to obtaining antisera ultimately specific to a single hormone. This may be ill founded since the β-subunit, in which apparently resides the biological specificity, has greater immunogenicity than the α-subunit[96,97]. Further, the author is not aware of any specific instance where removal of an antibody contaminant to a hormone has not been possible (namely LH antibody from an antiserum to TSH or FSH can be removed resulting in antisera specific to TSH or FSH and not having any effect on LH action), leading to a situation where the use of an antiserum results in the neutralisation of more than one hormonal activity. The steroid hormone antibodies, hitherto used mainly in radioimmunoassay, are beginning to be used in physiological and biochemical studies. The major hindrance in this area was one of obtaining antibodies specific to the individual steroids. While this was achieved early for the oestrogen series, it is only recently that we have antibodies specific to progesterone and testosterone. The exploitation of these hormone antibodies, which have proved their worth as powerful laboratory investigative tools, in clinical medicine, particularly in the field of fertility control, is not far fetched, the only drawback holding its immediate application being the fear of serum sensitivity.

Acknowledgement

The author wishes to express his sincere thanks to his colleagues—Miss C. Mrinalini Rao, Dr A. Jagannadha Rao and Messrs K. Muralidhar, P. Pasupathy and R. Srinivasan for help rendered in the preparation of this manuscript. Aided by a grant from the Indian Council of Medical Research, New Delhi, and Ford Foundation, New York.

References

1. Li, C. H., Moudgal, N. R., Trenkle, A., Bourdel, G. and Sadri, K. (1962). Some aspects of immunochemical methods for the characterization of protein hormones, *Ciba Foundation Colloquia on Endocrinology*, 14, 20 (G. E. W. Wolstenholme and Margaret P. Cameron editors) (London: Churchill)

2. Hayashida, T. (1966). Immunological reactions of pituitary hormones, *The Pituitary Gland*, II, 613 (G. W. Harris and B. T. Donovan editors) (London: Butterworths)
3. Midgley, A. R. Jr (1969). Immunological characterization of the gonadotropins, *Reproduction in Domestic Animals* (Second Edn.), 47 (H. H. Cole and P. T. Cupps, editors) New York: Academic Press)
4. Diczfalusy, E. (1960). Immunoassay of gonadotropins, *Karolinska Symposia on Research methods in Reproductive Endocrinology*, 1st Symposium pp. 1–381 (Karolinska Institutet: Stockholm)
5. Edwards, R. G. (1969). *Immunology and Reproduction*, Proceedings of the First Symposium of the International Coordination Committee for the Immunology of Reproduction, pp. 1–293. (London: International Planned Parenthood Federation)
6. Parkes, A. S. and Rowlands, I. W. (1936). Inhibition of ovulation in the rabbit by anti-gonadotrophic serum, *J. Physiol. (London)* **88,** 305
7. Kabat, E. A. and Meyer, M. M. (1964). *Experimental Immunochemistry* (Springfield: Charles C. Thomas)
8. Moudgal, N. R. and Li, C. H. (1961). An immunochemical study of sheep pituitary interstitial cell stimulating hormone, *Arch. Biochem. Biophys.*, **95,** 93
9. Vaitukaitis, J., Robbins, J. B., Nieschlag, E. and Ross, G. T. (1971). A method for producing specific antisera with small doses of immunogen, *J. Clin. Endocrinol. Metab.*, **33,** 988
10. Lieberman, S., Erlanger, B. F., Beiser, S. M. and Agate, F. J. (1959). Steroid–protein conjugates: Their chemical, immunochemical and endocrinological properties, *Rect. Progr. Hormone Res.*, **15,** 165
11. Goodfriend, L. and Sehon, A. (1970). Early approaches to production, analysis and use of steroid–specific antisera, *Immunologic Methods in Steroid Determination*, 15 (F. G. Peron and B. V. Caldwell, editors) (New York: Appleton–Century–Crofts)
12. Gross, S. J. (1970). Specificities of steroid antibodies, *Immunologic methods in Steroid Determination*, 41 (F. G. Peron and B. V. Caldwell, editors) (New York: Appleton–Century–Crofts)
13. Thorneycroft, I. H. and Stone, S. C. (1972). Radioimmunoassay of serum progesterone in women receiving oral contraceptive steroids, *Contraception*, **5,** 129
14. Niswender, G. D. and Midgley, A. R., Jr (1970). Hapten-radioimmunoassay for steroid hormones, *Immunologic methods in Steroid Determination*, 149 (F. G. Peron and B. V. Caldwell, editors) (New York: Appleton–Century–Crofts)
15. Ferin, M., Zimmering, P. E. and Vande Wiele, R. L. (1969). Effects of antibodies to estradiol-17β on PMS-induced ovulation in immature rats, *Endocrinology*, **84,** 893
16. Caldwell, B. V., Scaramuzzi, R. J., Tillson, S. A. and Thorneycroft, I. G. (1970). Physiologic studies using antibodies to steroids, *Immunologic methods in Steroid Determination*, 183 (F. G. Peron and B. V. Caldwell, editors) (New York: Appleton–Century–Crofts)
17. Jagannadha Rao, A. and Moudgal, N. R. (1970). An Immunochemical study of ovine pituitary Follicle Stimulating Hormone (FSH), *Arch. Biochem. Biophys.*, **138,** 189
18. Madhwa Raj, H. G. and Moudgal, N. R. (1970). Hormonal control of gestation in the intact rat, *Endocrinology*, **86,** 874
19. Jagannadha Rao, A., Madhwa Raj, H. G. and Moudgal, N. R. (1972). Effect of LH, FSH and their antisera on gestation in the hamster (*Mesocricetus auratus*), *J. Reprod. Fert.*, **29,** 279
20. Bourdel, G. and Li, C. H. (1963). Effect of rabbit antiserum to sheep pituitary interstitial–cell stimulating hormone in adult female rats, *Acta Endocrinol.*, **42,** 473
21. Hayashida, T. (1963). Inhibition of spermiogenesis, prostate and seminal vesicle development in normal animals with antigonadotropic hormone serum, J. *Endocrinol.*, **26,** 75
22. Laurence, K. A. and Ichikawa, S. (1968). Effects of active immunization with bovine luteinizing hormone on reproduction in the female rat, *Endocrinology*, **82,** 1190
23. Talaat, M. and Laurence, K. A. (1969). Effects of active immunization with ovine FSH on the reproductive capacity of female rats and rabbits, *Endocrinology*, **84,** 185
24. Talaat, M. and Laurence, K. A. (1971). Impairment of spermatogenesis and libido through antibodies to luteinizing hormone, *Fert. Steril.*, **22,** 118
25. Wakabayashi, K. and Tamaoki, B. I. (1966). Influence of immunization upon the anterior pituitary–gonad system of rats and rabbits with special reference to

histological changes and biosynthesis of luteinising hormone and steroids, *Endocrinology*, **79**, 477

26. Midgley, A. R. Jr (1963). Immunofluorescent localization of human pituitary luteinising hormone, *Exp. Cell Res.*, **32**, 606
27. Midgley, A. R. Jr and Pierce, G. B. Jr (1962). Immunohistochemical localisation of human chorionic gonadotropin, *J. Exp. Med.*, **115**, 289
28. Behrman, H. R., Moudgal, N. R. and Greep, R. O. (1972). Studies with LH antisera *in vivo* and *in vitro* on luteal steroidogenesis and enzyme regulation of cholesteryl ester turnover, *J. Endocrinol.*, **52**, 419
29. Moyle, W. R., Moudgal, N. R. and Greep, R. O. (1971). Cessation of steroidogenesis in leydig cell tumors after removal of luteinising hormone and adenosine cyclic 3′,5′ monophosphate, *J. Biol. Chem.*, **246**, 4978
30. Moudgal, N. R., Moyle, W. R. and Greep, R. O. (1971). Specific binding of luteinising hormone to leydig tumor cells, *J. Biol. Chem.*, **246**, 4983
31. Bindon, B. M. (1970). Prolonged activity *in vivo* of rabbit antisera to placental gonadotropins, *J. Endocrinol.*, **46**, 221
32. Moudgal, N. R., Behrman, H. R. and Greep, R. O. (1972). Effect of LH antiserum (A/S) on progestin secretion from the pregnant rat ovary, *J. Endocrinol.*, **52**, 413
33. Madhwa Raj, H. G., Sairam, M. R. and Moudgal, N. R. (1968). Involvement of luteinising hormone in the implantation process of the rat, *J. Reprod. Fert.*, **17**, 335
34. Schwartz, N. B. (1969). A model for the regulation of ovulation in the rat, *Rect. Progr. Hormone Res.*, **25**, 1
35. Quadri, S. K., Harbers, L. H. and Spies, H. G. (1966). Inhibition of spermatogenesis and ovulation in rabbits with antiovine LH rabbit serum, *Proc. Soc. Exp. Biol. Med.*, **123**, 809
36. Ely, C. A., Tuercke, R. and Chen Bei-Loo (1966). Comparison of antisera to various gonadotropins as they effect the mouse vaginal cycle, *Proc. Soc. Exp. Biol. Med.*, **122**, 601
37. Munshi, S. R. and Rao, S. S. (1967). Biological specificity of antigens present in ovine luteinizing hormone, *Ind. J. Exp. Biol.*, **5**, 135
38. Schwartz, N. B. and Ely, C. A. (1970). Comparison of effects of hypophysectomy, antiserum to ovine LH and ovariectomy on estrogen secretion during the rat estrous cycle, *Endocrinology*, **86**, 1420
39. Kelly, W. A., Robertson, H. A. and Stansfield, D. A. (1963). The suppression of ovulation in the rat by rabbit antiovine LH serum, *J. Endocrinol.*, **27**, 127
40. Schwartz, N. B. and Gold, J. J. (1967). Effect of a single dose of anti LH serum at proestrus on the rat estrus cycle, *Anat. Rec.*, **157**, 137
41. Jagannadha Rao, A., Madhwa Raj, H. G. and Moudgal, N. R. (1971). Effect of gonadotropin antisera on ovulation in the intact hamster, Abstr. **67**, 25 *Proceedings of IVth Annual meeting of the Society for the Study of Reproduction, Boston, USA*
42. Spies, H. G., Slyter, A. L. and Quadri, S. K. (1967). Regression of corpora lutea in pregnant gilts administered antiovine LH rabbit serum, *J. Anim. Sci.*, **26**, 768
43. Vande Wiele, B. L., Bogumil, J., Dyrenfurth, I., Ferin, M., Jewelewicz, R., Warren, M., Rizkallah, T. and Mikhair, G. (1970). Mechanisms regulating the menstrual cycle in women, *Rect. Progr. Hormone. Res.*, **26**, 63
44. Lostroh, A. J. and Johnson, R. E. (1966). Amounts of Interstitial Cell-Stimulating Hormone and Follicle Stimulating Hormone required for follicular development, uterine growth and ovulation in the hypophysectomized rat, *Endocrinology*, **79**, 991
45. Macdonald, G. (1970). Inhibition of FSH induced ovulation by anti-luteinising hormone (LH A/S), Abstr. **49**, 25 *Proceedings of the III Annual meeting of the Society for the Study of Reproduction, Columbus, USA*
46. Harrington, F. E. and Elton, R. L. (1969). Induction of ovulation in adult rats with Follicle Stimulating Hormone, *Proc. Soc. Exp. Biol. Med.*, **132**, 841
47. Harrington, F. E., Bex, F. J., Elton, R. L. and Roach, J. B. (1970). The ovulatory effects of Follicle Stimulating Hormone treated with chymotrypsin in chloropromazine treated rats, *Acta Endocrinol.*, **65**, 222
48. Jagannadha, Rao, A., Moudgal, N. R., Lipner, H. and Greep, R. O. (1972). Role of FSH and LH in initiation of ovulation in rats and hamsters—A study using rabbit antisera to ovine FSH and LH (communicated to *J. Reprod. & Fertil.*)
49. Goldman, B. D. and Mahesh, V. B. (1969). A possible role of acute FSH-release in ovulation in the hamster as demonstrated by utilisation of antibodies to LH and FSH, *Endocrinology*, **84**, 236

50. Moudgal, N. R., Macdonald, G. J. and Greep, R. O. (1971). Effect of HCG antiserum on ovulation and corpus luteum formation in the monkey (*Macaca fascicularis*), *J. Clin. Endocrinol. Metab.*, **32,** 579
51. Sasamoto, S. (1969). Inhibition of HCG-induced ovulation by anti-HCG serum in immature mice pre-treated with PMSG, *J. Reprod. Fert.*, **20,** 271
52. Madhwa Raj, H. G. and Moudgal, N. R. (1970). Effect of anti-luteinising hormone serum on the ovulation of rats, *Nature* (*London*), **227,** 1344.
53. Madhwa Raj, H. G., Sairam, M. R. and Moudgal, N. R. (1967). Role of gonadotropins in implantation: A study using specific antigonadotropins, *Indian. J. Exp. Biol.*, **5,** 123
54. Jagannadha Rao, A., Madhwa Raj, H. G. and Moudgal, N. R. (1970). Need of luteinizing hormone for early pregnancy in the golden hamster (*Mesocricetus auratus*), *J. Reprod. Fert.*, **23,** 353
55. Macdonald, G. J., Armstrong, D. T. and Greep, R. O. (1967). Initiation of blastocyst implantation by luteinizing hormone, *Endocrino'ogy*, **80,** 172
56. Yoshinaga, K., Hawkins, R. A. and Stocker, J. F. (1969). Estrogen secretion by rat ovary *in vivo* during the estrous cycle and pregnancy, *Endocrinology*, **85,** 103
57. Raziano, J., Ferin, M. and Vande Wiele, R. L. (1972). Effect of antibodies to estradiol-17-β and to progesterone on nidation and pregnancy in rats, *Endocrinology*, **90,** 1133
58. Meyer, G. (1963). *Delayed Implantation*, 213 (A. C. Enders, editor) (Chicago: University of Chicago Press)
59. Linkie, D. M. and Niswender, G. D. (1972). Serum levels of prolactin, Luteinizing Hormone and Follicle Stimulating Hormone during pregnancy in the rat, *Endocrinology* **90,** 632
60. Munshi, S. R., Purandarae, T. V. and Rao, S. S. (1972). Effect of antiserum to ovine luteinizing hormone on corpus luteum function in mice, *J. Reprod. Fert.*, **30,** 7
61. Ahmad, N., Lyons, W. R. and Papkoff, H. (1969). Maintenance of gestation in hypophysectomized rats with highly purified pituitary hormones, *Anat. Rec.*, **164,** 291
62. Greenwald, G. S. (1967). Luteotropic complex of the Hamster, *Endocrinology*, **80,** 118
63. Loewit, K. K., Badawy, S. and Laurence, K. A. (1969). Alterations of corpus luteum function in the pregnant rat by antiluteinizing serum, *Endocrinology*, **84,** 244
64. Spies, H. G. and Quadri, S. K. (1967). Regression of Corpora lutea and interruption of pregnancy in rabbits following treatment with rabbit serum to ovine LH, *Endocrinology*, **80,** 1127
65. Wiest, W. G., Kidwell, K. R. and Balogh, K. (1968). Progesterone catabolism in the rat ovary. A regulatory mechanism for progestational potency during pregnancy, *Endocrinology*, **82,** 844
66. El Tomi, A. E. F., Boots, L. and Stevens, V. C. (1971). Effect of antibodies to human placental lactogen on reproduction in pregnant rats, *Endocrinology*, **88,** 805
67. Evans, H. M., Simpson, M. E., Lyons, W. R. and Turpeinen, K. (1941). Anterior pituitary hormones which favour the production of traumatic uterine placentomata, *Endocrinology*, **28,** 933
68. Kiracofe, G. H., Singh, A. R. and Nghiem, N. D. (1969). Depression of decidual growth in the rat with LH antiserum and pituitary autotransplantation, *J. Reprod. Fert.*, **20,** 473
69. Moudgal, N. R., Rao, A. J., Raj, H. G. M. and Maneckjee, R. (1970). Effect of sheep LH and FSH antisera (A/S) in reproductive processes of the rat and hamster, Abst. **85,** 79 *Proceedings of the 52nd meeting of the Endocrine Society, St Louis, USA*
70. Maneckjee, R., Madhwa Raj, H. G. and Moudgal, N. R. (1973). Comparative effects of antiserum to luteinising hormone on pseudopregnancy and pregnancy induced in the same rat, *Biol. Reprod.*, **8,** 43
71. Chang, C. C., Badawy, S. and Laurence, K. A. (1971). Decidual cell reponse (DCR) and enzyme activity of the ovary in pseudopregnant rats after administration of anti-luteinising hormone (LH) serum, *Fert. Steril.* **22,** 663
72. Yoshinaga, K., Moudgal, N. R. and Greep, R. O. (1971). Progestin secretion by the ovary in lactating rats: Effect of LH antiserum, LH and prolactin, *Endocrinology*, **88,** 1126
73. Brumley, L. E. and De Feo, V. J. (1964). Quantitative studies on deciduoma formation and implantation in the lactating rat, *Endocrinology*, **75,** 883
74. Moudgal, N. R., Macdonald, G. J. and Greep, R. O. (1972). Role of endogenous primate LH in maintaining corpus luteum function of the monkey, *J. Clin. Endocrinol. Metab.*, **35,** 113

75. Snook, R. B., Brunner, M. A., Saatman, R. R. and Hansel, W. (1969). The effect of antisera to bovine LH in hysterectomized and intact heifers, *Biol. Reprod.*, **1,** 49
76. Behrman, H. R. and Armstrong, D. T. (1969). Cholesterol esterase stimulation by luteinising hormone in luteinised rat ovaries, *Endocrinology*, **85,** 474
77. Behrman, H. R., Orczyk, G. P., Macdonald, G. J. and Greep, R. O. (1970). Prolactin induction of enzymes controlling luteal cholesterol ester turnover, *Endocrinology*, **87,** 1251
78. Rothchild, I. (1972). Hormonal requirements for maintenance of the corpus luteum. *Proceedings of the NIH conference on the Regulation of Mammalian Reproduction, Bethesda, 1970*, (S. J. Segal, R. Crozier, P. Corfman and P. Condliffe, editors) (Springfield: Charles C. Thomas)
79. Yoshinaga, K. (1972). Effect of antiserum to rat placenta on the maintenance of embryo at various stages of pregnancy, Abst. **125,** 89 *Proceedings of the Fifth Annual meeting of the Society for the Study of Reproduction. East Lansing, Michigan, 1972*
80. Marsh, J. M. (1969). The role of adenosine 3′,5′-monophosphate in the action of luteinising hormone steroidogenesis, *Progress in Endocrinology*, 83 (C. Guol, editor) (Amsterdam: Excerpta Medica)
81. Greep, R. O. (1971). Regulation of luteal cell function *Hormonal Steroids*, 670 (V. H. T. James and L. Martini, editors) (Amsterdam: Excerpta Medica)
82. Rothchild, I. (1964). An explanation for the cause of luteolysis in the rat and its possible explanation to other species, *Proceedings of the Second International Congress of Endocrinology, London*, 686 (Amsterdam: Excerpta Medica)
83. Choudary, J. B. and Greenwald, G. S. (1968). Comparison of the luteolytic action of LH and estrogen in the hamster, *Endocrinology*, **83,** 129
84. Jagannadha Rao, A. and Moudgal, N. R. (1972). Effect of specific LH and FSH antisera on testicular function in hamsters and guinea pigs (communicated)
85. Greep, R. O., Ferold, W. L. and Hisaw, F. L. (1936). Effects of two hypophyseal gonadotropic hormones on the reproductive system of the male rat, *Anat. Res.*, **65,** 261
86. Means, A. R. and Vaitukaitis, J. (1972). Peptide hormone 'receptors': specific binding of ^{3}H-FSH to testis, *Endocrinology*, **90,** 38
87. Dorrington, J. H., Vernon, R. G. and Fritz, I. B. (1972). Effect of FSH on the 3′,5′-AMP content of seminiferous tubules, Abst. **537,** 214 *Proceedings of the IV International Congress of Endocrinology, Washington, D.C. USA*
88. Lunenfeld, B. and Eshkol, A. (1968). The role of gonad-stimulating hormone on the development of the infantile ovary, *Gonadotropins*, 197 (E. Rosemberg, editor) (Los Altos, Calif, Geron-X)
89. Goldman, B. D. and Mahesh, V. B. (1970). Induction of infertility in male rats by treatment with gonadotropin antiserum during neonatal life. *Biol. Reprod.*, **2,** 444
90. Jagannadha Rao, A. and Moudgal, N. R. (1972). Effect of purified LH and FSH antisera on the development of gonads in male and female rat neonates (communicated)
91. Goldman, A. S., Baker, M. K., Chen, J. C. and Wieland, R. G. (1972). Blockade of masculine differentiation in male rat foetuses by maternal injection of antibodies to testosterone-3-bovine serum albumin, *Endocrinology*, **90** ,716
92. Makino, T., Fang, V. S., Yoshinaga, K. and Greep, R. O. (1972). Effect of implantation of anti LH into median eminence on rat pituitary and serum LH, *Proc. Soc. Exp. Biol. Med.*, **140,** 703
93. Muralidhar, K., Samy, T. S. A. and Moudgal, N. R. (1972). Preparation of luteinising hormone (LH) and follicle stimulating hormone (FSH) from pituitaries of different species using polymerised antisera, Abst. **244,** 98 *Proceedings of the IV International Congress of Endocrinology, Washington, D.C. USA.*
94. Pierce, J. (1971). Eli Lilly lecture: The subunits of pituitary thyrotropin—their relationship to other glycoprotein hormones, *Endocrinology*, **89,** 1331
95. Sairam, M. R., Papkoff, H. and Li, C. H. (1972). Human pituitary interstitial cell stimulating hormone: Primary structure of α-subunit, *Biochem. Biophys. Res. Commun.*, **45,** 530
96. Papkoff, H., Solis-Wallekermann, J., Martin, M. and Li, C. H. (1971). Immunochemical properties of ovine interstitial cell stimulating hormone (ICSH) subunit, *Arch. Biochem. Biophys.*, **143,** 226
97. Gospodarowicz, D. (1971). Immunologic and steroidogenic activity of luteinising hormone (LH) compared to its two subunits: CI and CII, *Endocrinology*, **89,** 669

3
Role of Prostaglandins in Reproduction

H. R. BEHRMAN*
Harvard Medical School, Boston
AND
B. V. CALDWELL
Yale School of Medicine, New Haven

3.1 INTRODUCTION

Although the prostaglandins were first discovered and isolated from sheep seminal vesicles in 1934 by von Euler[1] and considerable work on their

*Present address: Merck Institute for Therapeutic Research, Rahway, New Jersey 07065 (U.S.A.)

chemistry and characterisation accomplished by Bergstrom and his associates over the next 35 years[2–4], it was really the re-discovery of prostaglandins in 1968–1969 that stimulated the current excitement in these compounds as potent mediators of physiological processes. Karim[5] and Wiqvist *et al.*[6] almost simultaneously initiated studies on the action of prostaglandins on the pregnant uterus and opened the door to potential use of these compounds for regulation of fertility control. Other areas of research have also been stimulated to a great degree, but the field of reproductive physiology was influenced most significantly by the emergence of prostaglandins as possible regulators of cellular processes.

Much of the research that is presently planned and conducted with prostaglandins in reproduction is exploratory in nature. However, there is a current trend towards studies into the mechanism of prostaglandin action and advances in this area hold considerable promise. Credit must go to the Upjohn Co. for their early and continued supply of prostaglandin to most investigators and with the ready availability of synthesis inhibitors, antagonists and agonists, progress should be even more rapid. Indeed, the findings of Vane and his colleagues[7] that the most common of drugs, aspirin, may have as its mechanism of action the inhibition of prostaglandin synthesis began a frenzy of studies in many different fields. These fields include gastrointestinal, renal, pulmonary and reproductive physiology, skin disorders, inflammation, headaches and so on, but they are linked by a common thread, prostaglandins. The excitement in prostaglandin research is thus self-regenerating since each day seems to bring a new finding. But of even greater importance is the necessity to extend our knowledge of prostaglandins because of the recognised therapeutic potential use of these compounds.

It will be the primary purpose of this chapter to describe the studies which have been conducted which demonstrate that prostaglandins play a role in regulating many processes in reproductive physiology, including ovulation, oviduct and uterine motility, corpus luteum function, pregnancy and the initiation of labour. In addition to merely reviewing the copious literature, it will be our purpose to present and discuss the current concepts on how prostaglandins exert their actions at the organ, tissue and cellular levels.

3.2 SYNTHESIS, METABOLISM AND ASSAY

Prostaglandins (Table 3.1) are derived from the essential fatty acids and synthesised by a complex, membrane-bound enzyme located in the microsomal fraction of most mammalian cells. Little information is available on the physiological regulation of prostaglandin synthesis, but availability of substrate appears to be a rate-limiting step. In addition, the availability of co-factors such as reduced glutathione, oxygen, nucleotides and antioxidants, as well as the amount of enzyme present to catalyse the synthesis of prostaglandins, is also probably involved in the control of prostaglandin synthesis. For more detailed discussion, the reader is directed to published reviews[8,9].

A characteristic of prostaglandins is that the concentration present within cells and tissue is governed by:

Table 3.1 Structure and nomenclature of the naturally occurring prostaglandins and their derivatives (after Ramwell *et al.*[104])

Formula	*Name*	*Older abbreviation*	*Present abbreviation*
	Prostaglandin E_1: 11α, 15(*S*)-dihydroxy-9-oxo-13-*trans*-prostenoic acid	PGE	PGE_1
	Prostaglandin E_2: 11α, 15(*S*)-dihydroxy-9-oxo-5-*cis*-13-*trans*-prostadienoic acid		PGE_2
	Prostaglandin E_3: 11α, 15(*S*)-dihydroxy-9-oxo-5-*cis*-13-*trans*-17-*cis*-prostatrienoic acid		PGE_3
	Prostaglandin $F_{1\alpha}$: 9α,11α, 15(*S*)-trihydroxy-13-*trans*-prostenoic acid	PGF_{1-1}; PGF_1	$PGF_{1\alpha}$
	Prostaglandin $F_{2\alpha}$: 9α,11α, 15(*S*)-trihydroxy-5-*cis*-13-*trans*-prostadienoic acid	PGF_{1-2}	$PGF_{2\alpha}$
	Prostaglandin $F_{3\alpha}$: 9α,11α, 15(*S*)-trihydroxy-5-*cis*-13-*trans*-17-*cis*-prostatrienoic acid	PGF_{1-3}	$PGF_{3\alpha}$
	Prostaglandin $F_{1\beta}$: 9_β,11α, 15(*S*)-trihydroxy-13-*trans*-prostenoic acid	PGF_{2-1}	$PGF_{1\beta}$
	Prostaglandin $F_{2\beta}$: 9_β,11α, 15(*S*)-trihydroxy-5-*cis*-13-*trans*-prostadienoic acid	PGF_{2-2}	$PGF_{2\beta}$
	Prostaglandin $F_{3\beta}$: 9_β,11α, 15(*S*)-trihydroxy-5-*cis*-13-*trans*-17-*cis*-prostatrienoic acid	PGF_{2-3}	$PGF_{3\beta}$

Table 3.1 (*continued*)

Formula	*Name*	*Older abbreviation*	*Present abbreviation*
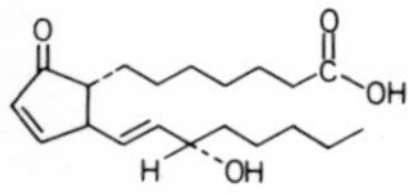	Prostaglandin A_1: 15(*S*)-hydroxy-9-oxo-10,13-*trans*-prostadienoic acid	PGE_1–220; PGE_1–218; PGE_1–217; Δ^{10}–PGE_1	PGA_1
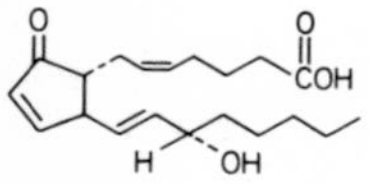	Prostaglandin A_2: 15(*S*)-hydroxy-9-oxo-5-*cis*-10,13-*trans*-prostatrienoic acid	PGE_2–217; Δ^{10}–PGE_2	PGA_2
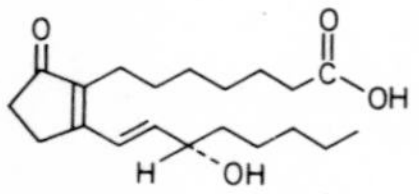	Prostaglandin B_1: 15(*S*)-hydroxy-9-oxo-8(12),13-*trans*-prostadienoic acid	PGE_1–278	PGB_1
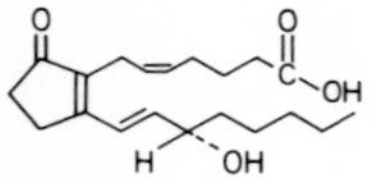	Prostaglandin B_2: 15(*S*)-hydroxy-9-oxo-5-*cis*-8(12),13-*trans*-prostatrienoic acid	PGE_2–278	PGB_2
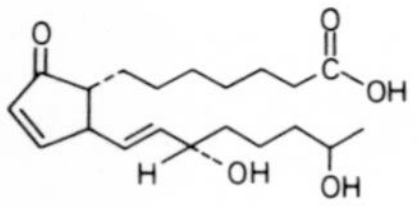	19-Hydroxyprostaglandin A_1: 15(*S*),19-dihydroxy-9-oxo-10,13-*trans*-prostadienoic acid	19-Hydroxy-PGE_1–217	19-Hydroxy-PGA_1
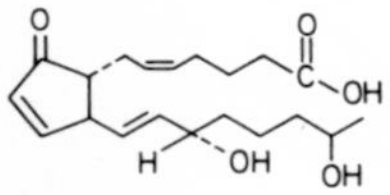	19-Hydroxyprostaglandin A_2: 15(*S*),19-dihydroxy-9-oxo-5-*cis*-10,13-*trans*-prostatrienoic acid	19-Hydroxy-PGE_2–217	19-Hydroxy-PGA_2
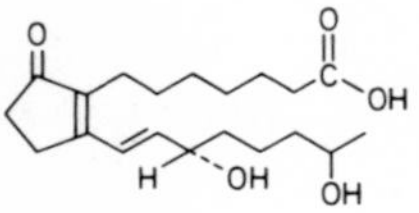	19-Hydroxyprostaglandin B_1: 15(*S*),19-dihydroxy-9-oxo-8(12),13-*trans*-prostadienoic acid	19-Hydroxy-PGE_1–278	19-Hydroxy-PGB_1
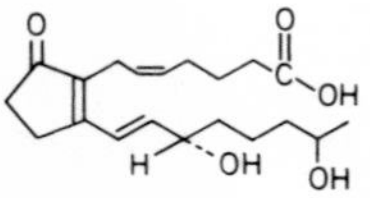	19-Hydroxyprostaglandin B_2: 15(*S*),19-dihydroxy-9-oxo-5-*cis*-8(12),13-*trans*-prostatrienoic acid	19-Hydroxy-PGE_2–278	19-Hydroxy-PGB_2
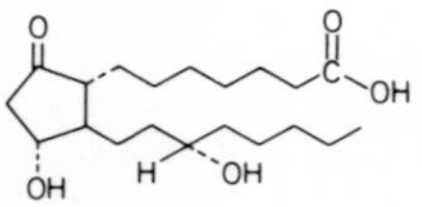	11α,15(*S*)-Dihydroxy-9-oxo prostanoic acid		Dihydro–PGE_1

Table 3.1 (*continued*)

Formula	*Name*	*Older abbreviation*	*Present abbreviation*
	15(*S*)-Hydroxy-9-oxo-8(12)-prostanoic acid	PGE_1–237	Dihydro–PGB_1
	9α,15(*S*)-Dihydroxy-11-oxo-13-*trans*-prostenoic acid		11–Dehydro–$PGF_{1\alpha}$
	11α-Hydroxy-9,15-di-ketoprost-13-enoic acid		15–Keto–PGE_1
	2,3-Dinorprostaglandin $F_{1\alpha}$: 2,3-Dinor-9α,11α,15(*S*)-trihydroxy-13-*trans*-prostenoic acid		Dinor–$PGF_{1\alpha}$

(a) rate of synthesis,
(b) rate of degradation, and
(c) rate of transport outside the cell and tissue into the vascular and lymphatic system.

Since the concentration of prostaglandin necessary to produce biological effects is extremely small, in some cases approaching $10^{-11}mol^{-1}$, each of the above factors must be considered as possible regulatory influences governing prostaglandin action. In addition, however, the nature of the prostaglandin synthesised adds a greater degree of specificity since it is known that different prostaglandins can produce opposite effects.

In general, it appears that prostaglandins produce their biological effects in the tissues where synthesis occurs, except where vascular and/or lymphatic connections permit a direct transport to a secondary tissue before entry into the systemic circulation. A well-known example of the latter case is the transport of prostaglandins from the uterus to the ovary, as will be discussed later. Prostaglandins have an extremely short half-life in circulating fluids due to their rapid degradation primarily in the liver and lungs and thus a great degree of localised action of prostaglandins in the tissue of origin can be achieved, since after one circulation little active prostaglandin remains. This is further augmented by the dilution which occurs in the blood producing a concentration which may be too low to elicit a biological action.

Although considerable research is necessary to gain a complete understanding of prostaglandin metabolism, significant advances have been made by Hamburg and Samuellson[10] in examining degradation of the E-prostaglandins in the human. A rapid and early step which occurs in prostaglandin degradation is oxidation of the hydroxyl group at position 15 to a keto group. This enzyme, 15-hydroxyprostaglandin dehydrogenase (NAD dependent), catalyses the conversion of prostaglandin to this inactive metabolite and occurs primarily in the lungs and liver but is also present in many other tissues such as the placenta[11] and testis[12]. The activity of this enzyme and its regulation may play a fundamental role in preventing build-up of prostaglandins in tissues where a deleterious action would result. For example, in the gravid uterus of the human, synthesis of prostaglandin is perhaps offset by the high activity of 15-hydroxyprostaglandin dehydrogenase[11]. Subsequent to 15-oxidation, reduction of the double bond at position 13–14 occurs and this is then followed by β-oxidation to tetranor prostaglandin. Metabolism of the other prostaglandin classes such as the F series, appears to follow a similar sequence[9].

A limitation to the rapidity with which we can understand the possible physiological role of new compounds is usually the development of methodology of sufficient sensitivity and precision to enable measurement in biological fluids. In the specific case of prostaglandin research, this has certainly been the situation. Early methods which were developed included various bio-assays with all of the classical problems associated with lack of specificity, reproducibility and sensitivity. In retrospect, it is also now clear that these methods gave generous over-estimations and could not always distinguish between the various classes of prostaglandins. The next group of methods reported were elegantly specific and included gas–liquid chromatography, often associated with mass spectroscopy, enzymatic and fluorescence procedures. These methods, however, were limited primarily because of the requirement of sophisticated apparatus, the relatively few samples which could be processed and also their lack of sensitivity.

With the recent developments in radioimmunoassay procedure, however, most of the limitations have been overcome. To be sure, g.l.c.–mass spectrometric methods will continue to be the yardstick against which all other procedures must be compared[13], however many of the recent advances in our knowledge of the role of prostaglandins in various physiological processes come from the use of these radioimmunoassay methods[14–17]. There will continue to be discussion over the relatively high base levels recorded by radioimmunoassay and over the possible limitations of its use to only certain groups of prostaglandins, but increasing use and refinements should soon make these procedures as reliable as the radioimmunoassay for steroids and other hormones. It might also be pointed out here that the original levels of prostaglandins reported with all of the above mentioned were equal to, if not higher than, those reported by radioimmunoassay.

3.3 PROSTAGLANDINS AND OVARIAN FUNCTION

3.3.1 Ovulation

The process of follicular growth, maturation and ovulation cannot be viewed as being autonomous but must be linked to the secretions of the pituitary under the influence of the hypothalamus. Although studies are still in the preliminary stages, recent reports have suggested that gonadotrophin synthesis and perhaps release from the pituitary in response to hypothalamic releasing factors is mediated by prostaglandins. In particular, Guillemin's group has demonstrated *in vitro* that a prostaglandin antagonist (7-oxa-13-prostynoic acid) can block LRF stimulation of LH secretion[18]. Both hypothalamic extracts and PGE_1 have been shown to increase adenyl cyclase activity, and CAMP concentrations in the pituitary[19] and other trophic hormones have been shown to increase greatly in the peripheral circulation following prostaglandin administration[20].

The most compelling evidence for the involvement of prostaglandins in the regulation of the hypothalamic–pituitary–gonadal axis has come from a different line of investigation. In the past year, Vane[7] and others[21] have shown that the non-steroidal, anti-inflammatory agents have in common the ability to inhibit prostaglandin synthesis. It is probable that these agents exert all or at least a major part of their effects by inhibiting the cellular synthesis of prostaglandins. It is interesting that their common side-effects may also be related to this same characteristic. This new pharmacological finding has had even wider implications however, and has provided an important tool for the study of the role of prostaglandins in physiological events.

The first use of this tool in the elucidation of reproductive processes was that reported by Orczyk and Behrman[16] and by Armstrong and Grinwich[22] who independently showed that indomethacin administration to rats could inhibit ovulation in the rat if given at the appropriate time prior to the expected release of LH. Orczyk and Behrman continued their investigations to show that indomethacin did reduce pituitary and hypothalamic concentrations of prostaglandins, and most significantly that a combination of $PGF_{2\alpha}$ and PGE_2 could overcome the indomethacin block of ovulation. To determine the site at which indomethacin or aspirin was blocking ovulation in the rat, Behrman *et al.*[23] used luteinising hormone and synthetic gonadotrophin releasing factor (GN–RF) and found that neither LH nor GN–RF could overcome the block of ovulation caused by indomethacin. Aspirin-induced block, however, was reversed with both LH or GN–RF which suggested that the primary site of action for aspirin was at the hypothalamic level, while the primary action of indomethacin might be at the ovarian level. Of course, it might just be that aspirin, known to be less potent than indomethacin in inhibiting prostaglandin synthesis, was more easily overcome by the exogenous or endogenous released LH. Both indomethacin and aspirin, however, reduced hypothalamic, pituitary and plasma levels of PGF[16].

A similar approach has also been employed to study ovulation in the rabbit, an induced ovulator in which ovulation occurs only after mating at a very precise time (10–12 h post-coitus). In this species, indomethacin given before, or at the time of, mating did not block ovulation[24]. However, if the injection was given 8 h after mating, a single intravenous dose of 20 mg kg^{-1} was sufficient to block ovulation in 100% of the animals. In this study also, the administration of 250 μg $PGF_{2\alpha}$ given at 6, 8 and 10 h after coitus, was able to overcome the block induced by indomethacin in about 50% of the animals[25]. In the rabbit, the stimulus for LH release is thought to be via a neural pathway with LH reaching peak levels within 90 min.

The fact that indomethacin could not block ovulation when given at the time or prior to mating means that the neural and humoral regulation of the hypothalamic–pituitary axis may not be mediated by prostaglandins, and the block of ovulation in this species was at the ovarian level. The timing also means that the block of ovulation is probably not of steroidogenesis *per se*, since the normal increase of 20α-OH progesterone induced by an endogenous LH release 2–6 h after mating has already reached a peak and is declining by the time of the indomethacin administration[26].

The characteristic findings upon histological observation in the rabbit and rat showed there to be a direct effect of indomethacin on the follicles. Luteinisation of the follicles was evident after at least two days following the expected time of ovulation, but in every ovary examined the egg was found to be retained within grossly haemorrhagenic follicles. Progesterone levels in the peripheral plasma of these indomethacin-treated rabbits were not significantly different from normal during the pseudopregnancy that ensued[25]. This showed that the action of LH on steroidogenesis was not blocked, and that a second role of LH, that of actually causing expulsion of the egg, can be said to be mediated by prostaglandins.

The finding of normal patterns of steroidogenesis following block of ovulation with indomethacin suggests the potentially exciting possibility that ovulation may be blocked, ovarian steroidogenesis maintained and a method of contraception available that would not require exogenous administration of steroids. The logical extension of this work to the primate is currently being performed. However, the single most important obstacle to the use of such an approach in the primate will obviously be the lack of a reliable means for predicting the expected time of ovulation.

In conclusion, it can be stated with some confidence that prostaglandins play a direct role in mediating the action of LH to produce the physical process of ovulation, and most likely also mediating the release of LH from the pituitary in the rat.

3.3.2 Corpus luteum function

The corpus luteum was first recognised as a possible regulator of ovarian periodicity in 1887 and although no subject in reproductive physiology received equal research attention, over 70 years was required before a mechanism could be described for the regulation of corpus luteum function. The process of ovulation and the subsequent formation of luteal tissue, with its primary function of progesterone secretion, was known to be under the control of pituitary hormones, primarily LH and to varying degrees, prolactin. Also, most investigators accepted the view that the corpus luteum was directly responsible for regulating the precise length of the oestrous and pseudopregnancy cycles and also, in some non-primate species, the length of gestation. The process for controlling the functional life span, and the induction of corpus luteum regression, 'luteolysis', however, remained a matter of much controversy.

Since 1923, it was clear that a product of the uterus was directly involved in luteolysis when Loeb[27] reported that hysterectomy in the guinea-pig prolonged the functional life span of the *corpus luteum* from a normal interval of 15 days to 60 or more days, a period approximating gestation in this species. The same observation was subsequently extended to include the rat, hamster, mouse, rabbit, sheep, pig, cow and horse. In all of these non-primate mammalian species, merely removing the uterus was able to lengthen corpus luteum life span[28]. In the primates, however, hysterectomy was shown to have no effect on corpus luteum function and regression of luteal tissue occurred at the same time, that is 8–10 days following ovulation in the presence or absence of the uterus[29].

Attention was therefore focused on the non-primate species and it was discovered that the uterus induced luteolysis in a local manner. That is, the 'luteolytic factor' was transmitted through a local tissue or vascular connection to regulate luteal tissue only on the ovary adjacent to the uterine horn in direct tissue continuity[30]. One uterine horn would result in the regression of corpora lutea in the adjacent ovary, while corpora lutea in the ovary with no remaining uterine connection would be maintained. The fact that the luteolytic factor was blood-borne was shown by Caldwell and Moor[31] and later by McCracken *et al.*[32] in reporting that blood, taken from the uterine vein of sheep at the appropriate time, could cause the premature regression of luteal tissue when perfused into other animals. Another clue that it was a uterine product came from the work showing that endometrial extracts could also induce luteal regression, and that transplantation of the uterus with an ovary into the neck of sheep resulted in normal oestrous cycle lengths.

Despite all of the above information, it was not until Pharriss and Wyngarden[33] reported that $PGF_{2\alpha}$ could cause regression of corpora lutea when administered during pseudopregnancy in the rat, that interest centred on whether or not this compound could be the long-sought luteolytic factor. Their observations were rapidly extended to include all of the species that had shown a uterine relationship to luteal function, and work proceeded to establish whether the uterus produced $PGF_{2\alpha}$, whether it was secreted at the appropriate time and whether it could get to the ovary in a local manner. Most of the definitive work has been performed in sheep by a number of workers; therefore, this species will be described in detail here. It is becoming increasingly clear that the same pattern would seem to be operating in most other non-primate mammalian species; however, the situation in the primate is considerably different and will be considered separately.

In sheep, ovulation occurs to initiate each new oestrous cycle during the breeding season, and is characterised by a sequence of hormonal events as shown in Figure 3.1. Hysterectomy in this species results in prolonged luteal function from a normal 16-day period to approximately 165 days, or about the length of normal gestation. Through a variety of experiments, we can now show that the normal mechanism for ending the oestrous cycle most likely begins with a rise in oestradiol on day 13–14 which causes a rise in $PGF_{2\alpha}$ from the uterus. Figure 3.2 shows some of the evidence for this assertion, showing that $PGF_{2\alpha}$ rises sharply after a single administration of oestradiol to ovariectomised ewes previously pre-treated with progesterone[34]. In animals immunised against oestradiol, or previously hysterectomised, no such increase in $PGF_{2\alpha}$ can be detected. In the normal

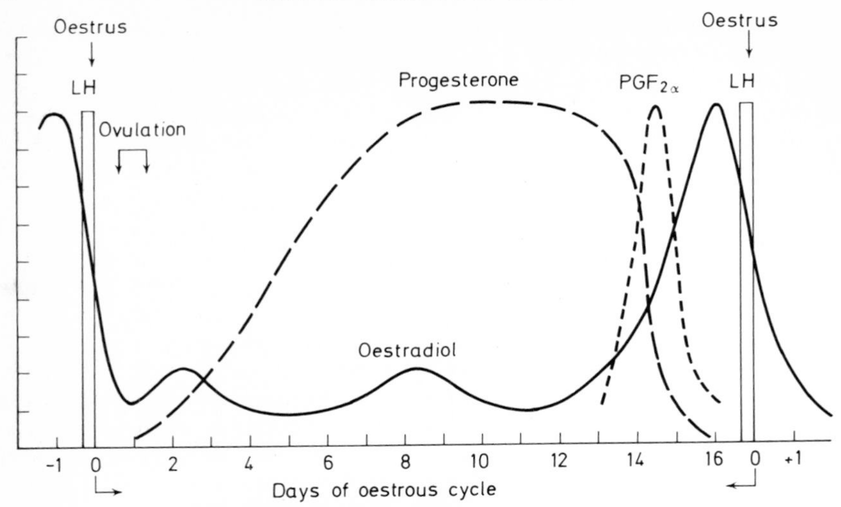

Figure 3.1 A model of the sheep oestrous cycle

oestrous cycle, the levels of $PGF_{2\alpha}$ in the peripheral circulation are highest just at the time of luteal regression. The $PGF_{2\alpha}$, produced in response to the oestradiol, then causes corpus luteum regression, an event which is essential for ovulation to take place since we have shown that if progesterone remains high in sheep, LH release and ovulation is blocked[35].

The sequence of hormonal events regulating ovarian periodicity in sheep is, therefore,

(a) oestradiol as the signal for $PGF_{2\alpha}$ release,
(b) progesterone decline,
(c) oestradiol continuing to rise and becoming the signal for
(d) LH release,
(e) ovulation and the end of one cycle and beginning of the next.

The most important new piece of evidence in support of this concept came from the work of Goddings' group[36] and McCracken *et al.*[32] in their experiments which demonstrated the manner in which $PGF_{2\alpha}$ could be transferred from the uterine vein into the ovarian artery. Using radioactive $PGF_{2\alpha}$, these workers clearly established that $PGF_{2\alpha}$ secreted into the uterine vein of sheep passed selectively through the wall of the vein into the ovarian artery which, in this and many other species, closely adheres to the surface of the uterine vein.

In the rabbit, a different line of investigation has been employed to demonstrate that in this species, like the sheep, $PGF_{2\alpha}$ is the uterine luteolysin[24,25]. Table 3.2 shows that hysterectomy, immunisation against $PGF_{2\alpha}$ and treatment with a prostaglandin synthesis inhibitor (indomethacin) all result in prolonged luteal function during pseudopregnancy. These procedures also result in low to undetectable levels of $PGF_{2\alpha}$ being found in the peripheral circulation.

Obviously many details of the precise control of the sequence of hormonal events which regulate corpus luteum function in all of the non-primate

Table 3.2 **The length of pseudopregnancy in rabbits immunised against $PGF_{2\alpha}$**

Group	*Cycles*	*Duration of pseudopregnancy (days)*
Control	10	16±0.5
Hysterectomised		25–29.0
Immunised	5	25±1.2
Indomethacin	4	26–30.0

mammalian species remain to be worked out. However, from the evidence available it is clear that $PGF_{2\alpha}$ plays a central role in many of these species, the primary function of which is to cause regression of corpora lutea at the end of an infertile cycle. The way in which an embryo is able to overcome

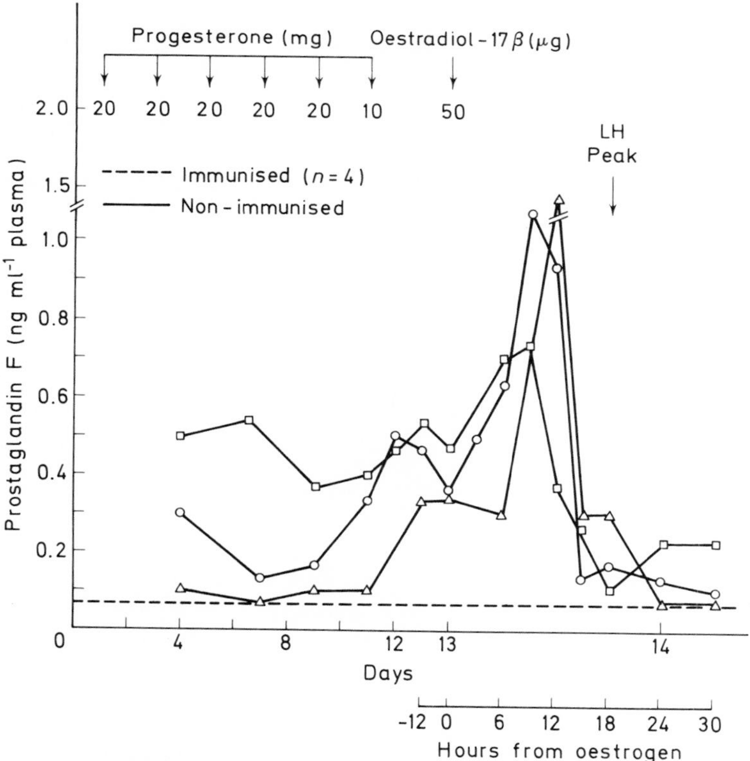

Figure 3.2 Prostaglandin F levels in ovariectomised ewes

this luteolytic influence so that progesterone levels may remain high enough to maintain pregnancy must be established, but possibly arises from a neutralisation of the prostaglandin effect at the level of the corpus luteum (gonadotrophins) or attenuation of uterine prostaglandin production.

The situation in the primate is clearly different since hysterectomy in the monkey and human has no effect on corpus luteum function. It is reasoned that a different mechanism has evolved in the primate species to accomplish the maintenance of the corpus luteum during pregnancy, that being the secretion of a placental trophic hormone. In pregnant women, human chorionic gonadotrophin (HCG) is produced 8–9 days after ovulation and is responsible for the continued production of progesterone from the ovary for the next critical few weeks. This 8–9-day period is exactly the same time that the corpus luteum would normally begin to regress in the absence of an embryo and thus it is felt that the dominance has shifted from a uterine luteolytic control over corpus luteum function in non-primates to a placental luteotrophic control in the primate. $PGF_{2\alpha}$ in the primate is probably still luteolytic as several studies have shown[37,38]. Figure 3.3 demonstrates that $PGF_{2\alpha}$ infused into a monkey caused a rapid decrease in progesterone production when given at high dose levels. Of interest also was the finding

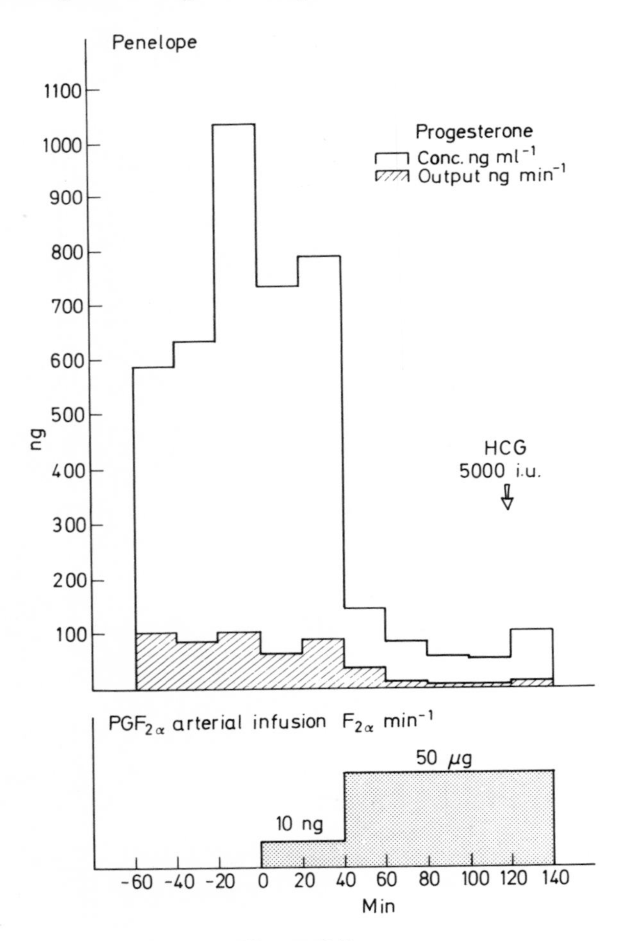

Figure 3.3

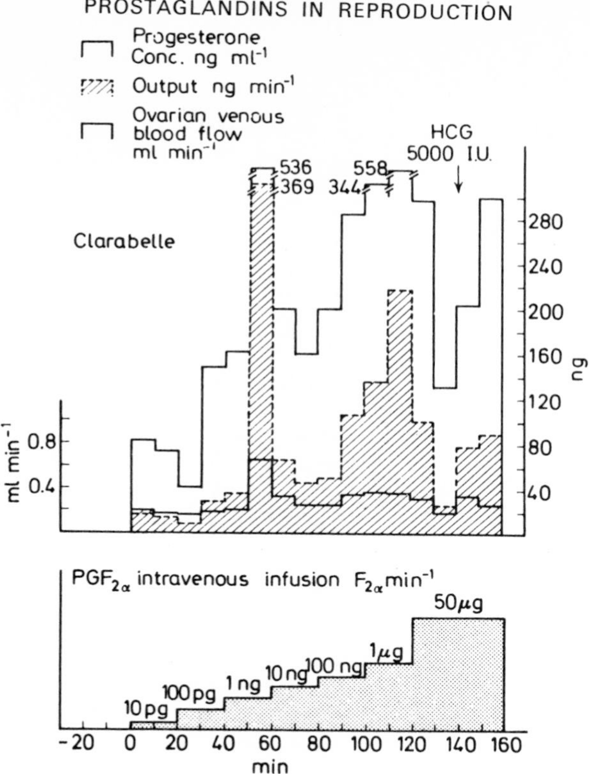

Figure 3.4 Effect of $PGF_{2\alpha}$ infusion on the primate *corpus luteum*

in another study (Figure 3.4) that low doses of $PGF_{2\alpha}$ stimulated progesterone secretion, although when the dose was increased progesterone production was severely depressed[38]. In most cases, it was noted that HCG administered after a luteolytic dose of $PGF_{2\alpha}$ was able to overcome the block on progesterone secretion and bring the levels back to the stimulated rate.

In the human, of studies reported to date in which $PGF_{2\alpha}$ was administered during the luteal phase of normal cycles only one worker has claimed that one cycle was shortened[39]. Two other recent reports do not substantiate this finding[40,41], however, and because of their much greater numbers, the current feeling is that $PGF_{2\alpha}$ has not yet been demonstrated to be luteolytic in the human, which may be due to the side-effects of $PGF_{2\alpha}$ prohibiting use of a concentration sufficient to induce luteolysis. Work currently in progress in several laboratories should answer this in the near future.

Several pieces of evidence have been reported that suggest that prostaglandin release may be under the control of other hormones. In particular, the work on sheep which showed that $PGF_{2\alpha}$ was released under a single injection of oestradiol[34], and the work of Liggins which showed that cortisol, and also oestradiol, could induce the formation of $PGF_{2\alpha}$ in the maternal cotyledons of sheep at the late stages of pregnancy[42]. Since oestrogens have been reported to be 'luteolytic' in a number of non-primate species, the recent findings of Auletta *et al.*[43] and Karsh *et al.*[44] were not surprising. They independently reported that oestrogen administered to monkeys caused a decline in progesterone, and Auletta *et al.* suggested that the blood levels of $PGF_{2\alpha}$ increased significantly in this species after the injection of diethylstilboestrol (Figure 3.5). Continuing these studies, we have also been

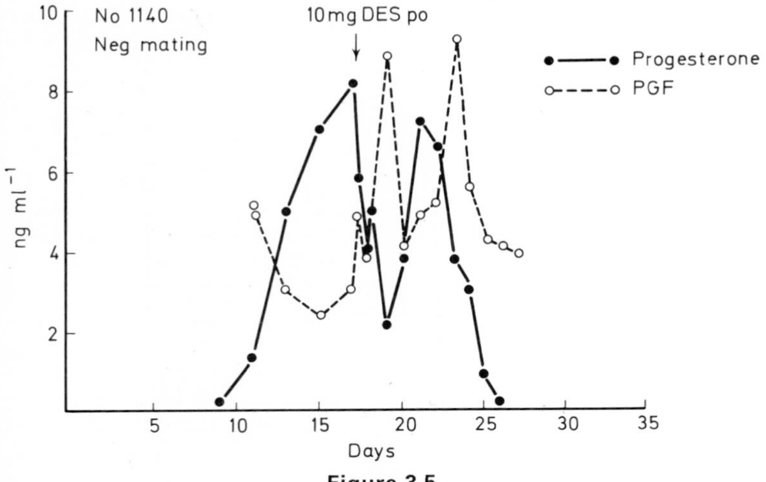

Figure 3.5

able to show that diethystilboestrol is luteolytic in women, although no detectable changes in peripheral levels of $PGF_{2\alpha}$ have been noticed. It is possible, in light of the above findings, that $PGF_{2\alpha}$ may actually be the mediator of the well-known post-coital contraceptive effect of oestrogens. If such is the case, the possibility of finding a compound which would be able to release $PGF_{2\alpha}$ in a manner similar to oestrogens, but be devoid of the side-effects, may be one approach to finding a more acceptable means of post-coital birth control. It is interesting to note here that the common major side-effect of diethylstilboestrol and $PGF_{2\alpha}$ administration is nausea and vomiting, again suggesting that oestrogens may be acting by promoting prostaglandin release.

No discussion of the role of prostaglandins on corpus luteum function would be complete without a mention of the rather confusing data that has been reported using various *in vitro* preparations. Until recently, most workers who tested $PGF_{2\alpha}$ or PGE_2 on corpora lutea tissue slices, homogenates or pieces reported a stimulation of progesterone production[45]. The same prostaglandin, including PGE_2[46], when tested *in vivo*, as mentioned above, were primarily luteolytic. Two recent reports, however, using modified organ culture systems, have described *in vitro* systems in which $PGF_{2\alpha}$ has for the first time had a luteolytic influence[47,48]. The reason for this paradox has certainly not been satisfactorily answered and will be discussed in more detail in Section 3.3.3.

In conclusion, it can be stated that $PGF_{2\alpha}$ is most likely luteolytic in all species, and is the 'uterine luteolytic factor' responsible for regulating the functional life span of corpora lutea in most non-primate mammalian species. Whether or not this activity of $PGF_{2\alpha}$ may ever be put to practical use in humans as a contraceptive device remains to be established. However, other pharmacological approaches using this knowledge may be fundamental to the development of an effective 'once-a-month' pill.

3.3.3 Cellular mechanisms of prostaglandin action

At present, an important area of prostaglandin research is emerging directed towards gaining an understanding of the mechanism of prostaglandin action at the cellular level. Prostaglandins are present and produced in all mammalian cells and produce effects in practically all biological systems tested. An outstanding characteristic of these compounds is their ability to mimic the action of many hormones. Since prostaglandins seem to be active in all endocrine systems, the general belief arose that they perhaps act as intracellular messengers. This interpretation is supported by their ability to cause an alteration of intracellular cyclic 3′5′-AMP levels, a response which probably explains their non-specificity of target organ action in the various endocrine systems.

Recently, an interesting hypothesis was forwarded by Kuehl *et al.*[49] on the role of prostaglandins as mediators of hormone action. Kuehl and Humes[50] suggested, in particular, that the action of LH on the ovary is mediated at the cellular level by the ability of prostaglandins to increase cyclic 3′5′-AMP. The basis of this hypothesis rests on:

(a) The ability of LH to elevate cyclic 3′5′-AMP levels and progesterone secretion.

(b) The ability of PGE_2 to mimic this effect of LH.

(c) The non-additivity at maximal concentrations of PGE_2 and LH with respect to elevation of cyclic 3′5′-AMP levels.

(d) The ability of an antagonist of prostaglandin action (7-oxa-13-prostynoic acid) to competitively block the action of LH to elevate cyclic 3′5′-AMP levels.

These data and their formulation into a workable hypothesis have added a new fundamental dimension to the role of prostaglandins in biological processes and receive support from similar observations made in other endocrine glands such as the thyroid[51] and the pituitary[19].

The above hypothesis is, however, not completely substantiated since Kuehl (personal communication) finds that the prostaglandin antagonists also inhibit phosphodiesterase and cyclic AMP-dependent protein kinase activity and, therefore, also probably inhibit protein synthesis. In addition, Kuehl (personal communication) and Grinwich *et al.*[26] have shown the inhibition of prostaglandin synthesis with indomethacin does not prevent the steroidogenic action of LH. However, it is possible that prostaglandin synthesis is not a pre-requisite for the acute steroidogenic action of LH on the ovary which may be mediated by bound endogenous prostaglandin. It has been shown by Chasalow and Pharriss[52], Behrman[53] and Demers *et al.*[48] that the ovary synthesises prostaglandin, and these same authors have reported a stimulation of prostaglandin synthesis by LH which would be consistent with the Kuehl hypothesis. The data point to the possibility that prostaglandins may be necessary at low concentrations for continued ovarian function with retained ability to respond to gonadotrophins. The observation that chronic treatment of rats with indomethacin did not prevent steroidogenesis[26] and extended luteal function in the rabbit[25] is not consistent with the hypothesis of a prerequisite for prostaglandins by the ovary for continued function. However, these data are negative in character and do not preclude the possibility of continued prostaglandin synthesis (albeit much reduced by indomethacin) at levels sufficient to maintain cellular integrity and responsiveness to gonadotrophins.

Perhaps the most specific action of $PGF_{2\alpha}$ so far described is its ability to cause corpus luteum regression. However, the mechanism of prostaglandin-induced luteolysis is not completely understood at present. The initial hypothesis for this action was forwarded by Pharriss *et al.*[54], who first suggested that $PGF_{2\alpha}$ produced constriction of the ovarian vein thereby causing engorgement of the ovary and loss of luteal function. This hypothesis was later modified to a $PGF_{2\alpha}$-induced decrease in blood flow to the ovary substantiated with data derived from an indirect method (hydrogen desaturation technique) for determining blood flow[55]. On the other hand, Behrman *et al.* did not observe a decrease in rat ovarian venous flow rate within 30 min[56], 6 h or 12 h after $PGF_{2\alpha}$ administration[57], although progesterone secretion was markedly decreased in each case. More recently, Novy[58], using labelled latex beads, demonstrated that $PGF_{2\alpha}$ induced a decrease in blood flow to and from the corpus luteum. At the moment, the effect of $PGF_{2\alpha}$ on ovarian blood flow is not completely resolved. It must be considered that possibly no effect on blood flow to or from the corpus luteum is produced by $PGF_{2\alpha}$, but rather an induced increase in ateriovenous shunting of blood flow, thereby reducing exchange at the site of the corpus luteum.

There is evidence that luteolysis may be induced by a mechanism completely independent from any vascular effects. O'Grady *et al.*[47] and Demers *et al.*[48] have cultured rabbit and rat corpora lutea explants, respectively. Both laboratories independently observed a decrease in progesterone biosynthesis when $PGF_{2\alpha}$ was added to the media. In addition, Demers *et al.*[48] found a $PGF_{2\alpha}$-induced decrease in luteal protein synthesis. These data argue strongly against luteolysis arising from any change in blood flow produced by $PGF_{2\alpha}$.

An alternate possibility for the mechanism of luteolysis induced by $PGF_{2\alpha}$ is an attenuation of gonadotrophin action through the rather complex interrelationship of hormones with $PGF_{2\alpha}$ as shown in Figure 3.6. For example, in the rat both LH and prolactin are necessary for continued corpus luteum function. The action of LH was shown by Armstrong *et al.*[59] to cause an acute increase in progesterone biosynthesis and prolactin has been shown to maintain the ability of the corpus luteum to secrete progesterone[60]. Indeed, Armstrong *et al.*[61] have demonstrated that pre-treatment with prolactin may be necessary in order to demonstrate an acute stimulation of progesterone synthesis by LH. Behrman and Armstrong[62] described an early action of LH on the rat corpus luteum to be activation of cholesterol ester hydrolase, thereby providing a ready source of cholesterol for conversion to progesterone.

In addition, Behrman *et al.*[60] demonstrated that the synthesis of cholesterol esters is catalysed within the corpus luteum by cholesterol ester synthetase, an enzyme which is specifically maintained by prolactin. Thus, the active turnover of cholesterol ester appears to be regulated by both gonadotrophins, a process which is fundamentally linked to steroidogenesis. Pharriss[63] demonstrated that in hypophysectomised rats $PGF_{2\alpha}$ prevented maintenance of corpus luteum function by prolactin. Later, Behrman *et al.*[64] showed that $PGF_{2\alpha}$ prevented that trophic expression of prolactin to maintain the enzymes

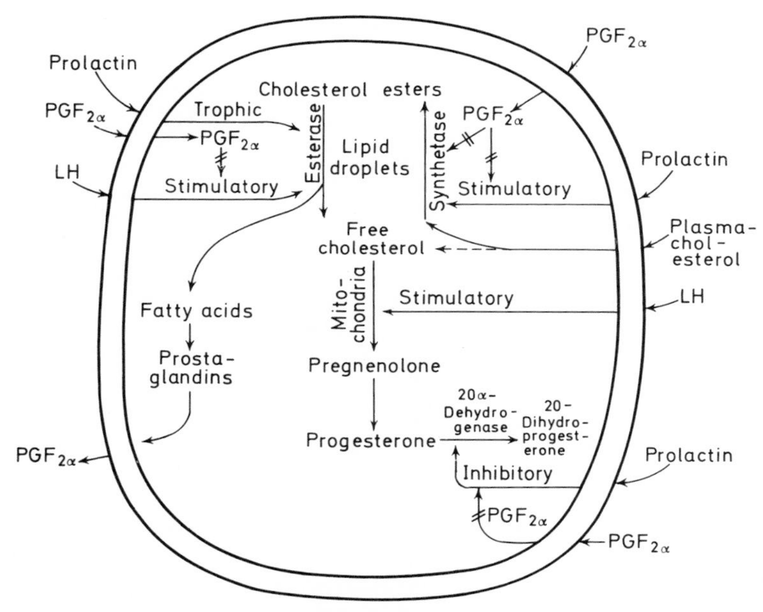

Figure 3.6

required for cholesterol ester turnover with a resultant loss in progesterone biosynthesis, and suggested that the action of $PGF_{2\alpha}$ may be to neutralise gonadotrophin expression. This hypothesis gains support from a similar action of prostaglandin reported in other endocrine systems[65].

The mechanism of prostaglandin-induced luteolysis is not known completely and further research at the cellular level is necessary. At first it appears paradoxical that, on the one hand, prostaglandin is necessary to mediate the action of a hormone and, on the other hand, appears to attenuate the action of the same hormone. However, it is possible that the type of action is dependent upon the nature of the prostaglandin.

For example, Labhsetwar[46] has shown that $PGF_{2\alpha}$ is a more potent luteolytic agent than PGE_2. Kuehl *et al.*[66] find that the E-prostaglandins are more potent than the F-prostaglandins in stimulating cyclic AMP accumulation and have reported that $PGF_{2\alpha}$ at low doses is a very potent stimulator of cyclic GMP formation. The possibility arises that PGE_2 may stimulate cyclic AMP and $PGF_{2\alpha}$ stimulate cyclic GMP, each of which result in different responses.

3.4 PROSTAGLANDINS AND THE REPRODUCTIVE TRACT

3.4.1 The action of prostaglandins on the uterus and oviduct

In reviewing the literature, an immediate paradox is evident between the action of the various prostaglandins on the uterus when tested *in vitro* and *in vivo*. Further confusion also resulted from the discrepancies in action when the prostaglandins were applied to the non-pregnant or pregnant uterus. Although it is difficult to reconcile some of the early studies with those of more recent vintage, the following statements about the action of prostaglandins under the variety of circumstances in which they have been used seems to be warranted.

3.4.1.1 In vitro *studies*

(a) PGE compounds (E_1, E_2 and E_3, in order of potency) when applied to isolated strips of non-pregnant uteri cause an inhibition of motility and a decrease in the amplitude of contraction. PGA, PGB and their 19-OH derivatives, all of which are present in human seminal fluid, also cause this decrease in activity but are much less potent.

(b) PGF compounds have a marked stimulatory action on the non-pregnant uterus *in vitro*[69].

(c) Extracts of human seminal fluid, containing at least 13 different prostaglandins when tested on the same type of human uterine muscle strips, are always inhibitory to about the same extent as PGE compounds, suggesting that it is these compounds in seminal fluid which cause the decrease in muscle tone, overcoming the effects of the PGF present in seminal fluid in large amounts[70].

(d) The same studies, when conducted on strips of uterine muscle taken from pregnant women, show considerably different results. Both PGE and PGF compounds stimulate uterine motility of the pregnant uterus as is also true of extracts of seminal fluid, which were inhibitory in test systems using muscle from non-pregnant women[71].

3.4.1.2 In vivo *studies*

(a) All studies in which various prostaglandins (PGE_1, PGE_2, $PGF_{2\alpha}$, $PGF_{1\alpha}$) were administered by various routes to pregnant or non-pregnant women give a marked stimulation of uterine activity. Although there were qualitative and quantitative differences in the form and amplitude of the contractions induced with the various prostaglandins and via the various routes, in some cases the uterine tone increased and fairly regular patterns of contractions were recorded[72].

(b) The main differences in response of the pregnant uterus *in vivo* to the various prostaglandins is one of relative potency, with PGE_1 and PGE_2 being approximately 10 times more active than $PGF_{2\alpha}$; $PGF_{1\alpha}$ was the least active. The routes of administration did not seem to be very important;

however, for various reasons, intra-amniotic injections have been employed with the most success as will be discussed later in considering the clinical uses of the prostaglandins. A single, intra-amniotic injection of 40 mg $PGF_{2\alpha}$ or 5 mg PGE induces contractions in the mid-trimester patient for a period of up to 24 h. During this time, the uterine tone increases to 40–50 mmHg and the amplitude of the contractions may be consistently above 100 mmHg.

In summary, the predominant influence of all prostaglandins tested *in vivo* in pregnant or cycling women is one of stimulation of myometrial activity. The reasons for the discrepancies between *in vivo* and *in vitro* results remain to be explained, and the importance of prostaglandin effects *in vitro* to any physiological evaluation of the role on myometrial activity cannot be assessed at this time.

Because of the technical difficulties in measuring the response of the fallopian tube *in vivo*, most studies using prostaglandins have been performed with isolated segments of the oviduct tested *in vitro*. The results of such studies are difficult to interpret because of the previously described discrepancies between the action of the prostaglandins on the uterine musculature *in vitro* when compared to *in vivo*. Nevertheless, much attention has been given to the finding that PGE_1 and PGE_2 cause contraction of the proximal end of the tube in human tissues. $PGF_{2\alpha}$ showed a stimulatory action on all portions of the tube and PGE_3 caused an inhibition of all segments. In the only *in vivo* study in man, it was noted that $PGF_{2\alpha}$ had a stimulatory action on the ovary, tube and uterus, whereas PGE_2 relaxed the tube but caused contractions of the ovary and uterus. Since this technique could not differentiate between the action of the prostaglandins on the various portions of the tube, it could not be stated whether or not PGE_2 had any different effects on the various segments of the tube that has been shown *in vitro*[73].

All of the PGs are present in large concentrations in human semen, but PGEs predominate. Sanberg *et al.*[74] have noted the possibility that PGE in seminal fluid could aid in egg transport through its differential action on the fallopian tube, and cause its retention at the ampullary–isthmal junction. The contraction of the proximal (isthmal region) and relaxation of the distal portions could cause a suction in the tube aiding in transport of the egg into the tube, and also the constriction at the ampullary–isthmal junction could be a result of the differential activity of PGE on these two regions. Chang and Hunt[75] have reported several studies in which they have shown the rapid expulsion of eggs from the rabbit oviduct with exogenous prostaglandin. However, as with many of the areas of possible prostaglandin control, oviduct motility and egg transport still remain a matter for more detailed investigation.

There is very little direct evidence that prostaglandins play any role in the process of implantation. Certainly the weight of circumstantial evidence suggests that prostaglandins can influence the transport of the egg through the oviduct, and in this way interfere with the normal process of implantation by altering the precise timing of the arrival of the blastocyst into the uterine lumen. However, when one considers the actual process of blastocyst invasion into maternal tissue, no information on the action of prostaglandins in this process is known.

What can be discussed are the various experiments which have been conducted for other reasons but which have provided some information in this regard. For example, indomethacin, a prostaglandin synthesis inhibitor, when used in the rabbit to block ovulation did not interrupt the establishment of pregnancy when given during the period of tubal transport and implantation[24]. However, without more detailed studies we cannot make any conclusions from this finding.

For some years, the mechanism of action of intrauterine devices (IUD) has been the subject of much study with speculation centring on its effect on tubal transportation and implantation. Recently, it was found that these devices cause a local inflammatory reaction in the endometrium and result in a considerable rise in prostaglandin levels in both uterine tissue and uterine vein blood[76]. Animals treated with indomethacin do not show this increase in prostaglandin levels, and one preliminary study has indicated that implantation can occur in these animals whereas it is blocked in control-IUD treated animals. The possibility suggested by these studies is that the IUD owes its antifertility activity to increase in prostaglandins and also indicates that further investigation in this use[8]of prostaglandins, their analogues or inhibitors may provide a new approach to fertility control.

3.4.2 Placental function

A number of reports have appeared in the last year describing the effects on hormone levels of prostaglandins administered at various times throughout pregnancy in the woman. As with other areas of prostaglandin research, different investigations have yielded quite different results and conclusions which cannot be entirely reconciled at this point. The majority of studies which have been done during the induction of labour at term with $PGF_{2\alpha}$ have shown no marked effect of the drug on plasma progesterone levels. LeMaire *et al.*[77], in a recent study, showed that there was no significant drop in progesterone levels when either $PGF_{2\alpha}$ or oxytocin were used unsuccessfully to induce labour. When, however, either drug was successful in initiating labour, there was a marked drop in progesterone at high drug infusion rates (20 μg min^{-1}) but there was no difference between the patients receiving $PGF_{2\alpha}$ or oxytocin. The distinct possibility in these studies is that progesterone levels decline only after disruption of the vasculature. These authors concluded that neither oxytocin nor $PGF_{2\alpha}$ given intravenously to term-pregnant women significantly altered plasma levels of progesterone. A directly contrasting view is held by Csapo and his colleagues[78] who, in preliminary data, have shown $PGF_{2\alpha}$ infusions do cause a decline in progesterone levels when infused at term and theorise that this fall is essential to the initiation of labour.

The use of prostaglandins for the termination of pregnancy during the first and second trimester has not gained wide acceptance as an alternate choice to surgical approaches, but has provided additional evidence on the possible effects of prostaglandins on placental function. Speroff *et al.*[79] recently showed that the infusion of up to 200 μg min^{-1} of $PGF_{2\alpha}$ had no significant effect on plasma progesterone levels in women receiving the drug between 7.5 and 20 weeks of gestation. Most significant was the finding that in seven patients treated with $PGF_{2\alpha}$ at 7 weeks gestation in which abortion was successfully induced, no significant changes in progesterone or 17-hydroxyprogesterone were noted until just prior to, and following delivery of, the conceptus (Figure 3.7). It was, therefore, concluded that neither

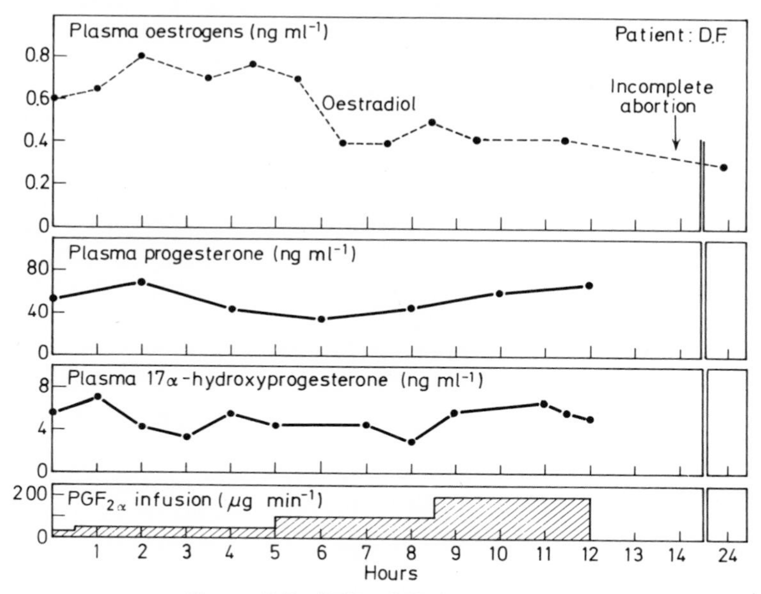

Figure 3.7 $PGF_{2\alpha}$ infusion at 7 weeks

placental nor ovarian steroidogenesis was markedly influenced by $PGF_{2\alpha}$ since 17-hydroxyprogesterone is uniquely an ovarian product in the human.

The oestrogen levels in the Speroff study showed a differential response to $PGF_{2\alpha}$. Oestradiol levels decline gradually throughout the infusion, but oestriol levels fell significantly prior to any change in progesterone or oestradiol concentrations. Oestriol levels were not shown by LeMaire *et al.*[77] to be affected by either oxytocin or $PGF_{2\alpha}$ administered for induction of labour at term. One possible explanation for this discrepancy might be the differences in the amount of drug infused.

One study is particularly notable since it is in marked contrast to those cited above. Jewelewicz *et al.*[41] have reported that $PGF_{2\alpha}$ infused for termination of pregnancy between 12–16 weeks gestation had significant effects in reducing progesterone, oestradiol and HCG. FSH and rennin levels, on the other hand, remained constant. Speroff *et al.*[79] had also measured HCG in their series of patients and could find no significant decline in this hormone prior to expulsion of the foetus. They did, however, find a drop in human placental lactogen (HPL) levels during the course of the infusion. Once again, the discrepancies are hard to explain, especially since the protocols followed by both groups were essentially the same. However, since the Jewelewicz study now stands alone against several others, some caution must be taken in reviewing their work.

In an attempt to summarise the information available on the effect of $PGF_{2\alpha}$ infused during pregnancy, it must be re-stated that some uncertainty still exists, but most workers have shown little or no effect of this drug on progesterone levels, some gradual effects on oestrogens, primarily oestriol and conflicting results on protein hormones of placental origin. Most investigators attempt to explain any effects on these hormones by citing the possibility that the increased uterine tone and contractions induced by $PGF_{2\alpha}$ may alter the blood flow to and from the placental tissue, possibly producing anoxia and thus alter the levels of the hormones in a mechanical way rather than by acting directly on the synthesis on these hormones. It would seem safe to state that the primary mechanism of action of $PGF_{2\alpha}$ in inducing abortion or stimulation of labour at term is probably not through any 'luteolytic' action, but rather through its well-established oxytocic action.

Recently another more direct approach has been employed to study the effects of $PGF_{2\alpha}$ on placental function. Satoh and Ryan[81] have reported a significant increase in cyclic AMP production in placental tissue incubated with prostaglandins. However, Alsat and Cedard[82] have demonstrated that $PGF_{2\alpha}$ and PGE_2 increased aromatisation of testosterone by human placenta and these authors postulate a possible mediation of hormone stimulation of aromatisation by prostaglandin.

3.4.3 Parturition

The mechanism(s) which regulate the onset of labour have been studied in a great number of species and with many different experimental designs. Until recently, these studies have met with little success for, in all cases, some important mediator or regulator has been lacking. When prostaglandins were first shown to be potent stimulators of myometrial activity, many investigators were quick to include them in their experimental approaches and theories. Although we are a long way from describing the sequence of events which lead to parturition in all species, the work of Liggins *et al.*[83], Bedford *et al.*[84] and Thorburn *et al.*[85] have provided the basis for a comprehensive theory on the initiation of labour in the sheep. The degree to which it may be extended to other species remains to be seen, but striking parallels in the human, cow, goat and pig, to name a few, certainly suggest that this hypothesis may have general applicability. Recently the proceedings of a conference on the control of parturition have been published to which the reader is referred for specific details of this work[86]. Only the essentials of the scheme will be presented here, recognising that it is a working hypothesis open to further investigation.

It is now generally believed that the stimulation of the foetal adrenal by ACTH from the foetal pituitary causes an increase in glucocorticoid secretion which then initiates the chain of events leading to the maturity of the lung (induced surfactant production) and activation of various enzymes in the placenta and uterus. Most importantly, oestradiol levels in the blood increase shortly after cortisol and, at the same time, $PGF_{2\alpha}$ concentrations in maternal cotyledons and in uterine vein blood also begin to rise. Progesterone levels, on the other hand, begin to fall at this time, although no direct causitive relationship has been established between this finding and the rise in cortisol, oestradiol, or $PGF_{2\alpha}$. It is known that ACTH or dexamethasone infusions in sheep initiate labour, cause progesterone levels to decline and $PGF_{2\alpha}$ levels to increase. The previously discussed role of $PGF_{2\alpha}$ in causing a decline in ovarian progesterone production makes the speculation that this may be the cause of progesterone decline in placental tissue rather attractive. Liggins *et al.*[83], however, do not feel that the decline in progesterone concentration is essential to initiate labour since exogenous progesterone did not block labour, although the amplitude of the contractions was reduced and the duration of labour was increased.

Oestradiol has been shown by others to cause $PGF_{2\alpha}$ release in several species, while $PGF_{2\alpha}$', on the other hand, has recently been shown to increase aromatisation in the human placenta[82] and, therefore, increase oestrogen production. Once cortisol starts the chain of events, it is possible that oestradiol and $PGF_{2\alpha}$ then feed-back positively on each other and cause continued high levels of both oestradiol and $PGF_{2\alpha}$ to maintain labour. Also compatible with the view that oestradiol may be a trigger in this sequence are the findings that infusions of oestradiol cause an enhancement of rhythmic contractions, increased vascularity and permiability of the uterus, while overcoming the progesterone dominance and increasing the myometrial response to oxytocin[87]. It should also be pointed out that oestrogen infusions in rats and humans have been shown to stimulate uterine contractions at term.

Although $PGF_{2\alpha}$ levels in placental tissue and maternal blood increase dramatically just prior to the initiation of labour, $PGF_{2\alpha}$ infusion into sheep near term did not induce labour. This is in direct contrast to the same studies conducted in humans and other species, and imply that $PGF_{2\alpha}$ was not the

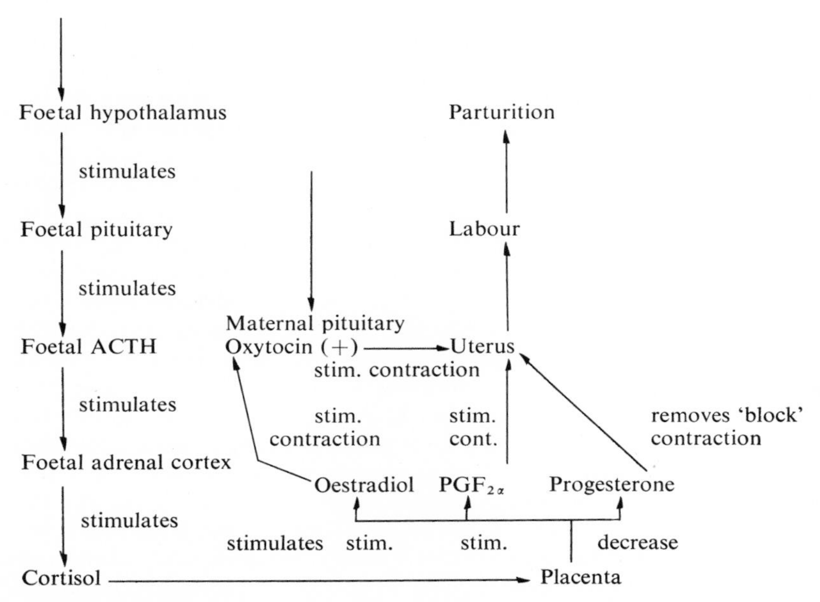

Figure 3.8 A model for the initiation of labour in sheep (modified from Liggins[42].)

final stimulus for uterine contractions in sheep. Liggins and others did note that $PGF_{2\alpha}$ seemed to enhance the activity of oxytocin, and Roberts and Share[88] had previously suggested that oestrogen injections to normally cycling ewes led to marked elevations in oxytocin. Since a decline in progesterone levels is known to increase the response of myometrium to oxytocin and $PGF_{2\alpha}$, perhaps it is the integration of these events which leads to labour rather than any single occurrence.

In summary, the following sequence of events extracted from the work of the several investigators seems to be the most reasonable hypothesis for control of parturition in sheep (see Figure 3.8):

(a) Increase in ACTH from foetal pituitary causes an increase in size and activity of the foetal adrenal with a subsequent increase in glucocorticoid secretion 3–4 days prior to delivery.

(b) Increased glucocorticoid secretion causes an increase in oestrogen secretion and/or $PGF_{2\alpha}$ production.

(c) Progesterone levels decline, perhaps through the action of $PGF_{2\alpha}$ and removal of a 'physiological block' of myometrial activity. This may or may not be important to the whole process.

(d) $PGF_{2\alpha}$, and/or oestradiol, sensitises the myometrium to oxytocin or promotes its secretion, thereby initiating labour.

3.5 CLINICAL USE OF PROSTAGLANDINS IN REPRODUCTIVE PHYSIOLOGY

3.5.1 Prostaglandins for the induction of labour at term

Karim *et al.*, in 1968, first reported clinical results using prostaglandin for the successful induction of labour at term in 10 women[89]. For the next 2 years, Karim and several others continued these studies with both $PGF_{2\alpha}$ and PGE_2 given mainly by intravenous routes and reported that the prostaglandins were superior to oxytocin in initiating labour. Since that time, however, at the suggestion of Anderson *et al.*[90], the design of further studies included a Bishop scoring index to assess the inducibility of patients on the various drugs. Using this scoring system, many workers conducted double blind studies and reported at the recent Upjohn Brook Lodge meeting[91] that a careful analysis of the data showed no significant differences between the efficacy of $PGF_{2\alpha}$, PGE_2 or oxytocin in inducing labour at term. Neither the success rate nor the treatment–delivery interval was different when all groups were compared using the inducibility scoring system. The general feeling is that the results of early findings may have been due to different experimental design, or differences in patients and, at the moment, oxytocin remains the method of choice for labour induction at term.

One possible clinical use for prostaglandins in this area may be for the induction of labour in cases of intrauterine death, missed abortion or anencephalia where oxytocin infusions are not very successful unless continued for extremely long periods of time. Recent reports have been rather optimistic in this limited but important area[92].

The recent findings that the oral administration of PGE_2 in doses of less than 1 mg h^{-1} or $PGF_{2\alpha}$ at 5 mg h^{-1} could induce labour at term have re-stimulated interest in these compounds for this purpose[91]. A large number of patients must be treated using a proper scoring system and double blind design before any attempt is made at comparing these methods with the presently acceptable oxytocin infusions. The clinical future, however, continues to look bright for the prostaglandins despite these early setbacks. Newer analogues, dosage forms, or routes of administration may yet be developed to take advantage of their remarkable potency in stimulating myometrial activity for use as abortifacients and in induction of labour at term.

3.5.2 Prostaglandins for induction of therapeutic abortion

The rationale for using prostaglandins to terminate early pregnancy came from the *in vitro* and *in vivo* findings, as mentioned previously, which showed that $PGF_{2\alpha}$ and PGE_2 were both potent stimulators of uterine contractions. Initial clinical research was carried out independently in Uganda and Stockholm and centred on the response of the pregnant human uterus to intravenous infusions of the drugs[5,6]. Clinical trials were undertaken shortly thereafter since both groups found that, unlike oxytocin, prostaglandin had a stimulating effect on the early pregnant uterus.

The first two publications reporting the use of prostaglandins for therapeutic abortion appeared in the same issue of Lancet in January, 1970[93,94] and aroused considerable interest in both the medical profession and lay public. In the 2½ years since these early results, thousands of abortions have been performed using various prostaglandins given by various routes and with various results being reported. In June 1972, The Upjohn Company, which sponsored most of these studies, conducted a meeting at which the results were reported[91].

In summary, on the past and potential clinical use of prostaglandins the following conclusions were generally accepted:

(a) $PGF_{2\alpha}$ or PGE_2 are effective means for interrupting pregnancy at any stage if given intravenously at rather high doses. The side-effects of this route of administration, however, were generally unacceptable and make their use in this manner questionable.

(b) Intravaginal application of the prostaglandins is also effective but causes moderate to severe side-effects using the presently available compounds and is also, therefore, an unacceptable method.

(c) In general, any method which caused blood levels of prostaglandins to rise in excess of 2–3 ng ml^{-1} of plasma was effective, but produced a high incidence of side-effects.

(d) The intra-amniotic approach in which a large single dose of $PGF_{2\alpha}$ (40 mg) is administered or PGE_2 (10 mg) into the amniotic cavity during the mid-trimester is a very satisfactory method of interrupting pregnancy up to 20–22 weeks (95–100% effective), comparing favourably to the introduction of hypertonic saline for the same purpose. In fact, it was felt that prostaglandins offer some advantage over saline in this rather difficult procedure.

(e) The use of the prostaglandin analogue 15-methyl-$PGF_{2\alpha}$ or 15-methyl-PGE_2 did not seem to offer any major advantage over the parent compound, except that a smaller dose was effective.

(f) Prostaglandins given by any route to interrupt pregnancy in the first 8–9 weeks of pregnancy do not compare favourably with vacuum curetage.

(g) The interruption of pregnancy earlier than 3–4 weeks of gestation by intravaginal or intravenous prostaglandin administration has not been a satisfactory method, yielding too low a success rate with unacceptable side-effects.

(h) Combinations of oxytocin and prostaglandins may offer some advantage over prostaglandins alone, but it is still uncertain how effectively this enhancement may be.

The results, therefore, of approximately 2000–3000 cases suggest that prostaglandins will not gain the overnight success as abortifacients as predicted following the early studies. As a means for interrupting the mid-trimester pregnancy, an intra-amniotic injection may yet prove to be the most reliable method since abortion is relatively fast (16–24 h as an average) and side-effects are less than those seen with hypertonic saline. An example of the recent studies is shown in Table 3.3 comparing the use of $PGF_{2\alpha}$ with and without oxytocin[95].

Table 3.3

	Complete abortion	*Partial abortion*	*Total*
$F_{2\alpha}$	30	10	40
$F_{2\alpha}$ (corrected)*	36	4	40
$F_{2\alpha}$+syntocinon	37	3	40

*Corrected results comparing intra-amniotic $PGF_{2\alpha}$ for induction of mid-trimester abortion with and without a syntocinon infusion (80 patients)[95].

3.6 MALE REPRODUCTION

Prostaglandins were first isolated from the male but relatively little information on the role of prostaglandins is available in this sex. An early clue to a possible role was noted in 1933 by Goldblat and in 1935 by von Euler who independently observed that human semen contained a substance which stimulated uterine smooth muscle contraction (see uterus and oviduct for further details). However, relatively few species contain the high quantities of seminal prostaglandins characteristic of the human[96] and some species such as the rabbit are thought to have little or none present.

Recent evidence by Ellis[97] indicates that the testis of the rat contains the enzymatic machinery necessary for prostaglandin synthesis. The ability of the rat testis to synthesise prostaglandin was directly related to androgen synthesis, as prostaglandin synthesis decreased after hypophysectomy and increased following administration of luteinising hormone. However, in these same experiments androgen synthesis was maintained when the animals were adrenalectomised, but prostaglandin synthesis was reduced. Thus it appears that steroidogenesis may be independent from prostaglandin biosynthesis in the testis. On the other hand, recent reports have shown that $PGF_{2\alpha}$ injected into male mice decreased testosterone levels in circulating plasma.

Although no direct evidence for a role of prostaglandin in spermatogenesis has been shown, indirect evidence indicates that prostaglandin may play a role in this process. PGE levels were reported by Bygdeman[99] to be possibly associated with infertility in human males and Sturde[100] has reported that androgen therapy increased sperm numbers and prostaglandin levels in infertile males. The action of androgen in improving fertility in these cases may be directly related to a role of prostaglandin in spermatogenesis but no definite conclusions can be reached until further evidence is available.

Direct addition of $PGF_{2\alpha}$ has been shown to produce contractions of the rabbit testis whereas PGE caused relaxation[101]. However, concrete evidence for the physiological importance of controlled testicular contractions has not yet been presented, although Davis *et al.*[102] suggest that this phenomenon may be necessary for transport of non-motile sperm from the seminiferous tubules towards the epididymis. Recently, $PGF_{2\alpha}$ has been shown to inhibit adrenalin-stimulated seminal vesicle contractions[103]. These authors postulate that PGE may play a role in delaying seminal vesicle secretion during emission, but the physiological importance of this process is unknown.

The level of prostaglandin in tissues of the male reproductive system appear to be controlled by synthesis, but the importance of degrading enzymes may also play a predominant role. For example, Nalcano and Prancan[12] have demonstrated that the testis has a capacity similar to that of the lung for conversion of PGE_1 to the inactive metabolite 15-keto-PGE_1, and thus the necessity for controlling prostaglandin levels through degradation provides a means for regulation. Clearly, considerable research must be conducted to understand the role of prostaglandins in male reproduction since much of the present evidence, although promising, is inconclusive. The fact that prostaglandins are synthesised in the testis, change during different functional states and modify function when added *in vitro* imply an important but unknown role.

References

1. Euler, U. S. von (1934). Zur kenntnis der pharmakologischen wirkungen von nativsekreten und extrackten männlicher accessorischer geschlechtsdrusen, *Arch. Exp. Pathol. Pharmakol.*, **175,** 78
2. Bergstrom, S. and Sjovall, J. (1957). The isolation of prostaglandin, *Acta Chem. Scand.*, **11,** 1086
3. Bergstrom, S. (1967). Prostaglandins: Members of a new hormonal system, *Science*, **157,** 382
4. Bergstrom, S. and Samuelsson, B. (1962). Isolation of prostaglandin E_1 from human seminal plasma, *J. Biol. Chem.*, **237,** 3005
5. Karim, S. M. M. (1968). Appearance of prostaglandin $F_{2\alpha}$ in human blood during labour, *Brit. Med. J.*, **4,** 618

6. Wiqvist, N., Bygdeman, M., Kwon, S. U., Mukherjee, T. and Roth-Brandel, U. (1968). Effect of prostaglandin E_1 on the midpregnant human uterus, *Amer. J. Obstet. Gynecol.*, **102,** 327
7. Vane, J. R. (1971). Inhibition of prostaglandin synthesis as a mechanism of action for aspirin-like drugs, *Nature* (*New Biol.*) (*London*), **231,** 232
8. Hamburg, M. and Samuelsson, B. (1966). Novel biological transformations of 8,11,14-eicosatrienoic acid, *J. Amer. Chem. Soc.*, **88,** 2349
9. Samuelsson, B. (1972). Biosynthesis of prostaglandins, *Fed. Proc.* (*Fed. Amer. Soc. Exp. Biol.*), **31,** 1442
10. Hamberg, M. and Samuelsson, B. (1971). On the metabolism of prostaglandins E_1 and E_2 in man, *J. Biol. Chem.*, **246,** 6713
11. Jarabak, J. (1972). Human placental 15-hydroxyprostaglandin dehydrogenase, *Proc. Nat. Acad. Sci.* (*USA*), **69,** 533
12. Nalcano, J. and Prancan, A. V. (1971). Metabolic degradation of prostaglandin E_1 in the rat plasma and in rat brain, heart, lung, kidney and testicle homogenates, *J. Pharm. Pharmacol.*, **23,** 231
13. Shaw, J. E. and Ramwell, P. W. (1969). Separation, identification and estimation of prostaglandins. *Method. Biochem. Anal.*, **17,** 325
14. Jaffe, B. M., Smith, J. W., Newton, W. T. and Parker, C. W. (1971). Radioimmunoassay of the F-prostaglandins, *J. Clin. Endocrinol. Metab.*, **33,** 171
15. Caldwell, B. V., Burstein, S., Brock, W. A., and Speroff, L. (1971). Radioimmunoassay of the F-prostaglandins, *J. Clin. Endocrinal. Metab.*, **33,** 171
16. Orczyk, G. P. and Behrman, H. R. (1972). Ovulation blockade by aspirin or indomethacin. *In vivo* evidence for a role of prostaglandin in gonadotrophin secretion, *Prostaglandins*, **1,** 3
17. Levine, L. and Vunakis, H. von (1970). Antigenic activity of prostaglandins, *Biochem. Biophys. Res. Commun.*, **41,** 1171
18. Amos, M., Blackwell, R., Vale, W., Burgus, R. and Guillemin, R. (1971). Stimulation of concommitant secretion *in vitro* of LH and FSH of highly purified hypothalamic LRF; evidence for a prostaglandin receptor for the release of LH, *15th Int. Cong. Phys. Sci.*, **9,** 17
19. Zor, U., Kaneko, T., Schneider, H. P. G., McCann, S. M. and Field, J. (1970). Further studies on stimulation of anterior pituitary cyclic adenosine 3′,5′-monophosphate formation by hypothalamic extract and prostaglandins, *J. Biol. Chem.*, **245,** 2883
20. Hertelendy, F., Todd, H., Erhart, K. and Blute, R. (1972). Studies on growth hormone secretion. IV *In vivo* effects of prostaglandin E_1, *Prostaglandins*, **2,** 79
21. Smith, J. B. and Willis, A. L. (1971). Aspirin selectivity inhibits prostaglandin production in human platelets, *Nature* (*New Biol.*) (*London*), **231,** 235
22. Armstrong, D. T. and Grinwich, D. L. (1972). Blockade of spontaneous and LH-induced ovulation in rats by indomethacin, an inhibitor of prostaglandin biosynthesis, *Prostaglandins*, **1,** 21
23. Behrman, H. R., Orczyk, G. P. and Greep, R. O. (1972). Effect of synthetic gonadotrophin-releasing hormone on ovulation blockade by aspirin and indomethacin, *Prostaglandins*, **1,** 245
24. O'Grady, J. P., Caldwell, B. V., Auletta, F. J. and Speroff, L. (1972). The effects of an inhibitor of prostaglandin synthesis (indomethacin) on ovulation, pregnancy and pseudopregnancy in the rabbit, *Prostaglandins*, **1,** 97
25. Caldwell, B. V., Speroff, L., Brock, W. A., Auletta, F. J., Gordon, J. W., Anderson, E. G. and Hubbins, J. C. (1972). Development and application of a radioimmunoassay for F-prostaglandins, *Brooklodge Symp. Prostaglandins* (E. Southern and N. Patel, editors) (Kalamazoo: Upjohn Co.)
26. Grinwich, D. L., Kennedy, T. G. and Armstrong, D. T. (1972). Dissociation of ovulatory and steroidogenic actions of luteinizing hormone in rabbits with indomethacin, an inhibitor of prostaglandin synthesis, *Prostaglandins*, **1,** 89
27. Loeb, L. (1923). The effect of extirpation of the uterus on the life and function of the *corpus luteum* in the guinea-pig, *Proc. Soc. Exp. Biol. Med.*, **20,** 441
28. Pharriss, B. B., Tillson, S. A. and Erickson, R. R. (1972). Prostaglandins and ovarian function, *Record Progr. Hormone Res.*, **28,** 51

29. Beling, C. G., Marcus, S. L. and Markham, S. M. (1970). Functional activity of the *corpus luteum* following hysterectomy, *J. Clin. Endocrinol. Metab.*, **30,** 30
30. Calwell, B. V., Rowson, L. E. A., Moor, R. M. and Hay, M. F. (1969). The utero–ovarian relationship and its possible role in infertility, *J. Reprod. Fert., Suppl.*, **8,** 59
31. Caldwell, B. V. and Moor, R. M. (1971). Further studies on the role of the uterus in the regulation of *corpus luteum* function in sheep, *J. Reprod. Fert.*, **26,** 133
32. McCracken, J. A., Baird, D. T. and Goding, J. R. (1971). Factors affecting the secretion of steroids from the transplanted ovary in the sheep, *Record Prog. Hormone Res.*, **27,** 537
33. Pharriss, B. B. and Wyngarden, L. J. (1969). The effect of prostaglandin $F_{2\alpha}$ on the progestogen content of ovaries from pseudopregnant rats, *Proc. Soc. Exp. Biol. Med.*, **130,** 92
34. Caldwell, B. V., Tillson, S. A., Brock, W. A. and Speroff, L. (1972). The effect of exogenous progesterone and oestradiol on prostaglandin F levels in ovariectomized ewes, *Prostaglandins*, **1,** 217
35. Scaramuzzi, R. J., Tillson, S. A., Thorneycroft, I. H. and Caldwell, B. V. (1971). Action of exogenous progesterone and oestrogen on behavioral oestrus and luteinizing hormone levels in the ovariectomized ewe, *Endocrinology*, **88,** 1184
36. Barret, S., Blockey, M. A., Brown, J. M., Catt, K. J., Cumming, I. A., Goding, J. R., Mole, B. J. and Obst, J. M. (1971). Initiation of the oestrus cycle in the ewe by infusions of prostaglandin $F_{2\alpha}$ to the autotransplanted ovary, *J. Reprod. Fert.*, **24,** 136
37. Kirton, K. T., Pharriss, B. B. and Forbes, A. D. (1970). Luteolytic effects of prostaglandin $F_{2\alpha}$ in primates, *Proc. Exp. Biol. Med.*, **133,** 314
38. Auletta, F. J., Speroff, L. and Caldwell, B. V. (1973). Prostaglandin $F_{2\alpha}$ induced steroidogenesis and luteolysis in the primate *corpus luteum*, *J. Clin. Endocrinol. Metab.*, in press
39. Lehmann, F., Peters, F., Breckwoldt, M. and Bettendorf, G. (1972). Plasma progesterone levels during infusion of prostaglandin $F_{2\alpha}$ in the human, *Prostaglandins*, **1,** 269
40. LeMaire, W. J. and Shapiro, A. G. (1972). Prostaglandin $F_{2\alpha}$: its effect on the *corpus luteum* of the menstrual cycle, *Prostaglandins*, **1,** 259
41. Jewelewicz, R., Cantor, B., Dyrenfurth, I., Warren, W. P. and Van de Wiele, R. L. (1972). Intravenous infusion of prostaglandin $F_{2\alpha}$ in the mid-luteal phase of the normal human menstrual cycle, *Prostaglandins*, **1,** 443
42. Liggins, G. C. (1972). Endocrine factors in the initiation of parturition. *Record Prog. Hormone Res.*, in press
43. Auletta, F. J., Caldwell, B. V., Van Wagenen, G. and Morris, J. Mcl (1972). Effects of postovulatory oestrogen on progesterone and prostaglandin F levels in the monkey, *Contraception*, **6,** 411
44. Karsh, F. J., Weick, R. F., Dierschke, D. J., Krey, L. C., Hotchkiss, J., Yamaji, T. and Knobil, E. (1972). Oestrogen-induced luteolysis in the rhesus monkey, *Proc. 4th Int. Cong. Endocrinol. Washington (D.C.)*, in press
45. Speroff, L. and Ramwell, P. W. (1970). Prostaglandin stimulation of *in vitro* progesterone synthesis, *J. Clin. Endocrinol. Metab.*, **30,** 345
46. Labhsetwar, A. R. (1972). Prostaglandin E_2: Evidence for luteolytic effects, *Prostaglandins*, **2,** 23
47. O'Grady, J. P., Kohurn, E. I., Glass, R. H., Caldwell, B. V., Brock, W. A. and Speroff, L. (1972). Inhibition of progesterone synthesis *in vivo* by prostaglandin $F_{2\alpha}$, *J. Reprod. Fert.*, **30,** 153
48. Demers, L., Behrman, H. R. and Greep, R. O. (1972). Effects of prostaglandins and gonadotrophins on luteal prostaglandin and steroid biosynthesis, *Advan. Biosci.*, **9,** in press
49. Kuehl, F. A., Humes, J. L., Tarnoff, J., Cirillo, V. J. and Ham, E. A. (1970). Prostaglandin receptor site: evidence for an essential role in the action of luteinizing hormone, *Science*, **169,** 883
50. Kuehl, F. A. and Humes, J. L. (1972). Direct evidence for a prostaglandin receptor and its application to prostaglandin measurements, *Proc. Nat. Acad. Sci. (USA)*, **69,** 480

51. Sato, S., Szabo, M., Kowalski, K. and Burke, G. (1972). Role of prostaglandin in thyrotropin action on thyroid, *Endocrinology*, **90**, 343
52. Chasalow, F. I. and Pharriss, B. B. (1972). Luteinizing hormone stimulation of ovarian prostaglandin biosynthesis, *Prostaglandins*, **1**, 107
53. Behrman, H. R. (1972). Regulation of ovarian steroid secretion, *Proc. 4th Int. Cong. Endocrinol. Washington (D.C.)*, in press
54. Pharriss, B. B., Cornette, J. C. and Gutknecht, G. D. (1970). Vascular control of luteal steroidogenesis, *J. Reprod. Fert., Suppl.*, **10**, 97
55. Gutknecht, G. D., Duncan, G. W. and Wyngarden, L. J. (1970). Effect of prostaglandin $F_{2\alpha}$ on ovarian blood flow in the rabbit as measured by hydrogen desaturation, *Physiologist*, **13**, 214
56. Behrman, H. R., Yoshinaga, K. and Greep, R. O. (1971). Extra-luteal effects of prostaglandins, *Ann. N.Y. Acad. Sci.*, **180**, 426
57. Behrman, H. R., Yoshinaga, K., Wyman, H. and Greep, R. O. (1971). Effects of prostaglandin on ovarian steroid secretion and biosynthesis during pregnancy, *Amer. J. Physiol.*, **221**, 189
58. Novy, M. J. (1972). Distribution of ovarian blood flow in rabbits as measured by radioactive microspheres, *Proc. Soc. Study Reprod.*, **A17**, 24
59. Armstrong, D. T., O'Brien, J. and Greep, R. O. (1964). Effects of luteinizing hormone on progestin biosynthesis in the luteinized ovary, *Endocrinology*, **75**, 488
60. Behrman, H. R., Orcyzk, G. P., Macdonald, G. J. and Greep, R. O. (1970). Prolactin induction of enzymes controlling luteal cholesterol ester turnover, *Endocrinology*, **86**, 1251
61. Armstrong, D. T., Miller, L. S. and Knudsen, K. A. (1969). Regulation of lipid metabolism and progesterone production in rat *corpora lutea* and ovarian interstitial elements by prolactin and luteinizing hormone, *Endocrinology*, **85**, 393
62. Behrman, H. R. and Armstrong, D. T. (1969). Cholesterol esterase stimulation by luteinizing hormone in luteinized rat ovaries, *Endocrinology*, **85**, 574
63. Pharriss, B. B. (1970). The possible vascular regulation of luteal function, *Perspect. Biol. Med.*, **13**, 434
64. Behrman, H. R., Macdonald, G. J. and Greep, R. O. (1971). Regulation of ovarian cholesterol esters: evidence for the enzymatic sites of prostaglandin-induced loss of *corpus luteum* function, *Lipids*, **6**, 791
65. Ozer, A. and Sharp, G. W. G. (1972). Effects of prostaglandins and their inhibitors on osmotic water flow in the toad bladder, *Amer. J. Physiol.*, **222**, 674
66. Kuehl, F. A., Cirillo, V. J., Ham, E. A. and Humes, J. L. (1972). The regulatory role of the prostaglandins on the cyclic 3′,5′-AMP system, *Advan. Biosci.*, **9**, in press
67. Pickles, V. R. (1967). The prostaglandins, *Biol. Rev.*, **42**, 614
68. Bydgeman, M. (1964). The effect of different prostaglandins on the human myometrium *in vitro*, *Acta Physiol. Scand.*, **63**, (Suppl. 242), 1
69. Bygdeman, M. (1967). Studies on the effects of prostaglandins in seminal plasma on human myometrium *in vitro* 93, *Prostaglandins*, *Proc. 2nd Nobel Symposium* (S. Bergstrom and B. Sammuelsson, editors) (Stockholm: Almqvist and Wiksell; New York: Interscience)
70. Bygdeman, M. and Eliasson, R. (1963). The effect of prostaglandin from human seminal fluid on the motility of the non-pregnant human uterus *in vitro*, *Acta Physiol. Scand.*, **59**, 43
71. Bygdeman, M., Kwon, S. V., Mutherjee, T., Roth-Brandel, U. and Wiqvist, N. (1970). The effect of the prostaglandin F compounds on the contractility of the pregnant human uterus, *Amer. J. Obstet. Gynecol.*, **106**, 567
72. Roth-Brandel, V., Bygdeman, M. and Wiqvist, N. (1970). Effect of intravenous administration of prostaglandin E_2 and $F_{2\alpha}$ on the contractility of the non-pregnant human uterus *in vivo*, *Acta Obstet. Gynec. Scand.* **49**, Suppl. 5, 19
73. Karim, S. M. M. (1972). *The Prostaglandins* (S. M. M. Karim, editor) (New York: John Wiley & Sons)
74. Sandberg, F., Ingelman-Sundberg, A. and Ryden, G. (1965). The effect of prostaglandin $F_{1\alpha}$, $F_{1\beta}$, $F_{2\alpha}$ and $F_{2\beta}$ on the human uterus and the fallopian tubes *in vitro*, *Acta Obstet. Gynecol. Scand.*, **44**, 585
75. Chang, M. C. and Hunt, D. M. (1972). Effect of prostaglandin $F_{2\alpha}$ on the early pregnancy of rabbits, *Nature* (*London*), **236**, 120

76. Stillman, C. H. and Duby, R. T. (1972). Prostaglandin mediated luteolytic effect of an intrauterine device in sheep, *Prostaglandins*, **2,** 159
77. LeMaire, W. J., Spellacy, W. N., Shevach, A. B. and Gall, S. A. (1972). Changes in plasma estriol and progesterone during labor induced with prostaglandin $F_{2\alpha}$ or oxytocin, *Prostaglandins*, **2,** 93
78. Caspo, A. I. (1972). On the mechanism of the abortifacient action of prostaglandin $F_{2\alpha}$, *J. Reprod. Med.*, **9,** 400
79. Speroff, L., Caldwell, B. V., Brock, W. A., Anderson, G. G. and Hobbins, J. C. (1972). Hormone levels during prostaglandin $F_{2\alpha}$ infusions for therapeutic abortion, *J. Clin. Endocrinol. Metab.*, **34,** 531
80. Cantor, B., Jewelewicz, R., Warren, M., Dyrenfurth, I., Patner, A. and Vande Wiele, R. (1972). Hormonal changes during induction of mid-trimester abortion by prostaglandin $F_{2\alpha}$, *Amer. J. Obstet. Gynecol.*, **113,** 607
81. Satoh, K. and Ryan, K. J. (1972). Prostaglandins and their effects on human placental adenyl cyclase, *J. Clin. Invest.*, **51,** 456
82. Alsat, E. and Cedard, L. (1972). Action stimulatrice des prostaglandines sur la production d'oestrogenes par le placenta humaine perfuse *in vitro*, *C. R. Acad. Sci.*, in press
83. Liggins, G. C., Grieves, S. A., Kendall, J. Z. and Knox, B. S. (1972). The physiological roles of progesterone, oestradiol-17B and prostaglandin $F_{2\alpha}$ in the control of ovine parturition, *J. Reprod. Fert.*, *Suppl.*, **16,** 85
84. Bedford, C. A., Challis, J. R. G., Harrison, F. A. and Heap, R. B. (1972). The role of oestrogens and progesterone in the onset of parturition in various species. *J. Reprod. Fert.*, *Suppl.*, **16,** 1
85. Thorburn, G. D., Nicol, D. H., Kendall, J. Z., and Knox, B. S. (1972). Parturition in the goat and sheep; changes in corticosteroids, progesterone, oestrogens and prostaglandin F, *J. Reprod. Fert.*, *Suppl.*, **16,** 61
86. Perry, J. S. (1972). Control of Parturition, *J. Reprod. Fert.*, *Suppl.*, **16,** 1–136
87. Csapo, A. and Wood, C. (1968). The endocrine control of the initiation of labor in the human, *Recent Advances in Endocrinology*, 8th ed. (V. H. T. James, editor) (London: Churchill)
88. Roberts, J. S. and Share, L. (1969). Effects of progesterone and oestrogen on blood levels of oxytocin during vaginal distention, *Endocrinology*, **84,** 1076
89. Karim, S. M. M., Trussell, R. R., Patel, R. D. and Hillier, K. (1968). Response of pregnant human uterus to prostaglandin $F_{2\alpha}$-induction of labor, *Brit. Med. J.*, **4,** 621
90. Anderson, G., Hobbins, J. C., Codero, L. and Speroff, L. (1971). Clinical use of prostaglandins as oxytocin substances, *Ann. N.Y. Acad. Sci.*, **180,** 499
91. Southern, E. M. (1973). *Brooklodge Symp. The Prostaglandins*, Clinical application in human reproduction (E. Southern and N. Patel, editors) Futura Pub. Co. New York
92. Pedersen, P. H., Larsen, J. F. and Sorensen, B. (1972). Induction of labor with prostaglandin $F_{2\alpha}$ in missed abortion, fetus mortures and anencephalia, *Prostaglandins*, **2,** 135
93. Roth-Brandel, U., Bygdeman, M., Wiqvist, N. and Bergstrom, S. (1970). Prostaglandins for induction of therapeutic abortion, *Lancet*, **1,** 190
94. Karim, S. M. and Filshie, G. M. (1970). Therapeutic abortion using prostaglandin $F_{2\alpha}$, *Lancet*, **1,** 157
95. Anderson, G. G., Hobbins, J. C., Speroff, L. and Caldwell, B. V. (1972). Midtrimester abortion using intra-amniotic $PGF_{2\alpha}$, *Brooklodge Symp. Prostaglandins* (E. Southern and N. Patel, editors) (Kalamazoo: Upjohn Co.)
96. Eliasson, R. (1965). Effect of frequent ejaculations on the composition of human seminal plasma, *J. Reprod. Fert.*, **9,** 331
97. Ellis, L. C. (1972). Rat testicular prostaglandin synthesis and its relationship to androgen synthesis, *Fed. Proc.* (*Fed. Amer. Soc. Exp. Biol.*), **31,** A295
98. Bartke, A., Musto, M., Behrman, H. R. and Caldwell, B. V. (1973). Effects of a cholesterol esterase inhibitor and of prostaglandin $F_{2\alpha}$ on testis cholesterol and on plasma testoslerone in mice, *Endocrinology*, **92,** 1223
99. Bygdeman, M. (1969). Prostaglandins in human seminal fluid and their correlation to fertility, *Int. J. Fert.*, **14,** 228

100. Sturde, H. C. (1971). Das Verhalten der sperma-prostaglandine unter androgen therapie, *Arzneimittel-Forsch.*, **21,** 1293
101. Free, M. J. and Jaffe, R. A. (1972). Effect of prostaglandins on blood flow and pressure in the conscious rat, *Prostaglandins*, **1,** 483
102. Davis, J. R., Langford, G. A. and Kirby, P. J. (1970). The testicular capsule, *The Testis*, Vol 1, 281. (A. D. Johnson, W. R. Gomes, N. L. Vandemark, editors) (New York: Academic Press)
103. Ito, H., Katayama, T., Takagishi, H. and Momose, G. (1972). Contraction of seminal vesicle and prostaglandin E_1, *Prostaglandins*, **1,** 327
104. Ramwell, P., Shaw, J. E., Clarke, G. B., Grostie, M. F., Kaiser, D. G. and Pike, J. E. (1968). *Prostaglandins. In* Prog. in the chemistry of fats and other lipids, vol. 9, p. 231. (R. J. Holman, editor) New York: Pergamon Press

4
Recent Progress in the Study of Eggs and Spermatozoa: Insemination and Ovulation to Implantation

C. R. AUSTIN
Physiological Laboratory, Cambridge

4.1 SPERM TRANSPORT IN THE FEMALE TRACT

The consensus of opinion continues to be that in mammals the spermatozoa are not directionally orientated during their passage through the vagina, uterus and oviduct (Fallopian tube)—their individual tracks tend to be circular rather than straight, and there is no evidence that they can show chemotactic attraction in the way that is well known, in, for instance, fern spermatozoids. On deposition in the female tract at coitus—whether in uterus or vagina—the spermatozoa tend to disperse in all directions, rather like gas molecules diffusing from a focal point. General orientation is imposed by the structural or mechanical features of the female organs. Essentially, transport from the site of deposition to the site of fertilisation is effected by a process of mixing: soon after deposition the spermatozoa become suspended in female tract secretions which are then actively mingled by muscular contractions of the walls and by the beating action of cilia. As the suspension tends towards homogeneity, some of it passes into the next compartment where again mixing occurs. Recent reviews[1–4] conform with these ideas though not expressing precisely the same points.

Two phases of transport involve additional features: the passage of spermatozoa through the cervix uteri in those animals in which deposition occurs in the vagina (notably man, rabbit and sheep), and passage into and through the oviduct.

4.1.1 Sperm transport through the cervix

Spermatozoa have been found in the cervical canal in man a few seconds after sexual intercourse, and within the uterus a few minutes later. This rapid movement naturally raises the question as to how transport is effected; some authorities stress the importance of vigorous contractions of the vaginal wall, while others attribute transport to the development of negative pressure within the uterus. The latter idea, which envisaged the dipping of the external os uteri into the 'semen pool' in the fornix of the vagina, followed by insufflation of semen into the uterus, was a widespread view of a few years ago; in more recent times there has been a singular lack of support for the theory, and the current belief is that the vaginal contractions represent the main motive force. In this connection the observations of Fox, Wolff and Baker[5] are especially relevant. By means of pressure transducers and telemetry, they were able to make records during actual coitus. The intravaginal pressure was found to be negative on intromission and during male orgasm, and positive during female orgasm. Intrauterine pressures on the other hand increased greatly in female orgasm and showed a sharp fall thereafter. There seems little doubt that the effects would involve a vigorous to-and-fro movement of the column of cervical mucus, with a tendency to draw the sperm suspension a short distance into the canal, so that the spermatozoa would rapidly find their way into the mucus mass. There is, thus, no suggestion of a bulk movement of material from vagina to uterus, for the cervical mucus retains its composition and shows no dilution with vaginal secretions.

The composition of cervical mucus is such as to stimulate sperm motility, and so it is not long before spermatozoa achieve the uterine lumen through their own swimming movements. Clearly the composition of cervical mucus is a critical factor in human fertility: it has been studied by a number of investigators[6–11]. Nevertheless examination of post-coitus cervical mucus samples may offer no help in distinguishing between fertile and infertile couples[12].

Cervical mucus undergoes a distinct cycle of changes corresponding with the menstrual cycle. As the time of ovulation approaches, the quantity of mucus increases, its viscosity becomes less ('Spinnbarkeit' increases), the pH changes to reach a plateau at 8.0, 'ferning' becomes more distinct, and sperm penetrability improves. These changes begin about the 9th day of the cycle, and after the ovulatory LH peak drop again to previous levels within the next 2 days or so. Cervical mucus has been found to contain a variety of proteins, including blood proteins, but one of the most important from a functional viewpoint is mucoid, consisting of two glycoproteins with a very long molecular structure. Under normal circumstances the long mucoid threads would be oriented away from the mucosa that produces them, curving posteriorly until they take up a more or less parallel orientation.

As already noted sperm movement through cervical mucus is most rapid at about the ovulatory phase; this movement would be facilitated by the fact that the glycoproteins can be digested by proteases present in human semen which would of course reduce viscosity further[13,14]. Although human semen contains also another hydrolytic enzyme in abundance, namely hyaluronidase, and despite the fact that this enzyme was thought to have the same liquifying action on cervical mucus, it is almost certainly only the proteases that exert a significant effect. The viscosity of cervical mucus appears normally to be under the control of the hormone balance that underlies the menstrual cycle, and in addition can be influenced by hormone therapy—oestrogen tending to reduce viscosity and therefore favour sperm penetrability, and progesterone to increase it and consequently to militate against sperm passage through the cervical canal[15]. (Bedford[16] has data on the rate of sperm passage into the rabbit cervix after coitus.)

The development of ideas on the transport of spermatozoa through the human cervix initially depended on inferences drawn from the earlier work of Krehbiel and Carstens[17] in the rabbit. These authors maintained that stimulation of the vulva inherent in the coital act was responsible for producing vigorous contractile movements of the vagina which projected bulk quantities of sperm suspension through the cervix into the uterus. Later the problem was reinvestigated by Akester and Inkster[18] who used x-ray cinematography, and their findings are often misinterpreted as providing support for the views of Krehbiel and Carstens. In fact, while the second pair of authors did confirm that stimulation of the vulva caused vigorous to-and-fro movements of contrast medium within the vagina, they were definitely of the opinion that the function here was merely to transfer material to the cranial end of the vagina and not project any through into the uterus. The latter movement took place in the form of minute jets of medium associated with each of a continuous series of local contractions that involved the cranial end of the vagina and which proceeded independently of mating. The sperm numbers in the rabbit cervix after coitus have been found to be much higher than in the uterus, uterotubal junction or oviduct[19]. Indications are that the rabbit provides a useful but not exact model for the human subject.

Extensive chemical and physical studies on cervical mucus have been made on ruminant material (cow, sheep and goat). The streaming orientation of mucoprotein molecules leaving the cervical mucosa and passing distally through the lumen at the cervix is well shown in these animals[20–25]. The role of cervical mucus in sperm transport, however, is possibly better understood in the sheep where conditions are very similar; the work of Mattner[26] especially has clearly demonstrated the way in which spermatozoa, once involved in the cervix, move along channels roughly defined by mucoprotein threads and so pass in a curvilinear course to gather at the mucusal surface and in crypts. (Spermatozoa in the sheep evidently depend mainly on their own motility to establish a cervical population—Lightfoot and Restall[27]). Spermatozoa in these locations retain their motility and are believed eventually to escape, finding their way into the uterus and thence becoming available for transport to the site of fertilisation. (Lightfoot and Salamon[28] especially stress the importance of early establishment of a sperm population in the cervix for good fertility with artificial insemination.) Thus the cervix can be thought of as fulfilling a storage function. (Very high concentrations of spermatozoa were found in cattle in the cranial part of the vagina, just caudal to the external os uteri, so that the storage region may well differ in different species[19].) Dead spermatozoa in the uterus and in the cranial reaches of the cervical canal are steadily swept out through the cervix by the flow of mucus. In this flow too, leucocytes entering the uterine cavity pass out into the vagina; since they occupy the centre of the mucus flow they are prevented from reaching and phagocytosing the spermatozoa arranged along the mucus surfaces—the arrangements could clearly have a protective value for the spermatozoa. Nevertheless, de Boer[29] reports that India ink particles placed in the human cervical canal were later recovered in the oviducts in 30% of cases. Dead pig spermatozoa were also transported to the oviducts after artificial insemination, but not as efficiently as live ones[30].

4.1.2 Sperm transport through the oviduct

A problem that has long intrigued gametologists has been how the oviduct was capable of passing eggs in one direction and spermatozoa in the other. The oviduct, particularly in its more ovarian region, is richly supplied with ciliated epithelium, the cilia of which for a long time were thought to make their driving strokes only in an abovarian direction, thus causing a steady flow of oviductal secretion into the uterus. Clearly such a movement would facilitate the passage of eggs from the infundibulum to the site of fertilisation, and the passage of embryos from there to the isthmus and thence into the uterus, the movement being aided by contractile activity of the oviductal walls. But how were spermatozoa constrained to move against this adverse tide? An explanation put forward by Merton[31] has long been vogue; in brief his idea was that the spermatozoa were carried towards the ovaries in eddy currents which characteristically are found along the 'banks' of any stream bounded by irregular sides. Passing from eddy to eddy the spermatozoa would eventually reach the site of fertilisation. Merton did in fact demonstrate eddy currents associated with the beating of oviductal cilia *in vitro*. Sperm movement occurring in this fashion would no doubt be augmented by the mixing process referred to at the beginning of this section, but the mechanism does not seem adequate to explain the remarkable rapidity of sperm transport. More recently Blandau[32] has reported that there exist in the oviduct narrow strips of ciliated epithelium in which the cilial beat is in the direction opposing that obtaining in the rest of the epithelium. With this addition the explanation of sperm transport through the oviduct seems reasonably satisfactory.

The ciliation of the oviduct is remarkably impermanent. The histological and electron microscopic observations of Brenner[33,34] have shown very clearly how most of the cilia in the primate oviduct disappear in pregnancy and during the luteal phase of the cycle, to be renewed once more in the follicular phase of the succeeding cycle. He has also described with great precision the restoration of cilia—the replication of granules to form new basal bodies that give rise to cilia, and the growth of the cilial shaft. The ciliogenic cycle is evidently under control of the oestrogen–progesterone balance; regression of the ciliary population can be produced by progesterone treatment or by ovariectomy, and restored by the injection of oestrogen. The precocious ciliation that occurs in the late foetus is probably due to a burst in production of foetal oestrogens. In the rabbit, cilia are reduced in size after ovariectomy, but the number is little affected.

That the genotype of the spermatozoon could affect its efficiency in reaching and fertilising an egg was shown originally by Braden[35] and Braden and Gluecksohn-Waelsch[36] in their studies with mice bearing certain T alleles. The problem has since been reinvestigated by Olds[37] who confirms that the spermatozoa of sterile males fail to enter the oviducts, though present in high concentration and motility in the uterus. She surmises that the effect could be attributable to the presence in the semen of a chemical agent of some kind which induces the closure of the uterotubal junction.

4.1.3 Sperm metabolism in the female tract

The metabolic processes of spermatozoa can be both anabolic and catabolic; anabolic pathways may involve protein or lipid synthesis, and the catabolic either endogenous or exogenous substrate. The endogenous substrate is utilised oxidatively and the exogenous either oxidatively or glycolytically. Probably all these activities take place in the female tract, the balance depending in greater or less part on the concentration of spermatozoa and other metabolically functioning cells such as leucocytes, and on the local partial pressures of oxygen and carbon dioxide. In general, conditions in the female tract are likely to support oxidative metabolism[38].

Special interest, however, lies in the relations between the compound glycerylphosphorylcholine (GPC) and the enzyme GPC diesterase[39]. GPC has been identified in the semen of several different mammals (being produced in the epididymis) and occurs in remarkably high concentrations—in ram semen the level may exceed 1 %—but it is not metabolisable by spermatozoa. GPC diesterase has the action of breaking down GPC to phosphoglycerol and choline, and the former compound can readily be utilised as an energy source by spermatozoa. The enzyme is elaborated in the female tract, evidently in the uterine epithelium; the level of activity in the secretions varies with the stage of the oestrous cycle, and in the rat and ewe the high peak correlates well with pro-oestrus and oestrus. Ovariectomised rats receiving a single injection of oestrogen show a large increase in uterine diesterase activity, an effect inhibited by simultaneous administration of progesterone. The system is clearly appropriate for providing spermatozoa with an energy source at the time of sperm transport through the female tract, but its significance is not obvious because of the co-existence of several other energy sources.

Both respiratory and glycolytic pathways of sperm metabolism become increasingly used during incubation in the uterus, presumably in association with the process of capacitation, which is discussed later.

4.2 SPERM CAPACITATION AND THE ACROSOME REACTION

As the spermatozoa pass along the female genital tract to the site of fertilisation, and before they can penetrate through the egg investments, they undergo the process known as capacitation. By this term is implied a physiological change in the spermatozoon that renders it capable of undergoing the morphological change that constitutes the acrosome reaction, and this alteration of sperm structure permits the release of enzymes required for penetration of egg investments. The first intimation of a need for capacitation was obtained more than 20 years ago but despite intensive investigation all the evidence that has been obtained remains circumstantial, and there is still no clear indication as to just what is involved.

4.2.1 Acrosome reaction

A change in the acrosome, visible by light microscopy in motile spermatozoa recovered from the oviducts in several mammalian species, was first described by Austin and Bishop[40]. They recognised this change or 'loss' as the mammalian equivalent of the acrosome reaction, well known in the spermatozoa of marine invertebrates from the work of Dan[41–43] and other investigators. The mammalian acrosome reaction was inferred to release hyaluronidase for sperm passage through the cumulus oophorus and to expose a specific sperm organelle, the perforatorium, which was thought to carry a lytic agent, capable of acting upon the zona pellucida (the 'zona lysin') and thus permitting sperm transit through this coat. Ultrastructural evidence for the 'loss' of the acrosome by spermatozoa penetrating rabbit and rat eggs was forthcoming from the observations of Moricard[44], Austin[45] and Piko and Tyler[46]. Piko and Tyler noted additionally that the perivitelline sperm head was beset by an array of membranous vesicles which evidently derived from fusion between the outer leaf of acrosome membrane and the overlying plasma membrane. Fuller details of the way in which acrosomal and plasma membranes fuse so as to produce vesicles, release acrosome contents and yet retain an intact limiting membrane about the sperm head, emerged in later reports[47–52].

Since the acrosome reaction takes place through fusion of sperm membranes covering the anterior part of the sperm head and this must take place to be effective in the near vicinity of the eggs, the spermatozoon as deposited in the female tract is inferred to have some property that stabilises its membranes and prevents precocious reaction. Such a property could reside in an extraneous coat, the removal of which could constitute capacitation, but as yet no such coat has been clearly demonstrated (though there is some suggestive information). The problem of capacitation has been investigated most assiduously in the rabbit in which species the change *in vitro* has only recently been reported for the first time[53]. In addition, in recent years much work has been done with hamster, human, and mouse spermatozoa, all of which can readily undergo capacitation *in vitro*.

4.2.2 'Decapacitation'

In the rabbit the idea that capacitation involves the removal of a surface coating received some support from observation on the 'decapacitation factor'. This is a substance present in seminal plasma and capable of reversing the state of capacitation in a sperm population, so that the population requires to be returned to the female tract for 'recapacitation'[54]. Purification studies showed that the factor can be concentrated by high-speed centrifugation, is non-dialysable and moderately heat stable[55]. Subsequent work suggested that this material is in the nature of a glycoprotein complex; treatment with pronase separated the active component, which has a molecular weight of 300–500[56–58]. The decapacitation factor is evidently taken up when capacitated spermatozoa are suspended in solutions of the substance. The fluorescent staining of non-capacitated spermatozoa induced with tetracycline hydrochloride may be attributable to the presence of the decapacitation factor, since capacitated spermatozoa do not fluoresce[59,60]. Capacitation of rabbit spermatozoa seems to be achieved, in part at least, by treating them with mule eosinophils[61] and such treatment removes tetracycline from stained spermatozoa. Soupart[62] also has reason to believe that leucocytes play an active role in capacitation in the rabbit uterus. The rate of stain removal in the uterus, however, is much more rapid than that of capacitation[63]. Perhaps this is consistent with the idea that removal of the decapacitation factor from rabbit spermatozoa is simply the first step in the more complex process of capacitation.

4.2.3 Possible nature of capacitation

In the work on the hamster, man and the mouse, matured ovarian oocytes can be fertilised *in vitro* with spermatozoa recovered from the epididymis or from an ejaculate[50,64]. An interval of a few hours is required for the spermatozoa to undergo capacitation, which clearly can take place in the absence of female secretions other than those accompanying the oocytes on release from the ovary. In man the oocytes are recovered from follicles about 4 h before anticipated ovulation (promoted by hormone treatment), and washed ejaculated spermatozoa are added to them. The preparation is incubated for several hours and then examined; penetrated eggs have not been discovered before 7 h of incubation and this may be taken as an estimate of the time for capacitation, but since very few observations were made at shorter intervals it must be considered a generous estimate. The work on mice is limited to some experiments in which mouse spermatozoa were incubated with bovine follicular fluid and then placed with mouse oocytes; penetration was observed within an hour (the very brief capacitation time implied conforms with estimates made *in vivo* at an earlier date[65]). More extensive investigations have been made with hamster gametes.

The work of Yanagimachi[66–68] and Barros and Garavagno[69] showed that spermatozoa could undergo capacitation by being incubated in bovine follicular fluid or even in blood serum from several species, provided that this had been heated to destroy a factor (complement?) that is inimical to spermatozoa. (Serum taken at the time of ovulation is the most effective[70].) By means of fractionation and purification experiments Yanagimachi was able to show that the media in which capacitation occurred contained two factors, one dialysable and heat stable which apparently favoured the continuance of sperm motility, and the other non-dialysable and heat labile which seemed to trigger off the acrosome reaction. He inferred that the two factors are together necessary for producing the state of capacitation in the hamster spermatozoon. The inference though logical tends to confuse terminology and it would be more consistent to conclude that the dialysable factor had to do with the direct induction of the acrosome reaction; this is probably more than merely a semantic issue. Yanagimachi[68] has also drawn attention to the greater amplitude of flagellar beat in spermatozoa after capacitation.

Also working with the hamster, Bavister[71,72] has shown that the culture conditions—especially the pH—are of critical importance for capacitation, and that provided the medium is appropriate for prolonged sperm motility, capacitation can sometimes be obtained in essentially a defined medium (namely one in which the macromolecular component is polyvinylpyrrolidone). Prior incubation of spermatozoa in such a medium made it possible for fertilisation to be obtained within an hour of the addition of the spermatozoa to an egg suspension. The residual interval is thought to be attributable to the time taken by the acrosome reaction to proceed and for the spermatozoa to encounter the eggs.

On the basis of these observations it is suggested that the acrosome reaction in the hamster spermatozoon is provoked by some substance emanating from the oocyte or from the follicle cells investing it (the idea is in fact supported by Pavlok and McLaren's[73] data on the mouse), and exerting its action by becoming attached to receptor sites on the sperm head surface. The agent could take the form of a steroid such as progesterone which could then become incorporated in the cell membrane altering its properties by producing a less stable molecular configuration. Such a destabilised membrane might well be considered capable of taking part in fusion events with a closely adjacent membrane: in the case of the sperm head, the cell membrane would fuse with the immediately underlying acrosome membrane—a well recognised step in the acrosome reaction. According to this theory the receptor sites in the spermatozoon before capacitation would either be blocked by some form of 'stabilising' agent, or else exist in an inactive form, or even not exist at all. The idea of blocking requires that capacitation involve removal of the blocking agent and it is difficult to propose how this might occur other than through the action of a lytic enzyme for which we have no evidence. On the other hand, the notion that the receptor sites—presumably receptor proteins—could be generated or undergo activation through endogenous change avoids the difficulty of requiring an external lysin; endogenous changes would require no more than a suitable environment (which conforms with observations on hamster spermatozoa), and corresponds to acquisition of the competence to respond to steroid hormones—a feature of many target cells.

In the theory thus set out a prime role is attributed to progesterone; this substance was named because it is known to be produced in high concentration by follicle cells and because it is recognised to have a destabilising effect on cell membranes. But there is no direct evidence that progesterone is in fact involved and so the statement of the theory must be left for the moment with progesterone merely representing a type compound. These ideas were put forward in the hope that they might provide a useful working hypothesis[74].

4.3 MATURATION, OVULATION AND THE TRANSPORT OF OOCYTES

4.3.1 Course of maturation

Following the multiplication of the last generation of primordial germ cells (the oogonia), the nuclear state passes into the prophase of the first meiotic division, a state that distinguishes the oocyte. This happens before birth in most species or just after birth in some (hamster: Weakley[75], rabbit: Teplitz and Ohno[76]). In the prosimians a few oogonia evidently go on dividing, for they can be seen in the adult ovary where they give rise very belatedly to oocytes[77]. At all events, once the oocyte has been formed no further multiplication occurs in that cell line.

The first meiotic division proceeds to the preleptotene stage, at about which DNA is replicated for the last time before fertilisation[78], and on through leptotene and zygotene to pachytene. At this point the chromosomes shorten and thicken, becoming recognised as constituting the synaptonemal complex, the structure of which has been closely studied by electron microscopy[79]. Progress continues to diplotene when, in the monkey, man, cow and rabbit, the chromosomes achieve a 'lampbrush' state similar to that familiar in amphibian oocytes[80]. The difference from the 'dictyate stage' in the rat and mouse (well described by Franchi and Mandl[81]), is thought to be simply that here the chromosomes are more highly diffuse and so not easily recognised as of the 'lampbrush' type.

The first meiotic division now pauses, while the follicle develops and both oocyte and follicle increase greatly in size; it does not resume until a few hours before ovulation—obeyance may be attributable to association with granulosa cells[82]. For most or all of this time the oocyte synthesises RNA and protein, and it is tempting to infer that this RNA is the messenger for protein synthesis in the cleavage embryo[80]. Under the action of the preovulatory LH surge the first meiotic division goes to completion and the second meiotic division reaches metaphase or early anaphase; it is at this stage that follicle rupture occurs in most mammals. In man the interval between LH injection and metaphase II is some 36–40 h[83,84].

Evidence produced by Henderson and Edwards[85] gives good support to their hypothesis that oocytes reaching the stage of ovulation do so in the same sequence in which they were originally formed from oogonia. Their data also suggest that the chromosomes in the later-formed oocytes develop chiasmata less often, and consequently are more prone to show non-disjunction at the first meiotic division. Therefore oocytes ovulated later in a woman's reproductive life are more likely to be responsible for 'mongolism' and other trisomies, the essential point being that such chromosomal defects occur because the oocytes were late formed and not because they have aged excessively.

4.3.2 Mechanism of ovulation

The mechanism of ovulation is still very much a subject for debate. Follicles about to ovulate can be recognised by their sudden increase in size, accompanied by accumulation of the contained liquor folliculi. Mucoproteins in this fluid have been shown to undergo progressive depolymerisation with consequent increase in colloid osmotic pressure; the pressure is absorbed by influx of fluid[86]. Consistently, the permeability of the barrier between blood and liquor folliculi increases close to the time of ovulation[87]. Active growth of the granulosa and theca evidently contributes further to follicular enlargement. Blood vessels over the protruding surface become more evident except in a restricted region at the apex (the 'stigma') which remains avascular. Here the wall of the follicle becomes everted as a conical elevation at or near the tip of which the wall eventually breaks down, allowing follicular contents to escape.

Generally the outward flow of follicular fluid is gradual; evacuation is aided by slow contraction of the walls. Some pressure, attributable perhaps to contraction of the smooth muscle fibres in the theca externa[88], is clearly involved in the immediately post-ovulatory changes; measurements in the rabbit have shown that there are frequent pressure changes in the ripe follicle, produced by contractions of the walls, but there is no evidence of pressure increase being directly involved in the act of ovulation[89,90].

The final thinning of the follicle wall seems to be due to deterioration of the connective tissue elements[91] and could be brought about by the action of proteases or collagenases within the follicle. An increase of ovarian proteinase has been demonstrated in oestrus and dioestrus (as compared with pregnancy and pseudopregnancy)[92], and precocious ovulation has been provoked by injecting collagenases into the follicle, with precautions against increase in pressure[93]. Moreover, ovulation has been inhibited by the intrafollicular injection of actinomycin D; the mechanism of action is thought to involve inhibition of mRNA synthesis, the messenger being normally formed under the action of gonadotrophins, and being responsible for mediating enzyme synthesis[94]. Ovulation has also been inhibited by systemic (intraperitoneal) injection of actinomycin D, the time relations suggesting involvement of a similar system[95].

The evidence so far cited for the existence of a collagenase in the follicle, and the increase in its concentration near ovulation, is largely inferential, direct evidence having been difficult to produce. Espey and Rondell[96,97], therefore, made tests with a synthetic substrate (CBZ–gly–pro–gly–gly–pro–ala) which contains the gly–gly linkage thought to be specifically broken by collagenase, and the specific inhibitor parachloromercuribenzoate. They noted a significant decrease in the amount of collagenase-like enzyme in the projecting follicle wall and an increase in the concentration in the liquor folliculi. There appears also to be some seepage of the enzyme onto the follicle surface. Rondell[98] believes these rather strange results to mean that a collagenase-like enzyme is in fact released from the cells of the follicle wall and acts on an intercellular, interfibrillar substrate to promote follicle-wall distensibility. Under constant internal pressure the follicle wall would then stretch to rupture point—a not unattractive hypothesis. But, of course, this still does not prove the natural involvement of collagenase in ovulation.

4.3.3 Oocyte transport

On release from the follicle, the oocyte with its investing mass of cumulus cells passes onto the surface of the ovary. Then it is either ‘picked up’ by the fimbria of the oviduct, as in rabbit, guinea-pig and probably the primates (including man), or it is wafted into the infundibulum by fluid currents, as in the rat and mouse. The essential mechanism in both situations resides in the cilia with which the fimbria and infundibulum are so richly provided. In animals like the rabbit the cilia appear to obtain quite a strong purchase on the cumulus-cell mass, as is evident when the mass is pulled manually away from the fimbrial surface. Once attached to this surface the follicle-cell mass with the contained oocyte is moved steadily by the cilia into the lumen of the oviduct and along to the distended part of the ampulla where fertilisation occurs. In the rat and mouse the fluid currents that carry the oocyte and cumulus investment into the opening of the ampulla are produced by the vigorous beating of cilia. Once within the ampulla, transport to the site of fertilisation is probably by much the same means in both groups of animals, and these means may well include also peristalsis-like contractions of the oviduct walls. Both cilial activity and oviductal contractility are increased after ovulation. Oocyte transport in the oviduct is discussed in detail by Blandau[1,32].

Features of the hormonal control of oocyte transport were investigated by Boling and Blandau[99,100]. In their view the more rapid transport through the ampulla of the rabbit at about the time of ovulation is undoubtedly attributable to hormonal changes, oestrogen accelerating the movement by increasing the contractility of the smooth muscle in the ampullary walls, and progesterone having the opposite effect. In the castrate animal, oestrogen restored contractile activity to the ampulla and hastened the transport of oocytes, more especially at about 30 h after the last injection than at 1–3 h after. Work on the physiological and pharmacological aspects of human oviductal contractility has been reviewed by Coutinho[101].

4.4 SPERM PENETRATION INTO THE OOCYTE

4.4.1 Passage through cumulus oophorus and zona pellucida

With the occurrence of the acrosome reaction the contents of the acrosome are released to the exterior and it has long been inferred that they included the enzyme hyaluronidase. Recent studies have shown quite clearly that in addition to this enzyme there are a number of others in the sperm acrosome, mostly of a hydrolytic nature and including a protease with a close resemblance to trypsin[102–110]. The acrosomal protease has also been demonstrated cytochemically in the spermatozoa of several mammals[111]. The existence of this battery of enzymes, together with a close similarity between acrosomes and lysosomes in staining reactions with the fluorescent dye acridine orange, and the fact that the acrosome develops in the spermatid in association with the Golgi apparatus, led to the postulate that the acrosome is in fact a form of lysosome[112–114].

Be that as it may, the acrosomal enzymes, at least those that diffuse readily, seem to have only one function and that is to digest a passage enabling the spermatozoon to traverse the matrix of the cumulus surrounding the egg and reach the surface of the zona pellucida. Electron microscope observations are consistent in showing that by the time the spermatozoon reaches this point it lacks an acrosome in the usual sense—the acrosome reaction has proceeded to the point where the acrosomal contents have been dissipated and the cell membrane covering the acrosome, together with the outer acrosomal membrane, have been discarded from the sperm head. The spermatozoon, however, has yet to negotiate the zona pellucida, and to explain how this could be effected (and account at the same time for the narrow slit which is all that the sperm leaves in passing through the zona pellucida), it was proposed some years ago that the active agent must be a lysin closely adherent to the inner acrosome membrane, in much the same way as proteases in other kinds of cells are known to have firm adherence to cell membranes. A claim has been made to have extracted the zona lysin from rabbit spermatozoa[115], but the enzyme extracted is probably another one of the acrosome battery. Electron micrographs of spermatozoa after extraction show an intact inner acrosome membrane.

Overstreet and Adams[116] have evidence that spermatozoa from 'superior' male rabbits (as judged by their siring more young in mixed inseminations) tend not only to predominate in the upper reaches of the female tract but appear capable also of more rapid penetration of egg investments. This observation is made possible by staining the spermatozoa of one of the males in a mixed insemination with the dye fluorescein isothiocyanate, which permits easy distinction between spermatozoa recovered from the female tract or in contact with eggs without affecting their motility of fertility.

4.4.2 Sperm–egg fusion

The final step in sperm entry involves close application of the sperm head to the vitelline surface, followed by fusion between the limiting membranes of the two cells[4,49,50,117]. The essential mechanism, involving vesiculation of the two membranes over the region of contact appears to be the same as that involved in the acrosome reaction. It is at this point that a genetic factor seems capable of operating, for Krzanowska[118] has evidence that the ease of sperm–egg fusion is reduced in certain mouse strains—the effect being on both egg and spermatozoa. In consequence, eggs recovered at the time of fertilisation have several to numerous extra spermatozoa in the perivitelline space. From the work of Hanada and Chang[119] and Yanagimachi[120], sperm–egg fusion is inferred to have low species specificity. With zona-free eggs, fusion occurred between rat and guinea pig spermatozoa and hamster eggs, rat and hamster spermatozoa and mouse eggs, hamster spermatozoa and rat eggs, and—with highest frequency—mouse spermatozoa and rat and hamster eggs. In the last case, pronuclei were formed.

4.4.3 Cortical granule response, zona reaction, block to polyspermy

Passing through the zona pellucida with the aid of the supposed zona lysin, the spermatozoon reaches the surface of the vitellus and fuses with it. Among the evident consequences of this act are the disappearance (breakdown) of the cortical granules, and the evocation of a block to polyspermy in the surface of the vitellus, or the zona reaction or both. Fraser, Dandekar and Gordon[121] have observed that the cortical granules in rabbit eggs treated *in vitro* with capacitated (oviductal) spermatozoa disappeared as early as 45 min after semination in corona-free eggs and 1 h in eggs with intact cumulus. These times are thought to give a good estimate of the times of sperm fusion with the vitellus. Cooper and Bedford[122] noted that fertilised rabbit eggs devoid of the zona show more negative charged groups at the surface than do unfertilised eggs, but how this difference is related to the other changes is obscure.

On the basis of the known role of sea urchin cortical granules in the formation of the fertilisation membrane, the cortical granules in mammalian eggs were thought to be responsible in some way for inducing the zona reaction. The notion was that the contents of the cortical granules passed across the perivitelline space and induced the necessary change in the material of the zona pellucida. Observations by Yanagimachi and Chang[123] seem to support this idea for they were able to demonstrate the release of mucoprotein-like material into the perivitelline space coincident with the disappearance of cortical granules and Barros and Yanagimachi[124] have apparently been able to induce the zona reaction artificially by treating eggs with cortical-granule material. Conrad, Buckley and Stambaugh[125] have been able to show in the rat egg that a substance released from the cortical granules is a powerful protease inhibitor. The substance permeated the zona pellucida and was evidently responsible for the observed reduction in protease digestibility of the zona pellucida consequent upon sperm penetration. These observations provide an alternative explanation for the nature of the zona reaction—namely that spermatozoa excluded from penetrating fully into the egg are held in the thickness of the zona pellucida because it is there that they meet the lysin inhibitor.

Hunter and Léglise[126] removed the isthmus from the oviducts of pigs and sutured the ampulla to the remaining uterine end of the oviduct. When the animals were subsequently mated, there was found to be a large increase in the incidence of polyspermy. This was interpreted as support for the idea that the restrictive action of the isthmus on the passage of spermatozoa is one of the factors responsible for the normally low incidence of polyspermy (as suggested by Braden and Austin[65]). Another protective device against polyspermy is, of course, the zona reaction in eggs of animals such as the hamster, and Barros, Vliegluthart and Franklin[127] surmise from their data that the reason why polyspermy is so frequently seen in some preparations for *in vitro* fertilisation (for example, those of Yanagimachi and Chang[128]), is that the mechanism of the zona reaction is interfered with by the culture conditions. They found that the reaction took at least 4 h to become operative in their *in vitro* experiments. A recent estimate for zona-reaction time *in vivo* is less than 15 min[129].

Penetration of the spermatozoon through the cumulus and the zona pellucida of the mammalian egg may take place *in vitro* provided the environmental conditions are appropriate. Fertilisation *in vitro* has now been described for the eggs of the rabbit, golden hamster, Chinese hamster, mouse, man and cat—as repeatable experimental procedures; fertilisation *in vitro* of a few eggs of the sheep, cow and pig have also been recorded[130,131]. A recent contribution on human eggs is that by Seitz, Rocha, Brackett and Mastroianni[132], who obtained oocytes from follicles and matured them *in vitro* before adding spermatozoa; some of the oocytes showed evidence of fertilisation and a few underwent apparently normal cleavage on culture. Most success is obtained with the mouse but even in this species there are clearly many intervening features, and the procedure remains a most exacting exercise in attention to details such as extreme purity of water supply, and the unexpected and unwarranted variation in the purity of 'high grade' reagents. With the technique of *in vitro* fertilisation it has been possible to show that the presence of the investment of cumulus is not an essential requirement for the fertilisability of rabbit eggs[133]. Consistently the frequency of cleavage observed in rabbit eggs by Brackett, Killen and Peace[134], who added spermatozoa after removing one or more investments from the eggs (cumulus-free, corona-free, zona-free), was essentially the same as with intact eggs. A notable omission from the list of species in which fertilisation *in vitro* has been observed is the rat—only since this book went to press has success been achieved[135].

4.5 PRONUCLEAR DEVELOPMENT AND SYNGAMY

Soon after fusion of sperm and egg membranes, the sperm nucleus becomes surrounded with egg cytoplasm. The anterior half of the nucleus is the last part to be engulfed, apparently because of the close adhesion of the inner acrosome membrane[50,117,136]. As a result the posterior part of the sperm nucleus begins to expand first, losing the nuclear envelope over the lateral aspects but retaining it for a while in the region of attachment of the neck. During expansion the sperm chromatin loses its dense packing and takes the form of a complex of filamentous material, the peripheral fibrils being strongly oriented into the egg cytoplasm[136–138]. Nucleoli appear within the area of expanding sperm nucleus and a new nuclear envelope is formed by the elongation and alignment of cytoplasmic vesicles. Soon the nuclear envelope of the male pronucleus presents the standard appearance of a double wall traversed by numerous pores. The female pronucleus develops along similar lines, the alignment of vesicles destined to form the nuclear envelope beoming evident about individual or small groups of chromosomes, before a single nucleus is constituted[49,139]. Ultrastructural changes in the nucleoli are described by Schuchner[140].

The pronuclei enlarge greatly and come together in the centre of the egg. Nuclear envelopes come into close apposition and their surface convolutions tend to interdigitate. At the approach of syngamy the nuclear envelopes of the two pronuclei begin to break up: the previously continuous envelope becomes resolved into numerous individual vesicles which move off into the egg cytoplasm. Nucleoli disappear and the chromosomes condense into granular or fibrillar aggregates[141]. The gathering of chromosomes deriving from the two pronuclei in the prophase of the first cleavage division marks the completion of fertilisation. (The time sequence of nuclear and cytoplasmic events of fertilisation in the pig have been described by Hunter[142].)

Replication of DNA occurs during the pronuclear stage in the mouse egg, as demonstrated by photometric measurements on Feulgen-stained sections[143]. By including tritiated thymidine in the medium surrounding the living eggs, and observing labelling subsequently by means of radioautography, Mintz[144] found that incorporation occurred relatively early, within about 8 h of sperm penetration. RNA synthesis, on the other hand, appears to be only at maintenance levels in the ovulated oocyte and the egg undergoing fertilisation[145–147]; this is in strong contrast with the active synthesis during oocyte development while in the ovary. Pronuclear nucleoli (in contrast to the nucleoli of primary oocytes and eight-cell and later cleavage stages) have very little RNA, as observed by ultraviolet absorption at 260 nm[148,149]. Protein synthesis evidently occurs in pronucleate eggs, as indicated by tritiated-leucine incorporation, and this (as Mintz[144] points out) presumably depends on maternal ribosomes. By contrast, protein synthesis from as early as the two-cell stage onwards is in part at least regulated by newly synthesised RNA[150].

The energy metabolism of the oocyte is best considered with that of the cleavage embryo.

4.6 THE CLEAVAGE EMBRYO

4.6.1 Transport

After fertilisation in the ampulla of the oviduct the eggs pass through the isthmus and enter the uterus, and in most species this movement occurs whether or not fertilisation has taken place. The horse, however, is a notable exception—in this species it has been found that unfertilised eggs tend to get trapped in the last part of the oviduct while fertilised eggs (developing embryos) pass them by and enter the uterus in the normal way[151]. The means whereby the transport mechanism is able to distinguish between the two classes of eggs is quite obscure.

Mouse embryos have been shown to require an intact zona pellucida for normal tubal transport[152]. Eight-cell embryos apparently disaggregate if deprived of the zona and placed in an oviduct, though they will continue to develop *in vitro*. Blastocyst survival is unaffected by zona removal, presumably because the trophoblast cells are held together by tight junctions.

An attempt has been made to estimate the interval from ovulation to the entry of the embryo into the uterus in man[153]. The time of ovulation was inferred from urinary LH levels, properties of the cervical mucus, the basal body temperature, and endometrial biopsies. Uterine embryos were sought by one or several flushings. Eight eggs were recovered: one unfertilised on day 2, three unfertilised on day 3, two fertilised on day 4 and one fertilised on day 5. The fertilised eggs included a morula of 30-odd cells (to judge from the photograph) and a blastocyst which on fixation and staining was shown to have 186 nuclei. Nuclear staining characteristics of the blastocyst suggested the presence of at least two cell populations; there were few mitotic figures.

4.6.2 Culture of eggs and embryos *in vitro*

Extensive growth of preimplantation embryos under tissue culture conditions has been obtained for the rabbit, mouse, man and ferret[130]. First success was achieved with rabbit embryos which could be raised from the one-cell or two-cell stages to the morula, or from the morula to the early blastocyst, in a culture medium consisting of serum diluted with physiological saline solution[154,155]. More recently, culture from two-cell to morula has been achieved under much more defined conditions[156]. But the rabbit blastocyst presents difficulties, for *in vivo* it undergoes large expansion (compared with the blastocysts of rodents and man) as a preliminary to implantation. Many attempts have been made to get the full normal expansion in culture, and still there is only partial success. Kane and Foote[157,158] and Maurer, Onuma and Foote[159] reported a small proportion of expanding blastocysts grown from two- and four-cell embryos in a medium containing amino acids and other simple components (including glucose), plus serum dialysate but lacking in serum protein. A large increase in the number of expanding blastocysts occurred with the addition of serum albumen. To obtain good development from the one-cell stage pyruvate had to be added to the medium[160].

For full expansion of rabbit blastocysts there appears to be a requirement for a specific protein. This seems first to have been clearly recognised by Beier[161,162] and Krishnan and Daniel[163,164] (see also reviews by Beier, Kühnel and Petry[165], and Daniel[166]), who named the protein in question'u teroglobin' and 'blastokinin,' respectively. Investigations hvae since proceeded intensively. Uterine secretions in the rabbit contain proteins of uterine origin and of blood origin. The proteins of uterine origin all help to promote blastocyst expansion, but one of these—uteroglobin or blastokinin—is more effective, and in fact also stimulates both RNA and protein synthesis in the rabbit blastocyst. The specific protein has a molecular weight of 27 000 and its production appears to be progesterone dependent. Although the protein spectrum differs in different animals, the particular balance in the tract at any one time seems to be related directly to the needs of the embryo at the corresponding stage. Uteroglobin has been shown by an immunofluorescence technique to be synthesised in the endometrium and to pass among the cells of the trophoblast[167]. Recent publications have also appeared on blastokinin, including its composition and molecular nature[168,169], its occurrence in the Northern fur seal[170], its immunological assay[171], its steroid-binding property[172], its synergism with progesterone in stimulating synthetic activity in rabbit blastocysts *in vitro*[173,174], its disappearance from the uterus following ovariectomy, and reappearance with progesterone, but not oestrogen, injection[175], and its changing levels in pregnancy and pseudopregnancy, but lack of change in the presence of an IUD[176].

Most attention has been given to the culture of mouse embryos *in vitro*, and many facts have been ascertained as to their metabolic requirements under these circumstances (there are several recent reviews[177–186]). For successful culture, attention must be given to the pH of the medium, the nature of the buffering agent used (a bicarbonate buffer is almost mandatory, and bicarbonate also seems to be a required nutrient[183]), the composition of the gaseous phase must be closely controlled, the balance of salts and the osmolarity of the solution are critical, and finally appropriate energy sources must be supplied. The medium must also contain a macromolecular compound; commonly this is crystalline bovine serum albumin, but substitution can be made with polyvinylpyrrolidone. Except under certain circumstances a fixed nitrogen source—commonly in the form of amino acids—is also necessary, but this requirement can be met with a single substance, namely glutathione. From the two-cell stage on, embryos can develop without any amino nitrogen, provided polyvinylpyrrolidone is incorporated in the medium[187]. Best results with mouse embryos in culture are obtained when the embryos are F_1 hybrids of two relatively unrelated inbred strains[188].

The energy requirements represent a point of special interest. From the eight-cell stage onwards mouse embryos are capable of utilising glucose as well as a variety of hydroxy acids and fatty acids as energy sources; by contrast the oocyte before and during fertilisation is capable of utilising only pyruvate or oxaloacetate, and the two-cell stage these two together with lactate and phosphoenolpyruvate[189]. These observations show that in the early cleavage embryo there is a progressive development of enzymic capabilities reflecting the ontogeny of gene expression. An intriguing feature encountered by Donahue and Stern[190] is that one-cell eggs can survive on glucose only, *provided* the follicle cells are still about them—the follicle cells metabolise the glucose to pyruvate and oxaloacetate.

Other mammalian embryos have also been grown in culture. Rat eggs are difficult subjects; the undivided and earlier cleavage stages will only advance one or two divisions, but 8- to 12-cell embryos have been grown to blastocysts with some frequency[177]. Somewhat poor results have been obtained with sheep, goat and cow embryos, and guinea pig and hamster embryos have proved even less amenable to culture. Whittingham[191] has succeeded in growing ferret embryos from the two-cell stage to the blastocyst. The human embryo is evidently as good culture material as the mouse embryo. Most of the human oocytes have been obtained from follicles just before ovulation (after hormone treatment), and then fertilised *in vitro* with washed ejaculated spermatozoa; after transfer of the ootids to fresh culture medium, embryonic development has advanced well into the blastocyst stage[192,193]. Clearly, striking species differences exist, which presumably reflect peculiar requirements in culture conditions and energy sources.

4.6.3 Macromolecular synthesis

Despite the fact that the total cytoplasmic mass of the cleavage embryo decreases during the course of cleavage, protein synthesis is evidently occurring, for labelled amino acids become incorporated into protein from the stage of fertilisation onwards[144]. The rate is initially low but increases steadily. This synthesis was not completely inhibited by actinomycin D and so is inferred to depend in part on pre-existing RNA, presumably that elaborated in the oocyte before fertilisation. After the four-cell stage, nucleolar RNA synthesis was more strongly suppressed by actinomycin D, and development was correspondingly inhibited, and one may surmise that new synthesis of RNA (arising from translation of the embryonic genome) becomes increasingly important[144]. Monesi, Molinaro, Spalletta and Davoli[150] maintain that protein synthesis is dependent largely upon a continuous synthesis of RNA from as early as the two-cell stage; the mouse embryo contrasts strongly with the sea urchin embryo in which protein synthesis is entirely dependent until gastrulation upon RNA synthesised in the oocyte. From the eight-cell stage onwards, the utilisation of nucleoside precursors increases steeply[194] although it is known that the mouse embryo does not depend upon such precursors being in the medium.

DNA synthesis occurs with each cleavage division, very shortly after the mitosis—indeed a G_1 phase appears to be lacking, up to the 8- to 16-cell stage; differentiation between embryonic cells in the rates of cell division also begins to become evident at this time[195]. Mitochondrial DNA does not seem to undergo much replication during cleavage[194].

4.6.4 Egg and embryo transfer to uterine foster-mothers

The technique of egg or embryo transfer in mammals is of some antiquity, the first report apparently being that by Heape[196], who took four-cell embryos from one variety of rabbit and placed them in the oviducts of another. In due course young were born that had the fur characteristics of the donor variety.

Since then very many transfer experiments have been carried out involving oocytes before and during fertilisation, and embryos of all stages to the blastocyst, with transfer not only to the oviduct and uterus, but also to ectopic sites such as the anterior chamber of the eye, the kidney capsule, the brain, and even the testis. Adams and Abbott[197] have listed 451 original papers dealing with transfer work in 14 different species distributed over six orders of mammals (rabbit, rat, mouse, hamster, guinea-pig, ferret, mink, sheep, goat, cow, pig, deer, monkey and wallaby).

The procedure has been employed to test the normality of oocytes matured *in vitro*, to demonstrate the viability of embryos deriving from the *in vitro* fertilisation of oocytes, to determine the relative importance of ageing in the egg as opposed to ageing in the uterus, to determine the maximum and minimum capacity of the uterus, to induce twins in cattle, to exploit the reproductive potential of genetically valuable agricultural animals, and so on. The procedure has been a singularly successful one both in the frequency of its satisfactory outcome and in the wealth of information gained. It has also highlighted the resistance of the cleavage embryo to environmental insult; trauma if sufficiently severe will of course destroy a sufficient number of the component cells to lead to the death of the embryo as a whole, but, with a critical cell number surviving, development proceeds in an apparently normal way.

The normality of mouse embryos arising from oocytes that were matured and fertilised *in vitro* was shown by the development of 15-day foetuses after embryo transfer[198]. Cow oocytes removed as primary oocytes from follicles have been placed in culture and found to mature as far as metaphase I; when these were transferred to oestrus heifers which had received an intrauterine injection of semen, more than half were later observed to be undergoing fertilisation (the device clearly has useful agricultural applications)[199]. The demonstration of embryo viability after *in vitro* fertilisation was effected in the mouse by Whittingham[200] (following Chang's[201] lead in the rabbit).

The work on twin induction in cattle was that of Rowson, Lawson and Moor[202]; they found that transfer of two embryos to one uterine horn led often to death through competition, whereas transfer to separate horns yielded a high proportion of twins. Rowson and his colleagues have also transferred sheep and cow embryos to rabbits for intermediate incubation, after which the embryos were retrieved and placed in the relevant ruminant species. Survival of sheep and cow embryos was good even in oestrus rabbits[203], and there was a high percentage of young born after retransfer[204,205].

A recent use of the method of assessing the relative importance of egg and uterus ageing by transfer was reported by Maurer and Foote[206] who found that fewer embryos developed in old than in young rabbits after receipt of eggs from young animals (see also Adams[207]).

4.6.5 Experimental parthenogenesis

The possibility of inducing parthenogenetic development in mammalian oocytes has attracted the attention of investigators for a number of years. The subject received early stimulus from the work of Pincus and his associates[208–210]. The impetus was continued by Thibault[211], Chang[212] and Austin and Braden[213]. In these experiments, though rudimentary parthenogenesis was frequently observed in the eggs of several species, the most promising results seemed to be obtainable with the rabbit. Pincus and Shapiro claimed that a few rabbits of parthenogenetic origin were actually born. Subsequent workers were not able to confirm this finding, however; the furthest well-established stage of development was the implanted blastocyst, and this at very low frequency.

After that, interest waned until the recent simultaneous publication of papers by Graham[214] and Tarkowski, Whitowska and Nowicka[215] in the same issue of *Nature*. Both papers reported superior results to those of earlier workers, and both sets of data were on mice, but different methods were used to activate the eggs—Graham found that removal of the cumulus oophorus with hyaluronidase and culture of the eggs was sufficient to provoke development to the blastocyst stage, while Tarkowski and his associates obtained the same degree of development after applying an electric shock to the oviduct. With neither procedure did development continue in a normal way beyond mid-term, but a number of implanted embryos reached the egg-cylinder stage and two continued up to 9–10 days. Two main points of interest attach to this work:

(a) If full development can eventually be achieved the device would be of special value in the production of animal strains with a very high degree of homozygosity—they would breed 'truer' than is possible with inbred strains.

(b) A challenging problem in embryonic mechanisms is presented by the as yet regular failure of development soon after implantation. The explanation most often put forward is that the homozygosity exposes recessive lethal or sub-vital genes, and this is considered plausible by Graham[216]. This would apply if parthenogenesis follows the haploid line, but would be less likely if diploid heterozygosity were restored by suppression of the second polar body (the mechanism on which successful parthenogenesis in turkeys probably depends[217]). So far there is no good evidence that 'regulation to diploidy' can reliably be induced in mammals, but some of the mouse parthenogenones were indeed diploid. A possible manoeuvre for avoiding the difficulty is to fuse together two blastomeres deriving from separate haploid parthenogenetic embryos by treating them with inactivated Sendai virus. Fusion of blastomeres from diploid (fertilised) embryos yielded on culture tetraploid blastocysts; on transfer to a uterine foster-mother further development occurred, but this did not exceed that of haploid blastocysts[218]. Graham attributes the failure in this instance to the existence of tetraploidy.

Tarkowski[219] pointed out that exposure of lethal genes could well be only part of the explanation for the early death of mouse parthenogenones. Additional factors might be that the activating stimulus either fails to produce a cytoplasmic change normally induced by the spermatozoon (for example, the breakdown of cortical granules which in fact does not seem to occur with artificial activation), or else it damages some constituents of the egg. Evidence of abnormality seen in his experiments included extrusion of cytoplasmic fragments during cleavage, and lack of cytokinesis in some blastomeres, leading to haplodiploid mosaicism. (See also review by Tarkowski[220].)

4.6.6 Low temperature preservation of embryos

In view of the striking results that have been obtained with longterm low-temperature preservation of spermatozoa, blood cells and various tissues (such as ovary, cornea and cartilage)[221], a natural ambition is to do the same with mammalian embryos. The main advantage to be gained would be the ability to store valuable animal genotypes with the minimum expenditure of effort, time and space; this could find immediate application for the maintenance of inbred mouse strains (now very numerous) and in livestock improvement. Among the more specialised short-term uses of such a preservation technique would be for the accumulation of sufficient early embryonic material for sophisticated biochemical studies.

Encouraging early results with eggs subjected to freezing temperatures were obtained by Smith[222,223], who took rabbit eggs during fertilisation down to temperatures from −18 to −190 °C for periods up to 168 h and subsequently observed cleavage in culture; and by Sherman and Lin[224,225], who recorded term foetuses after transferring to a uterine foster-mother unfertilised mouse eggs that had been kept at −10 °C for $3\frac{1}{2}$ h. Glycerol was used as a protective agent in both sets of experiments. Embryos, however, have proved less amenable to freezing and the first successful report appears to be that by Whittingham[226]. He suspended mouse eight-cell eggs and early blastocysts in a culture medium which he then chilled to about 0 °C; polyvinylpyrrolidone solution was added and cooling continued to −79 °C. This temperature was held for about 30 min and the preparation was thawed and warmed to 37 °C. Embryos that appeared normal were washed, placed in fresh culture medium and incubated. Ninety-six (69 %) of the eight-cell embryos and 31 (91 %) of the early blastocysts developed into expanded blastocysts. Eight-cell embryos were also transferred to uterine foster-mothers and 4 (21 %) were developing as normal foetuses when examined at 18 days. Among the expanded blastocysts arising from frozen early blastocysts and transferred to foster-mothers, nine (69 %) were born as normal young. Pursuing this line of investigation, Whittingham, Mazur and Leibo[227] have since succeeded in obtaining normal foetuses and young from two- and eight-cell eggs frozen to −179 °C and −268 °C. The problem would seem to be well on the way to solution.

4.7 MANIPULATIVE INVESTIGATIONS ON EMBRYOS

4.7.1 Separation of blastomeres

The developmental potential of individual blastomeres of cleavage embryos in the rabbit has been examined recently by Moore, Adams and Rowson[228], who found that the single blastomeres of two-cell, four-cell and eight-cell rabbit embryos were capable of proceeding to full development after transfer to recipient animals. Development depended upon replacement of the blastomere in the zona pellucida from which the embryo had been removed, and consequently it was not possible to obtain development of more than one blastomere from any one embryo. The frequency of successful development was inversely related to the age of the embryo, the actual figures being 55% for the two-cell embryo, 35% for the four-cell embryo, and 10% for the eight-cell embryo. Thus, although the experiment demonstrates that some at least of the blastomeres retained totipotentiality, the possibility remains that other blastomeres had undergone a degree of determination and had lost this capacity. In the pig, some (35%) single blastomeres of four- and six-cell embryos were found capable of developing to blastocysts on transfer[229]. (Investigations on the development of single blastomeres of mouse embryos have been reviewed by Tarkowski[230].)

4.7.2 Chimaeras

The artificial production of chimaeric mice was first reported by Tarkowski[231], who induced the fusion of morulae in pairs after removal of the zona pellucida by dissection. The fusion products developed in a normal manner as single embryos though consisting of twice the normal number of cells, and after transfer to recipient animals developed to birth and ultimately to full maturity. Chimaera formation was also achieved by Mintz[232,233] who used a slightly different technique including the removal of the zona with the aid of the enzyme pronase. Close examination of the fused morulae showed that complete intermingling of the two strains of cells occurred and when blastocyst formation took place, the distribution of the two cell types between inner cell mass and trophoblast appeared quite random. The chimaeric mice raised to adulthood similarly showed great variations in the type of mosaicism though the mixing of cell types was very evident in colour and nature of coat. As expected, a certain proportion of animals grew up with varying degrees of hermaphroditism; there was however a larger than expected proportion of males at the expense of hermaphrodites. It was inferred that when male and female morulae were fused the male cells tended to get the upper hand and to determine completely the sex of the offspring. The technique of chimaera production has been used for the investigation of several problems in embryology, including the mode of differentiation of the multinucleate skeletal muscle fibres which turned out to involve fusion of precursor cells and not nuclear division without cell division[234].

Chimaera production by the fusion of morulae is subject to the limitation that morula cells, at least in the mouse, have apparently not yet undergone determination, and so the method does not lend itself easily to the investigation of problems relating to early differentiation. In this connection and in other ways also, the technique of chimaera production devised by Gardner[235,236] has great potential value. Gardner dissected mouse blastocysts in various ways by micromanipulation and then transferred the products to recipient animals for further development. He found that the inner cell mass removed from a blastocyst was unable to implant in a recipient uterus; by contrast the trophoblast, from which an inner cell mass was removed, could implant but failed to undergo further development. When, however, the inner cell mass from one embryo and the trophoblast from another were put together so as to form a reconstituted blastocyst, development would proceed to birth and maturity. Evidently trophoblast cells although differentiated for the act of implantation are incapable of pursuing the full course of this process without the influence exerted by immediately adherent inner cell mass cells. This influence inhibits the transformation of trophoblast cells into primary giant cells, and promotes the development of the ectoplacental cone which gives rise to the beginnings of the embryonic placenta.

Full development of an apparently normal (though possibly hermaphrodite) animal could also be obtained when the inner cell mass from one blastocyst was introduced into the blastocoelic cavity of another (intact) blastocyst; in no instance were indentical twins formed. Such animals were, of course, chimaeric as to coat colour. Chimaerism affecting coat colour to the extent of about 20% was obtained when only one cell from a donor blastocyst was introduced into the blastocoelic cavity of a recipient blastocyst. In such a case the recipient blastocyst generally contained about 15–20 inner cell mass cells, and so to account for the relatively high proportion of donor tissue ultimately observed, it was necessary to infer that only some four or five of the inner cell mass cells were actually destined to form embryo, the balance being presumably precursors of embryonic membranes. The technique is currently being applied to the clonal analysis of developmental problems.

4.8 IMPLANTATION OF THE EMBRYO

Implantation denotes the attachment of the embryo to, or its embedding in, the uterine endometrium as a necessary preliminary to placenta formation. The less common term 'nidation' is generally understood as synonymous, though some would see a difference in meaning, making it more appropriate to the embryo–uterine relation in particular species; on the other hand 'nidation' could be held preferable in any event because it implies the establishment of a specialised repository for the embryo (nidus= nest).

Implantation has been the subject of intensive research now for many years; though many problems remain, a detailed and fairly consistent picture seems to be emerging, at least in the most investigated species, the mouse and rat. Progress has continued especially in the following two areas (reviewed[237–248]).

4.8.1 Initiation of implantation

The importance of synchrony between the stage of development of the embryo and the hormonally controlled status of the uterus has long been appreciated, and much of the definitive research involving embryo transfer was done in the 1950s and early 1960s[249]. More recently synchrony was shown to be critical also in the hamster[250], the sheep[251] and pig[252]. An extensive series of experiments with the cow showed that in this animal the requirements are also acute: the percentage pregnancy rates observed were 0, 30, 52.2, 91.1, 56.5, 40 and 20, for degrees of synchrony corresponding to the days −3, −2, −1, 0, +1, +2 and +3[253]. If embryos are placed in uteri several days more advanced than they are themselves, they suffer some lasting detriment, as evinced by failure to develop when returned to a uterus at the right stage; this was demonstrated in the rabbit by Adams[254]. Sato and Yanagimachi[255], working with the hamster, uncovered a further point wherein hamster embryos may be regarded as unusual. They found that when early embryos were transferred to uteri in comparable stages of asynchrony, 30, 48 and 50% of four-cell and eight-cell embryos and blastocysts developed to 14-day foetuses, but none of the one-cell and two-cell stages did, suggesting that these have a much greater sensitivity to manipulation.

The precise order of events leading up to implantation has been debated by several investigators. In the mouse McLaren[242] recognised three major phases in the preparation of the uterus, each involving several different changes. In the '*presensitisation*' phase the uterine lumen becomes simple and sinuous, as distinct from the complex branched state of the oestrous uterus. The epithelium shows corrugations of the surface, and by electron microscopy the microvilli appear distinct and regular in form. DNA replication and cell division that proceed actively in the proliferative period come to a halt. In the '*sensitisation*' phase the epithelial surfaces come closer together and the microvilli diminish in height. The contained blastocyst loses its zona which undergoes actual dissolution (lysis). Stromal oedema begins to be evident but may be more distinctly a feature of the next phase. Uterine pH decreases—a change that may have begun in the preceding phase. The '*decidual*' phase is characterised by increase in capillary permeability (pontamine blue reactivity) in the vicinity of each blastocyst, the appearance of distinctive W-bodies in the epithelium, stromal oedema is more in evidence, and cell division and DNA replication in the stroma begin.

The presensitisation and sensitisation phases seem to be mainly attributable to progesterone and oestrogen influence (see Psychoyos[239,240] and Finn[244]), but a more specific problem relates to the nature of the decidualising stimulus which evidently emenates from the blastocyst. The stimulus does not seem to require any mechanical action of trophoblast cells on the endometrium[256], and Finn[244] points out that the endocrine requirements for a trauma-induced decidual reaction differ from those for blastocyst implantation. Various substances injected into the uterine lumen are effective, but it was difficult to see how these could be of the nature of the specific blastocyst stimulus. Moreover, as Shelesnyak[258] asserts, the injected substances may well have a damaging influence on the endometrium and could therefore be classed as traumatic agents. Perhaps the most interesting lead was given by Orsini[258] when she showed that a bubble of air was a sufficient cause. Following this up, Hetherington[259] found that if CO_2 was absorbed from the air used for injection the decidual reaction was much less, and conversely that moderate concentrations of CO_2 (0.01–0.1%) were highly effective. It is tempting to think that the CO_2 produced by the blastocyst may be sufficient in itself to induce the decidual response[260]. Bitton-Casimiri, Brun and Psychoyos[261] reported seeing vesicles containing granular material passing through the zona of rat blastocysts and into the surrounding medium, at about the time of implantation; the significance of the observation is obscure.

The lysis of the zona pellucida is a matter that has received special attention from Mintz[144,248], who studied zona loss just before implantation in mice carrying the t^{12} gene. Heterozygous $+/t^{12}$ embryos 'hatch' from the zona in the normal way in the late blastocyst stage, while the homozygous t^{12}/t^{12} embryos, which die at the morula stage, lose their zona a little later by lysis attributable to a uterine agent. The lytic agent may aid 'hatching', but is not indispensible for this action; it does, however, exert also some influence on the blastocyst surface that promotes attachment to the uterine epithelium as a preliminary to implantation. It may conceivably be related to the uterine factor that seems to potentiate the blastocyst for further development[262].

Ultrastructural changes in blastocyst and endometrium during implantation have been subject to thorough study[243,256,263–267].

4.8.2 The immunology of implantation

The development of thought on the immunologically unique position of the implanted embryo, though given direction initially by Medawar[268], owes much to Kirby's inquiring mind[247,269–278]. Some of the suggestions put forward by various people to account for the persistence of the embryo have proved groundless—contrary to their ideas, the blastocyst *is* antigenic, the pregnant female *is* immunologically competent, foetal cells (trophoblast) *do* invade the maternal vascular system, the uterus *can* reject homografts. The decidualised uterus, however, seems to be prevented from exerting immunological rejections. Trophoblast cells have but low antigenic potency, and their numbers in the female system are relatively very small. A layer of 'fibrinoid' material (mucoprotein) between embryo and host after implantation could have a protective function. A popular explanation for embryo survival invokes immunological enhancement, described first by Breyere and Barrett[279] in accounting for the fact that grafts of paternal skin evoke a milder reaction in females after several pregnancies. Anderson[280], nevertheless, prefers an explanation based on low-dose tolerance. Currie[281] confirmed the finding of Breyere and Barrett, but later concluded that enhancement was not an adequate explanation and that some form of foetal-maternal barrier was responsible for the immunological isolation of the foetus[282]. This view was based on his finding that the rate of rejection of grafted paternal tumour (transplantable fibrosarcoma) by immunised females was not affected by pregnancy. The problem is still clearly a matter for debate.

References

1. Blandau, R. J. (1969). Gamete transport—comparative aspects. *The Mammalian Oviduct*, 129 (E. S. E. Hafez and R. J. Blandau, editors) (University of Chicago Press)
2. Bishop, D. W. (1971). Sperm transport in the fallopian tube. *Pathways to Conception: The Role of the Cervix and the Oviduct in Reproduction*, 99 (A. I. Sherman, editor) (Illinois: Thomas)
3. Moghissi, K. S. (1971). Sperm migration through the cervical mucus. *Pathways to Conception: The Role of the Cervix and the Oviduct in Reproduction*, 214 (A. I. Sherman, editor) (Illinois: Thomas)
4. Bedford, J. M. (1972). Sperm transport, capacitation and fertilization. *Reproductive Biology*, 338 (H. Balin and S. Glasser, editors) (Amsterdam: Excerpta Medica)
5. Fox, C. A. Wolff, H. S. and Baker, J. A. (1970). Measurement of intravaginal and intrauterine pressures during human coitus by radio-telemetry. *J. Reprod. Fert.*, **22,** 243
6. Moghissi, K. S. and Syner, F. N. (1970). Studies on human cervical mucus. *Fert. Steril.*, **21,** 234
7. Schumacher, G. F. B. (1970). Biochemistry of cervical mucus. *Fert. Steril.*, **21,** 697
8. Schumacher, G. F. B. (1971). Soluble proteins in cervical secretions. *Pathways to Conception: The Role of the Cervix and the Oviduct in Reproduction*, 168 (A. E. Sherman, editor) (Illinois: Thomas)
9. Syner, F. N. and Moghissi, K. S. (1971). Mucoids of cervical mucus. *Pathways to Conception: The Role of the Cervix and Oviduct in Reproduction*, 156 (A. I. Sherman, editor) (Illinois: Thomas)
10. Iacobelli, S., Garcea, M. and Angeloni, C. (1971). Biochemistry of cervical mucus: A comparative analysis of the secretion from preovulatory, postovulatory and pregnancy periods. *Fert. Steril.*, **22,** 727
11. Moghissi, K. S. (1972). The function of the cervix in fertility. *Fert. Steril.*, **23,** 295
12. Gibor, Y., Garcia, C. J., Cohen, M. R. and Scommegna, A. (1970). The cyclical changes in the physical properties of the cervical mucus and the results of the postcoital test. *Fert. Steril.*, **21,** 20
13. Moghissi, K. S., Dabich, D., Levine, J. and Neuhaus, O. W. (1964). Mechanism of sperm migration. *Fert. Steril.*, **15,** 15
14. Moghissi, K. S. and Syner. F. N. (1970). The effect of seminal protease on sperm migration through the cervical mucus. *Int. J. Fert.*, **15,** 43
15. Zanartu, J. (1966). Effect of natural and synthetic sex steroids in cervical mucus, penetration and ascent of spermatozoa. *Vth World Congr. Fert. Steril., Stockholm, June, 1966* (A. Ingelman-Sundberg and B. Westin, editors) Int. Congr. Series No. 109, Excerpta Medica Foundation
16. Bedford, J. M. (1971). The rate of sperm passage into the cervix after coitus in the rabbit. *J. Reprod. Fert.*, **25,** 211
17. Krehbiel, R. H. and Carstens, H. P. (1939). Roentgen studies of the mechanism involved in sperm transportation in the female rabbit. *Amer. J. Physiol.*, **125,** 571
18. Akester, A. R. and Inkster, I. J. (1961). Cine-radiographic studies of the genital tract of the rabbit. *J. Reprod. Fert.*, **2,** 507
19. El-Banna, A. A. and Hafez, E. S. E. (1970). Sperm transport and distribution in rabbit and cattle female tract. *Fert. Steril.*, **21,** 534
20. Gibbons, R. A. (1959). Physical and chemical properties of mucoids from bovine cervical mucin. *Biochem. J.*, **72,** 27P
21. Tampion, D. and Gibbons, R. A. (1962). Orientation of spermatozoa in mucus of the cervix uteri. *Nature* (*London*), **194,** 381
22. Mattner, P. E. (1963). Capacitation of ram spermatozoa and penetration of the ovine egg. *Nature* (*London*), **199,** 772
23. Mattner, P. E. (1968). The distribution of spermatozoa and leucocytes in the female genital tract in goats and cattle. *J. Reprod. Fert.*, **17,** 253
24. Gibbons, R. A. and Mattner, P. E. (1967). Some aspects of the chemistry of cervical mucus. p. 695. *Fert. Steril. Proc. Vth World Congr.* (Westin, B. and Wiqvist, N. editors). Excerpta Medica Foundation

25. Gibbons, R. A. and Mattner, P. E. (1971). The chemical and physical characteristics of the cervical secretion and its role in reproductive physiology. *Pathways to Conception: The Role of the Cervix and the Oviduct in Reproduction*, 143 (A. I. Sherman, editor) (Illinois: Thomas)
26. Mattner, P. E. (1963). Spermatozoa in the genital tract of the ewe. III. Role of spermatozoan motility and of uterine contractions in transport of spermatozoa. *Austr. J. Biol. Sci.*, **16,** 877
27. Lightfoot, R. J. and Restall, B. J. (1971). Effects of site of insemination, sperm motility and genital tract contractions on transport of spermatozoa in the ewe. *J. Reprod. Fert.*, **26,** 1
28. Lightfoot, R. J. and Salamon, S. (1970). Fertility of ram spermatozoa frozen by the pellet method. I. Transport and viability of spermatozoa within the genital tract of the ewe. *J. Reprod. Fert.*, **22,** 385
29. de Boer, C. H. (1972). Transport of particulate matter through the human female genital tract. *J. Reprod. Feɪt.*, **28,** 295
30. Baker, R. D. and Degen, A. A. (1972). Transport of live and dead boar spermatozoa within the reproductive tract of gilts. *J. Reprod. Fert.*, **28,** 369
31. Merton, M. (1939). Studies on reproduction in the albino mouse. III. The duration of life of spermatozoa in the female reproductive tract. *Proc. Roy. Soc., Edinburgh.* **59,** 207
32. Blandau, R. J. (1972). An experimental approach to the study of egg transport through the oviducts of mammals. *N.I.H. Conf.: The Regulation of Mammalian Reproduction*, 1970, Maryland, U.S.A. (in press)
33. Brenner, R. M. (1971). Ciliogenesis in the primate oviduct. *Pathways to Conception: The Role of the Cervix and the Oviduct in Reproduction*, 50 (A. I. Sherman, editor) (Illinois: Thomas)
34. Brenner, R. M. (1972). Hormonal Regulation of Oviductal Epithelium. *N.I.H. Conf.: The Regulation of Mammalian Reproduction*, 1970, Maryland, U.S.A. (in press)
35. Braden, A. W. H. (1958). Influence of time of mating on the segregation ratio of alleles at the T locus in the house mouse. *Nature (London)*, **181,** 786
36. Braden, A. W. H. and Gluecksohn-Waelsch, S. (1958). Further studies of the effect of the T locus in the house mouse on male fertility. *J. Exp. Zool.*, **138,** 431
37. Olds, P. J. (1970). Effect of the T locus on sperm distribution in the house mouse. *Biol. Reprod.*, **2,** 91
38. Mann, T. (1967). Sperm metabolism. *Fertilization, Vol. 1*, 99 (C. B. Metz and A. Monroy, editors) (New York, London: Academic Press)
39. White, I. G. (1972). Biochemistry of semen and interaction in the female reproductive tract. *Search*, **3,** 22
40. Austin, C. R. and Bishop, M. W. H. (1958). Some features of the acrosome and perforatorium in mammalian spermatozoa. *Proc. Roy. Soc. B*, **148,** 234
41. Dan, J. C. (1952). Studies on the acrosome. I. Reaction to egg-water and other stimuli. *Biol. Bull.*, **103,** 54
42. Dan, J. C. (1956). The acrosome reaction. *Int. Rev. Cytol.*, **5,** 365
43. Dan. J. C. (1967). Acrosome reaction and lysins. *Fertilization, Vol. 1*, 237 (C. B. Metz and A. Monroy, editors) (New York, London: Academic Press)
44. Moricard, R. (1960). Observations de microscopie electronique sur des modifications acrosomiques lors de la penetration spermatique dans l'oeuf des mammiferes. *C.R. Soc. Biol. (Paris)*, **154,** 2187
45. Austin, C. R. (1963). Acrosome loss from the rabbit spermatozoon in relation to entry into the egg. *J. Reprod. Fert.*, **6,** 313
46. Piko, L. and Tyler, A. (1964). Five structural studies on sperm penetration in the rat. *Proc. Vth Int. Congr. Anim. Reprod.*, Treinto, **2,** 372
47. Barros, C., Bedford, J. M., Franklin, M. C. and Austin, C. R. (1967). Membrane vesiculation as a feature of the mammalian acrosome reaction. *J. Cell. Biol.*, **34,** C1
48. Bedford, J. M. (1967). Experimental requirements for capacitation and observations on ultrastructural changes in rabbit spermatozoa during fertilisation. *J. Reprod. Fert.*, **Suppl., 2,** 35
49. Austin, C. R. (1968). *Ultrastructure of fertilization* (New York: Holt, Rinehart and Winston)

50. Bedford, J. M. (1970). Sperm capacitation and fertilization in mammals. *Biol. Reprod.*, **Suppl., 2,** 128
51. Franklin, L. E., Barros, C. and Fussell, E. N. (1970). The acrosomal region and the acrosome reaction in sperm of the golden hamster. *Biol. Reprod.*, **3,** 180
52. Yanagimachi, R. and Noda, Y. D. (1970). Electron microscope studies of sperm incorporation into the golden hamster egg. *Amer. J. Anat.*, **128,** 429
53. Okinaga, Y. (1972). Fertilization *in vitro* using rabbit tubal ova. *Jap. J. Fertil. Steril.*, **17,** 1
54. Chang, M. C. (1957). A detrimental effect of seminal plasma on the fertilizing capacity of sperm. *Nature* (*London*), **179,** 258
55. Bedford, J. M. and Chang, M. C. (1962). Removal of decapacitation factor from seminal plasma by high speed centrifugation. *Amer. J. Physiol.*, **202,** 179
56. Williams, W. L., Hamner, C. E., Weinman, D. E. and Brackett, B. G. (1964). Capacitation of rabbit spermatozoa and initial experiments on *in vitro* fertilization. *Proc. Vth Int. Congr. Reprod., Trento, VII*, 288
57. Hunter, R. H. F. and Nornes, H. O. (1968). Characterization and isolation of a sperm-coating antigen from rabbit seminal plasma with capacity to block fertilization. *J. Reprod. Fert.*, **20,** 419
58. Williams, W. L., Robertson, R. T. and Dukelow, W. R. (1970). Decapacitation factor and capacitation. *Schering Symposium on Mechanisms involved in Conception*, 61. (Berlin, 1969) (G. Raspe, editor) (Oxford, Braunschweig: Pergamon, Vieweg)
59. Ericsson, R. J. (1967). Technology, physiology and morphology of sperm capacitation (rabbit, bull, man). *J. Reprod. Fert.*, **Suppl., 2,** 65
60 Ericsson, R. J. (1967). Fluorometric method for the measurement of sperm capacitation. *Proc. Soc. Exp. Biol. Med.*, **125,** 1115
61. Ericsson, R. J. (1969). Capacitation *in vitro* of rabbit sperm with mule eosinophils. *Nature* (*London*), **221,** 568
62. Soupart, P. (1970). Leukocytes and sperm capacitation in the rabbit uterus. *Fert. Steril.*, **21,** 724
63. Vaidya, R. A., Bedford, J. M., Glass, R. H. and Morris, J. M. (1969). Evaluation of the removal of tetracycline fluorescence from spermatozoa as a test for capacitation in the rabbit. *J. Reprod. Fert.*, **19,** 483
64. Thibault, C. (1969). *In vitro* fertilization of the mammalian egg. *Fertilization, Vol. II*, 405 (C. B. Metz and A. Monroy, editors) (New York, London: Academic Press)
64a. Thibault, C. (1972). *In vitro* fertilization of *in vitro* matured oocytes. *N.I.H. Conf.: The Regulation of Mammalian Reproduction*, 1970. Maryland, U.S.A. (in press)
65. Braden, A. W. H. and Austin, C. R. (1954). Fertilization of the mouse and the effect of delayed coitus and of hot-shock treatment. *Austr. J. Biol. Sci.*, **7,** 552
66. Yanagimachi, R. (1969). *In vitro* capacitation of hamster spermatozoa by follicular fluid. *J. Reprod. Fert.*, **18,** 275
67. Yanagimachi, R. (1969). *In vitro* acrosome reaction and capacitation of golden hamster spermatozoa bovine follicular fluid and its fractions. *J. Exp. Zool.*, **170,** 269
68. Yanagimachi, R. (1970). *In vitro* capacitation of golden hamster spermatozoa by homologous and heterologous blood sera. *Biol. Reprod.*, **3,** 147
69. Barros, C. and Garavagno, A. (1970). Capacitation of rabbit spermatozoa with blood sera. *J. Reprod. Fert.*, **22,** 381
70. Barros, C., Arrau, J. and Herrera, E. (1972). Induction of the acrosome reaction of golden hamster spermatozoa with blood serum collected at different stages of the oestrous cycle. *J. Reprod. Fert.*, **28,** 67
71. Bavister, B. D. (1969). Environmental factors important for *in vitro* fertilization in the hamster. *J. Reprod. Fert.*, **18,** 544
72. Bavister, B. D. (1971). *A study of* in vitro *fertilization and capacitation in the hamster*. Ph.D. Thesis, University of Cambridge.
73. Pavlok, A. and McLaren, A. (1972). The role of cumulus cells and the zona pellucida in fertilization of mouse eggs *in vitro*. *J. Reprod. Fert.*, **29,** 91
74. Austin, C. R., Bavister, B. D. and Edwards, R. G. (1972). Components of capacitation. *N.I.H. Conf: The Regulation of Mammalian Reproduction*, 1970. Maryland. (in press)

75. Weakley, B. S. (1967). Electron microscopy of the oocyte and granulosa cells in the developing ovarian follicles of the golden hamster (*Mesocricetus auratus*). *J. Anat.*, **100,** 503
76. Teplitz, R. and Ohno, S. (1963). Postnatal induction of ovogenesis in the rabbit (*Oryctolagus cuniculus*). *Exp. Cell. Res.*, **31,** 183
77. Ioannu, J. M. (1967). Oogenesis in adult prosimians. *J. Embryol. Exp. Morphol.*, **17,** 139
78. Baker, T. G., Beaumont, H. M. and Franchi, L. L. (1969). The uptake of tritiated uridine and phenylalanine by the ovaries of rats and monkeys. *J. Cell. Sci.*, **4,** 655
79. Solari, A. J. (1972). The behaviour of chromosomal axes during male meiotic prophase. *N.I.H. Conf.: The Regulation of Mammalian Reproduction*, 1970, Maryland, U.S.A. (in press)
80. Baker, T. G. (1971). Electron microscopy of the primary and secondary oocyte. *Schering Symposium on Intrinsic and Extrinsic Factors in Early Mammalian Development* (Venice, 1970) (G. Raspe, editor) (Oxford, Braunschweig: Pergamon-Vieweg)
81. Franchi, L. L. and Mandl, A. M. (1962). The ultrastructure of oogonia and oocytes in the foetal and neonatal rat. *Proc. Roy. Soc. B*, **157,** 99
82. Ohno, S. and Smith, J. B. (1964). Role of fetal follicular cells in meiosis of mammalian oocytes. *Cytogenetics*, **3,** 324
83. Jagiello, G., Karnicki, J. and Ryan, R. J. (1968). Superovulation with pituitary gonadotrophins. Method for obtaining metaphase figures in human ova. *Lancet*, **1,** 179
84. Steptoe, P. C. and Edwards, R. G. (1970). Laparoscopic recovery of preovulatory human oocytes after priming of ovaries with gonadotrophins. *Lancet*, **4,** 683
85. Henderson, S. A. and Edwards, R. G. (1968). Chiasma frequency and maternal age in mammals. *Nature* (*London*), **218,** 22
86. Zachariae, F. and Jensen, C. E. (1958). Studies on the mechanism of ovulation. Histochemical and physico-chemical investigations on genuine follicular fluids. *Acta Endocrinol.*, **27,** 343
87. Zachariae, F. (1958). Studies on the mechanism of ovulation. *Acta Endocrinol.*, **27,** 339
88. Lipner, H. J. and Maxwell, B. A. (1960). Hypothesis concerning the role of follicular contractions in ovulation. *Science*, **131,** 1737
89. Espey, L. L. and Lipner, H. (1963). Measurements of intrafollicular pressures in the rabbit ovary. *Amer. J. Physiol.*, **205,** 1067
90. Blandau, R. J. and Rumery, R. F. (1963). Measurements of intrafollicular pressure in ovulatory and preovulatory follicles of the rat. *Fert. Steril.*, **14,** 330
91. Espey, L. L. (1967). Tenacity of porcine Graafian follicle as it approaches ovulation. *Amer. J. Physiol.*, **212,** 1397
92. Reichert, L. E. (1962). Further studies on proteinases of the rat ovary. *Endocrinology*, **71,** 838
93. Espey, L. L. and Lipner, H. J. (1965). Enzyme induced rupture of rabbit Graafian follicles. *Amer. J. Physiol.*, **208,** 208
94. Pool, W. R. and Lipner, H. J. (1964). Inhibition of ovulation in the rabbit by actinomycin D. *Nature* (*London*), **203,** 1385
95. Austin, C. R. and Barros, C. (1968). Inhibition of ovulation by systematically administered actinomycin D in the hamster. *Endocrinology*, **83,** 177
96. Espey, L. L. and Rondell, P. (1967). Estimation of mammalian collagenolytic activity with a synthetic substrate. *J. Appl. Physiol.*, **23,** 757
97. Espey, L. L. and Rondell, P. (1968). Collagenolytic activity in the rabbit and sow Graafian follicle during ovulation. *Amer. J. Physiol.*, **214,** 326
98. Rondell, P. (1970). Follicular processes in ovulation. *Fed. Proc.*, **29,** 1875
99. Boling, J. L. and Blandau, R. J. (1971). Egg transport through the ampullae of the oviducts of rabbits under various experimental conditions. *Biol. Reprod.*, **4,** 174
100. Boling, J. L. and Blandau, R. J. (1971). The role of estrogens in egg transport through the ampullae of oviducts of castrate rabbits. *Fert. Steril.*, **22,** 544
101. Coutinho, E. M. (1971). Physiologic and pharmacologic studies of the human oviduct. *Fert. Steril.*, **22,** 807

102. Stambaugh, R. and Buckley, J. (1968). Zona pellucida dissolution enzymes of the rabbit sperm head. *Science, N.Y.*, **161,** 585
103. Stambaugh, R. and Buckley, J. (1969). Indentification and subcellular localization of the enzymes effecting penetration of the zona pellucida by rabbit spermatozoa. *J. Reprod. Fert.*, **19,** 423
104. Stambaugh, R. and Buckley, J. (1970). Comparative studies of the acrosomal enzymes of rabbit, rhesus monkey, and human spermatozoa. *Biol. Reprod.*, **3,** 275
105. Stambaugh, R. and Buckley, J. (1971). Acrosomal enzymes of mammalian spermatozoa affecting fertilization. *Fed. Proc.*, **30,** 1184
106. Stambaugh, R. and Buckley, J. (1972). Histochemical subcellular localization of the acrosomal proteinase effecting dissolution of the zona pellucida using fluorescein-labelled inhibitors. *Fert. Steril.*, **23,** 348
107. Zaneveld, L. J. D. and Williams, W. L. (1970). A sperm enzyme that disperses the corona radiata and its inhibition by decapacitation factor. *Biol. Reprod.*, **2,** 363
108. Zaneveld, L. J. D., Srivastava, P. N. and Williams, W. L. (1969). Relationship of a trypsin-like enzyme to capacitation. *J. Reprod. Fert.*, **20,** 337
109. Polakoski, K. L., Zaneveld, L. J. D. and Williams, W. L. (1972). Purification of acrosin, a proteolytic enzyme from rabbit sperm acrosomes. *Biol. Reprod.*, **6,** 23
110. Zaneveld, L. J. D., Polakowski, K. L. and Williams, W. L. (1972). Properties of a proteolytic enzyme from rabbit sperm acrosomes. *Biol. Reprod.*, **6,** 30
111. Yanagimachi, R. and Teichman, R. J. (1972). Cytochemical demonstration of acrosomal proteinase in mammalian and avian spermatozoa by a silver proteinate method. *Biol. Reprod.*, **6,** 87
112. Allison, A. C. and Hartree, E. F. (1968). Lysosomal nature of the acrosomes of ram spermatozoa. *Biochem. J.*, **111,** 35P
113. Allison, A. C. and Hartree, E. F. (1970). Lysosomal enzymes in the acrosomes and their possible role in fertilization. *J. Reprod. Fert.*, **21,** 501
114. Allison, A. C. and Young, M. R. (1969). Vital staining and fluorescence microscopy of lysosomes. *The Role of Lysosomes in Biology and Pathology, Vol. 2,* 600 (J. T. Dingle and H. B. Fell, editors) (Amsterdam: North Holland Publishing Co.)
115. Srivastava, P. N., Adams, C. E. and Hartree, E. F. (1965). Enzymic action of acrosomal preparation on the rat ovum *in vitro*. *J. Reprod. Fert.*, **10,** 61
116. Overstreet, J. M. and Adams, C. E. (1971). Mechanism of selective fertilization in the rabbit: Sperm transport and viability. *J. Reprod. Fert.*, **26,** 219
117. Bedford. J. M. (1972). An electron microscopic study of sperm penetration into the rabbit egg after normal mating. *Amer. J. Anat.*, **133,** 213
118. Krazanowska, H. (1970). Relation between fertilization rate and penetration of eggs by supplementary spermatozoa in different mouse strains and crosses. *J. Reprod. Fert.*, **22,** 199
119. Hanada, A. and Chang, M. C. (1972). Penetration of zona-free eggs by spermatozoa of different species. *Biol. Reprod.*, **6,** 300
120. Yanagimachi, R. (1972). Penetration of guinea-pig spermatozoa into hamster eggs *in vitro*. *J. Reprod. Fert.*, **28,** 477
121. Fraser, L. R., Dandekar, P. V. and Gordon, M. K. (1972). Loss of cortical granules in rabbit eggs exposed to spermatozoa *in vitro*. *J. Reprod. Fert.*, **29,** 295
122. Cooper, G. W. and Bedford, J. M. (1971). Charge density change in the vitelline surface following fertilization of the rabbit egg. *J. Reprod. Fert.*, **25,** 431
123. Yanagimachi, R. and Chang, M. C. (1961). Fertilizable life of golden hamster ova and their morphological changes at the time of losing fertilizability. *J. Exp. Zool.*, **148,** 185
124. Barros, C. and Yanagimachi, R. (1971). Induction of zona reaction in golden hamster eggs by cortical granule material. *Nature* (*London*), **233,** 268
125. Conrad, K., Buckley, J. and Stambaugh, R. (1971). Studies on the nature of the block to polyspermy in rabbit ova. *J. Reprod. Fert.*, **26,** 133
126. Hunter, R. H. F. and Léglise, P. C. (1971). Polyspermic fertilization following tubal surgery in pigs with particular reference to the role of the isthmus. *J. Reprod. Fert.*, **24,** 233

127. Barros, C., Vliegluthart, A. M. and Franklin, L. E. (1972). Polyspermic fertilization of hamster eggs *in vitro*. *J. Reprod. Fert.*, **28,** 117
128. Yanagimachi, R. and Chang, M. C. (1964). *In vitro* fertilization of golden hamster ova. *J. Exp. Zool.*, **156,** 361
129. Barros, C. and Yanagimachi, R. (1972). Polyspermy-preventing mechanisms in the golden hamster egg. *J. Exp. Zool.*, **180,** 251
130. Edwards, R. G. and Whitten, W. K. (1970). *In vitro* culture of mammalian eggs and embryos (including *in vitro* fertilization) (Bibliography) *Biblio. Reprod.*, **16,** 585
131. Hamner, C. E., Jennings, L. L. and Sojka, N. J. (1970). Cat (*Felis catus* L.) spermatozoa require capacitation. *J. Reprod. Fert.*, **23,** 477
132. Seitz, H. M., Rocha, G., Brackett, B. G. and Mastroianni, L. (1971). Cleavage of human ova *in vitro*. *Fert. Steril.*, **22,** 255
133. Fraser, L. R., Dandekar, P. V. and Vaidya, R. A. (1971). *In vitro* fertilization of tubal rabbit ova partially or totally denuded of follicular cells. *Biol. Reprod.*, **4,** 229
134. Brackett, B. G., Killen, D. E. and Peace, M. D. (1971). Cleavage of rabbit ova inseminated *in vitro* after removal of follicular cells and zona pellucidae. *Fert. Steril.*, **22,** 816
135. Miyamoto, H. and Chang. M. C. (1973). *In vitro* fertilization of rat eggs *Nature (London)*, **241,** 50
136. Barros, C. and Franklin, M. C. (1968). Behaviour of the gamete membranes during sperm entry into the mammalian egg. *J. Cell. Biol.*, **37,** C13
137. Zamboni, L. (1971). *Fine Morphology of Mammalian Fertilization* (New York, Evanston, San Francisco, London: Harper and Row)
138. Austin, C. R. (1972). Fertilization. *Developmental Biology* (J. W. Lash and J. R. Whittaker, editors) (Stamford: Sinauer Assoc.) (in press)
139. Zamboni, L. and Mastroianni, L. J. (1966). Electron microscope studies on rabbit ova II. The penetrated tubal ovum. *J. Ultrastruct. Res.*, **14,** 118
140. Schuchner, E. B. (1970). Ultrastructural changes of nucleoli during early development of fertilized rat eggs. *Biol. Reprod.*, **3,** 265
141. Longo, F. J. and Anderson, E. (1969). Cytological events leading to the formation of the two-cell stage in the rabbit: association of the maternally and paternally derived genomes. *J. Uultrastruct. Res.*, **29,** 86
142. Hunter, R. H. F. (1972). Fertilization in the pig: Sequence of nuclear and cytoplasmic events. *J. Reprod. Fert.*, **29,** 395
143. Alfert, M. (1950). A cytochemical study of oogenesis and cleavage in the mouse. *J. Cell Comp. Physiol.*, **36,** 381
144. Mintz, B. (1965). Nucleic acid and protein synthesis in the developing mouse embryo. *Ciba Foundation Symposium: Preimplantation stages of pregnancy*, 145 (London, 1965: Churchill)
145. Flax, M. H. (1953). Ph.D. Thesis, Columbia University. Cited by Mintz[144]
146. Mintz, B. (1962). Incorporation of nucleic acid and protein precursors by developing mouse eggs. *Amer. Zool.*, **2,** 432 (Abs.)
147. Mintz, B. (1964). Formation of genetically mosaic mouse embryos and early development of 'lethal $(t/^{12}/t^{12})$-normal' mosaics. *J. Exp. Zool.*, **157,** 273
148. Austin, C. R. (1953). Nucleic acids associated with the nucleoli of living segmented rat eggs. *Exp. Cell Res.*, **4,** 249
149. Austin, C. R. and Braden, A. W. H. (1953). The distribution of nucleic acid in rat eggs in fertilization and early segmentation. I. Studies on living eggs by ultraviolet microscopy. *Austr. J. Biol. Sci.*, **6,** 324
150. Monesi, V., Molinaro, M., Spalletta, E. and Davoli, C. (1970). Effect of metabolic inhibitors on macromolecular synthesis and early development in the mouse embryo. *Exp. Cell Res.*, **59,** 197
151. Van Niekerk, C. N. and Gerneke, W. H. (1966). Persistence and parthenogenetic cleavage of tubal ova in the mare. *Onderstepoort J. Vet. Res.*, **33,** 195
152. Bronson, R. A. and McLaren, A. (1970). Transfer to the mouse oviduct of eggs with and without the zona pellucida. *J. Reprod. Fert.*, **22,** 129
153. Croxatto, H. B., Diaz, S., Fuentealba, B., Croxatto, H. D., Carrillo, D. and Fabres, C. (1972). Studies on the duration of egg transport in the human oviduct. I. The time interval between ovulation and egg recovery from the uterus in normal women. *Fert. Steril.*, **23,** 447

154. Lewis, W. H. and Gregory, P. M. (1929). Cinematographs of living developing rabbit eggs. *Science*, **69,** 226
155. Lewis, W. H. and Gregory, P. M. (1929). Moving pictures of developing rabbit eggs (Abstr.). *Anat. Rec.*, **42,** *Suppl.*, 27
156. Brinster, R. L. (1970). Culture of two-cell rabbit embryos to morulae. *J. Reprod. Fert.*, **21,** 17
157. Kane, M. T. and Foote, R. H. (1970). Culture of two- and four-cell rabbit embryos to the blastocyst stage in serum and serum extracts. *Biol. Reprod.*, **2,** 245
158. Kane, M. T. and Foote, R. H. (1970). Fractionated serum dialysate and synthetic media for culturing 2- and 4-cell rabbit embryos. *Biol. Reprod.*, **2,** 356
159. Maurer, R. R., Onuma, H. and Foote, R. H. (1970). Viability of cultured and transferred rabbit embryos. *J. Reprod. Fert.*, **21,** 417
160. Kane, M. T. and Foote, R. H. (1971). Factors affecting blastocyst expansion of rabbit zygotes and young embryos in defined media. *Biol. Reprod.*, **4,** 41
161. Beier, H. M. (1967). Veranderungen am Proteinmuster des Uterus bei dessen Ernahrungsfunktion fur die Blastocyste des Kaninchens. *Verh. Dtsch. Zool. Ges.*, **31,** Heidelberg, 139
162. Beier, H. M. (1968). Uteroglobin: a hormone sensitive endometrial protein involved in blastocyst development. *Biochem. Acta*, **160,** 289
163. Krishnan, R. S. and Daniel, J. C. (1967). 'Blastokinin'—An inducer and regulator of blastocyst development in the rabbit uterus. *Science*, **158,** 490
164. Krishnan, R. S. and Daniel, J. C. (1968). Composition of 'Blastokinin' from rabbit uterus. *Biochem. Biophys. Acta*, **168,** 579
165. Beier, H. M., Kühnel, W. and Petry, G. (1971). Uterine Secretion Proteins as Extrinsic Factors in Preimplantation Development. *Schering Symposium on Intrinsic and Extrinsic Factors in Early Mammalian Development*, 165 (Venice: 1970) (G. Raspe, editor) (Oxford, Braunschweig: Pergamon, Vieweg)
166. Daniel, J. C. (1971). Uterine Proteins and Embryonic Development. *Schering Symposium on Intrinsic and Extrinsic Factors in Early Mammalian Development*, 190 (Venice: 1970) (G. Raspe, editor) (Oxford, Braunschweig: Pergamon, Vieweg)
167. Kirschner, C. (1972). Immune histologic studies on the synthesis of a uterine-specific protein in the rabbit and its passage through the blastocyst coverings. *Fert. Steril.*, **23,** 131
168. McGaughey, R. W. and Murray, F. A. (1972). Properties of blastokinin: amino acid composition, evidence for subunits, and estimation of isoelectric points. *Fert. Steril.*, **23,** 399
169. Murray, F. A., McGaughey, R. W. and Yarus, M. J. (1972). Blastokinin: its size and shape, and an indication of the existence of subunits. *Fert. Steril.*, **23,** 69
170. Daniel, J. C. (1972). Blastokinin in the Northern fur seal. *Fert. Steril.*, **23,** 78
171. Johnson, M. H., Cowan, B. D. and Daniel, J. C. (1972). An immunologic assay for blastokinin. *Fert. Steril.*, **23,** 93
172. Arthur, A. T., Cowan, B. D. and Daniel, J. C. (1972). Steroid binding to blastokinin. *Fert. Steril.*, **23,** 85
173. El-Banna, A. A. and Daniel, J. C. (1972). Stimulation of rabbit blastocysts *in vitro* by progesterone and uterine proteins in combination. *Fertility*, **23,** 101
174. El-Banna, A. A. and Daniel, J. C. (1972). The effects of protein fractions from rabbit uterine fluids on embryo growth and uptake of nucleic acid and protein precursors. *Fert. Steril.*, **23,** 105
175. Arthur, A. T. and Daniel, J. C. (1972). Progesterone regulation of blastokinin production and maintenance of rabbit blastocysts transferred into uteri of castrate recipients. *Fert. Steril.*, **23,** 115
176. Johnson, M. H. (1972). The protein composition of secretions from pregnant and pseudopregnant rabbit uteri with and without a copper intrauterine device. *Fert. Steril.*, **23,** 123
177. Brinster, R. L. (1969). Mammalian embryo culture: *The Mammalian Oviduct*, 419 (E. S. E. Hafez and R. J. Blandau, editors) (The University of Chicago Press)
178. Brinster, R. L. (1970). *In vitro* cultivation of mammalian ova: *Schering Symposium on Mechanisms Involved in Conception*, 199 (Berlin: 1969) (G. Raspe, editor) (Oxford, Braunschweig: Pergamon, Vieweg)

179. Brinster, R. L. (1971). Mammalian embryo metabolism: *The Biology of the Blastocyst*, 303 (R. J. Blandau, editor) (University of Chicago Press)
180. Brinster, R. L. (1971). Biochemistry of the early mammalian embryo: *The Biochemistry of Development*, 161 (P. F. Benson, editor) (London, Philadelphia: William Heinemann Medical Books, J. B. Lippincott)
181. Brinster, R. L. (1971). *In vitro* culture of the embryo: *Pathways to Conception: The Role of the Cervix and the Oviduct in Reproduction*, 245 (A. I. Sherman, editor) (Illinois: Thomas)
182. Brinster, R. L. (1972). Developing zygote: *Reproductive Biology*, 748 (H. Balin and S. Glasser, editors) (Amsterdam: Excerpta Medica)
183. Whitten, W. K. (1971). Nutrient requirements for the culture of preimplantation embryos *in vitro*. *Schering Symposium on Intrinsic and Extrinsic Factors in Early Mammalian Development*, 129 (Venice, 1970) (G. Raspe, editor) (Oxford, Braunschweig: Pergamon, Vieweg)
184. Biggers, J. D. (1971). Metabolism of mouse embryos. *J. Reprod. Fert.*, **14,** 41
185. Whittingham, D. G. (1971). Culture of mouse ova. *J. Reprod. Fert., Suppl.*, **14,** 7
186. Biggers, J. D., Whitten, W. K. and Whittingham, D. G. (1971). The culture of mouse embryos *in vitro*. *Methods in Mammalian Embryology*, 86 (J. C. Daniel, editor) (San Francisco: Freeman and Co.)
187. Cholewa, J. and Whitten, W. K. (1970). Development of 2-cell mouse embryos in the absence of a fixed nitrogen source. *J. Reprod. Fert.*, **22,** 553
188. Whitten, W. K. and Biggers, J. D. (1968). Complete development *in vitro* of the preimplantation stages of the mouse in a simple chemically defined medium. *J. Reprod. Fert.*, **17,** 399
189. Biggers, J. D. (1971). New observations on the nutrition of the mammalian oocyte and the preimplantation embryo: *The Biology of the Blastocyst*, 319 (R. J. Blandau, editor) (University of Chicago Press)
190. Donahue, R. P. and Stern, S. (1968). Follicular cell support of oocyte maturation: production of pyruvate *in vitro*. *J. Reprod. Fert.*, **17,** 395
191. Whittingham, D. G. (1971). Survival of mouse embryos after freezing and thawing. *Nature (London)*, **233,** 125
192. Edwards, R. G., Bavister, B. D. and Steptoe, P. C. (1969). Early stages of fertilization *in vitro* of human oocytes matured *in vitro*. *Nature (London)*, **221,** 632
193. Steptoe, P. C., Edwards, R. G. and Purdy, J. M. (1971). Human blastocysts grown in culture. *Nature (London)*, **229,** 132
194. Piko, L. (1970). Synthesis of macromolecules in early mouse embryos cultured *in vitro:* RNA, DNA, and a polysaccharide component. *Developmental Biology*, **21,** 257
195. Graham, C. F. (1972). Nucleic acid metabolism during early mammalian development. *N.I.H. Conf.: The Regulation of Mammalian Reproduction* (Maryland, U.S.A.: 1970) (in press)
196. Heape, W. (1891). Preliminary note on the transplantation and growth of mammalian ova within uterine foster mother. *Proc. Roy. Soc. B.*, **48,** 457
197. Adams, C. E. and Abbott, M. (1971). Bibliography on Recovery and Transfer of Mammalian Eggs and Ovarian Transplantation. *Biblio. Reprod.*, **18,** 325
198. Cross, P. C. and Brinster, R. L. (1970). *In vitro* development of mouse oocytes. *Biol. Reprod.*, **3,** 298
199. Hunter, R. H. F., Lawson, R. A. S. and Rowson, L. E. A. (1972). Maturation, transplantation and fertilization of ovarian oocytes in cattle. *J. Reprod. Fert.*, **30,** 325
200. Whittingham, D. G. (1968). Fertilization of mouse eggs *in vitro*. *Nature (London)*, **220,** 592
201. Chang, M. C. (1959). Fertilization of rabbit ova *in vitro*. *Nature (London)*, **184,** 466
202. Rowson, L. E. A., Lawson, R. A. S. and Moor, R. M. (1971). Production of twins by egg transfer. *J. Reprod. Fert.*, **25,** 261
203. Adams, C. E., Moor, R. M. and Rowson, L. E. A. (1968). Survival of cow and sheep eggs in the rabbit oviduct. *Proc. VIth Int. Congr. Anim. Reprod., Paris*, **1,** 573

204. Lawson, R. A. S., Adams, C. E. and Rowson, L. E. A. (1972). The development of sheep eggs in the rabbit oviduct and their viability after re-transfer to ewes. *J. Reprod. Fert.*, **29,** 105
205. Lawson, R. A. S., Rowson, L. E. A. and Adams, C. E. (1972). The development of cow eggs in the rabbit oviduct and their viability after re-transfer to heifers. *J. Reprod. Fert.*, **28,** 313
206. Maurer, R. R. and Foote, R. H. (1971). Maternal ageing and embryonic mortality in the rabbit. 1. Repeated superovulation, embryo culture and transfer. *J. Reprod. Fert.*, **25,** 329
207. Adams, C. E. (1970). Ageing and reproduction in the female mammal with particular reference to the rabbit. *J. Reprod. Fert., Suppl.*, **12,** 1
208. Pincus, G. and Shapiro, H. (1940). The comparative behaviour of mammalian eggs *in vivo* and *in vitro*. *Proc. Amer. Phil. Soc.*, **83,** 631
209. Pincus, G. and Shapiro, H. (1940). Further studies on the parthenogenetic activation of rabbit eggs. *Proc. Nat. Acad. Sci., Wash.* **26,** 163
210. Shapiro, H. (1942). Parthenogenetic activation of rabbit eggs. *Nature (London)*, **177,** 1134
211. Thibault, C. (1949). L'oeuf des mammiferes. San developpement parthenogenetique. *Ann. Sci. Nat. Zool. 11th Ser.*, **11,** 136
212. Chang, M. C. (1954). Development of parthenogenetic rabbit blastocysts induced by low temperature storage of unfertilized ova. *J. Exp. Zool.*, **125,** 127
213. Austin, C. R. and Braden, A. W. H. (1954). Nucleus formation and cleavage induced in unfertilized rat eggs. *Nature (London)*, **173,** 999
214. Graham, C. F. (1970). Parthenogenetic mouse blastocysts. *Nature (London)*, **226,** 165
215. Tarkowski, A. K., Witkowska, A. and Nowicka, J. (1970). Experimental parthenogenesis in the mouse. *Nature (London)*, **226,** 162
216. Graham, C. F. (1971). Experimental early parthenogenesis in mammals. *Schering Symposium on Intrinsic and Extrinsic Factors in Early Mammalian Development*, 87 (Venice: 1970) (G. Raspe, editor) (Oxford, Braunschweig: Pergamon, Vieweg)
217. Olsen, M. W. (1960). Nine-year summary of parthenogenesis in turkeys. *Proc. Soc. Exp. Biol. Med.*, **105,** 279
218. Graham, C. F. (1971). Virus assisted fusion of embryonic cells. *Acta Endocrinol., Suppl.*, **153,** 154
219. Tarkowski, A. K. (1971). (In discussion after paper by Graham[216], p. 98)
220. Tarkowski, A. K. (1971). Recent studies on parthenogenesis in the mouse. *J. Reprod. Fert., Suppl.*, **14,** 31
221. Smith, A. U. (1961). *Biological effects of freezing and supercooling* (Baltimore: Williams and Wilkins)
222. Smith, A. U. (1952). Behaviour of fertilized rabbit eggs exposed to glycerol and to low temperature. *Nature (London)*, **170,** 374
223. Smith, A. U. (1953). *In vitro* experiments with rabbit eggs. *Mammalian Germ Cells*, 217 (G. E. W. Wolstenholme, M. P. Cameron and J. S. Freeman, editors) (London: Churchill)
224. Sherman, J. K. and Lin, T. P. (1958). Survival of unfertilized mouse eggs during freezing and thawing. *Proc. Soc. Exp. Biol., N.Y.*, **98,** 902
225. Sherman, J. K. and Lin, T. P. (1959). Temperature shock and cold-storage of unfertilized mouse eggs. *Fert. Steril.*, **10,** 384
226. Whittingham, D. G. (1968). Fertilization of mouse eggs *in vitro*. *Nature (London)*, **220,** 592
227. Whittingham, D. G., Leibo, S. P. and Mazur, P. (1972). Survival of mouse embryos frozen to $-196°$ and $-269°C$. *Science*, **178,** 411
228. Moore, N. W., Adams, C. E. and Rowson, L. E. A. (1968). Developmental potential of single blastomeres of the rabbit egg. *J. Reprod. Fert.*, **17,** 527
229. Moore, N., Polge, C. and Rowson, L. E. A. (1969). The survival of single blastomeres of pig eggs transferred to recipient gilts. *Austr. J. Eiol. Sci.*, **22,** 979
230. Tarkowski, A. K. (1971). Developments of single blastomeres. *Methods in Mammalian Embryology*, 172 (J. C. Daniel, editor) (San Francisco: Freeman and Co.)

231. Tarkowski, A. K. (1961). Mouse chimaeras developed from fused eggs. *Nature (London)*, **190,** 857
232. Mintz, B. (1962). Formation of genotypically mosaic mouse embryos. *Amer. Zool.*, **2,** 432 (Abstr.)
233. Mintz, B. (1971). Allophenic mice of multi-embryo origin. *Methods in Mammalian Embryology*, 186 (J. C. Daniel, editor) (San Francisco: Freeman and Co.)
234. Mintz, B. and Baker, W. W. (1967). Normal mammalian muscle differentiation and the gene control of isocitrate dehydrogenase synthesis. *Proc. Nat. Acad. Sci.*, **58,** 592
235. Gardner, R. L. (1971). Manipulation on the Blastocyst: *Schering Symposium on Intrinsic and Extrinsic Factors in Early Mammalian Development*, 280 (Venice: 1970) (G. Raspe, editor) (Oxford, Braunschweig: Pergamon, Vieweg)
236. Gardner, R. L. (1972). Manipulation of development: *Reproduction in Mammals*, **2,** 110 (C. R. Austin and R. V. Short, editors) (Cambridge University Press)
237. Shelesnyak, M. C. and Marcus, G. J. (1969). *The Study of Nidation: Ovum Implantation: Proc. Weizmann Inst. of Science* (Israel, 1967) (M. C. Shelesnyak and G. J. Marcus, editors) (New York, London, Paris: Gordon & Breach)
238. Shelesnyak, M. C. and Marcus, G. C. (1971). Steroidal Conditioning of the Endometrium for Nidation: *Schering Symposium on Intrinsic and Extrinsic Factors in Early Mammalian Development*, 303 (Venice: 1970) (G. Raspe, editor) (Oxford, Braunschweig: Pergamon, Vieweg)
239. Psychoyos, A. (1967). The Hormonal Interplay Controlling Egg Implantation in the Rat: *Advances in Reproductive Physiology*, 257 (Anne McLaren, editor) (London: Logos Press)
240. Psychoyos, A. (1970). Conditionment hormonal de la réceptivité endométriale pour la nidation: *2nd Int. Seminar on Reproductive Physiology and Sexual Endocrinology: Ovo-Implantation, Human Gonadotropins and Prolactin*, 101 (Brussels: 1968) (P. O. Hubinont, F. Leroy, C. Robyn and P. Leleux, editors) (Basel, Munchen, New York: S. Karger)
241. McLaren, A. (1965). Maternal factors in nidation: *Symposium on the Early Conceptus, Normal and Abnormal.* (W. W. Park, editor)
242. McLaren, A. (1970). Early Embryo-Endometrial Relationships: *2nd Int. Seminar on Reproductive Physiology and Sexual Endocrinology: Ovo-Implantation, Human Gonadotropins and Prolactin*, 18 (Brussels: 1968) (P. O. Hubinont, F. Leroy, C. Robyn and P. Leleux, editors) (Basel, Munchen, New York: S. Karger)
243. Potts, M. (1969). The ultrastructure of egg implantation: *Advances in Reproductive Physiology*, **4,** 241
244. Finn, C. A. (1971). The biology of decidual cells. *Advances in Reproductive Physiology*, **5,** 1
245. Larsen, J. F. (1970). Electron Microscopy of Nidation in the Rabbit and Observations on the Human Trophoblastic Invasion: *2nd Int. Seminar on Reproductive Physiology and Sexual Endocrinology: Ovo-Implantation, Human Gonadotropins and Prolactin*, 38 (Brussels: 1968) (P. O. Hubinont, F. Leroy, C. Robyn and P. Leleux, editors) (Basel, Munchen, New York: S. Karger)
246. Wilson, I. B. and Smith, M. S. R. (1970). Primary Trophoblastic Invasion at the Time of Nidation: *2nd Int. Seminar on Reproductive Physiology and Sexual Endocrinology: Ovo-Implantation, Human Gonadotropins and Prolactin*, 1 (Brussels: 1968) (P. O. Hubinont, F. Leroy, C. Robyn and P. Leleux, editors) (Basel, Munchen, New York: S. Karger)
247. Kirby, D. R. S. (1970). Immunological Aspects of Ovo-Implantation: *2nd Int. Seminar on Reproductive Physiology and Sexual Endocrinology: Ovo-Implantation, Human Gonadotropins and Prolactin*, 86 (Brussels: 1968) (P. O. Hubinont, F. Leroy, C. Robyn and P. Leleux, editors) (Basel, Munchen, New York: S. Karger)
248. Mintz, B. (1971). Control of Embryo Implantation and Survival. *Schering Symposium on Intrinsic and Extrinsic Factors in Early Mammalian Development*, 317 (Venice: 1970) (G. Raspe, editor) (Oxford, Braunschweig: Pergamon, Veiweg)
249. Noyes, R. W., Dickmann, Z., Doyle, L. L. and Gates, A. H. (1963). Ovum transfers, synchronous and asynchronous, in the study of implantation: *Delayed Implantation*, 197, (A. C. Enders, editor) (University of Chicago Press)

250. Chang, M. C. and Pickworth, S. (1969). Egg Transfer in the Laboratory Animal: *The Mammalian Oviduct*, 389 (E. S. E. Hafez and R. J. Blandau, editors) (The University of Chicago Press)
251. Rowson, L. E. A. and Moor, R. M. (1966). Embryo transfer in the sheep: The significance of synchronizing oestrus in the donor and recipient animal. *J. Reprod. Fert.*, **11,** 207
252. Webel, S. K., Peters, J. B. and Anderson, L. C. (1970). Synchronous and asynchronous transfer of embryos in the pig. *J. Anim. Sci.*, **30,** 565
253. Rowson, L. E. A., Lawson, R. A. S., Moor, R. M. and Baker, A. A. (1972). Egg transfer in the cow: synchronization requirements. *J. Reprod. Fert.*, **28,** 426
254. Adams, C. E. (1971). The fate of fertilized eggs transferred to the uterus or oviduct during advancing pseudopregnancy in the rabbit. *J. Reprod. Fert.*, **26,** 99
255. Sato, A. and Yanagimachi, R. (1972). Transplantation of preimplantation hamster embryos. *J. Reprod. Fert.*, **30,** 329
256. Enders, A. C. and Schlafke, S. (1967). A morphological analysis of the early implantation stages in the rat. *Amer. J. Anat.*, **120,** 185
257. Shelesnyak, M. C. (1970). In Discussion (p. 35) after paper by McLaren[242]
258. Orsini, M. W. (1963). Induction of deciduomata in hamster and rat by injected air. *J. Endocrinol.* **28,** 119
259. Hetherington, C. M. (1968). The development of deciduomata induced by two non-traumatic methods in the mouse. *J. Reprod. Fert.*, **17,** 391
260. McLaren, A. and Menke, T. M. (1971). CO_2 output of mouse blastocysts *in vitro*, in normal pregnancy and in delay. *J. Reprod. Fert., Suppl.*, **14,** 23
261. Bitton-Casimiri, V., Brun, J. L. and Psychoyos, A. (1971). Active release of material from rat blastocysts developing *in vitro*. *J. Reprod. Fert.*, **27,** 461
262. Kirby, D. R. S. (1965). The invasiveness of the trophoblast. *The Early Conceptus, Normal and Abnormal*, 68 (W. W. Park, editor) (Edinburgh, London: Livingstone)
263. Potts, M. (1966). The attachment phase of ovo-implantation. *Amer. J. Obstet. Gynec.*, **96,** 1122
264. Potts, M. (1968). The ultrastructure of implantation in the mouse. *J. Anat.*, **103,** 77
265. Reinius, S. (1969). Ultrastructure of blastocyst attachment in the mouse. *Z. Zellforsch*, **77,** 257
266. Nilsson, O. (1970). Some Ultrastructural Aspects of Ovo-Implantation: *2nd Int. Seminar on Reproductive Physiology and Sexual Endocrinology: Ovo-Implantation, Human Gonadotropins, and Prolactin*, 18 (Brussels: 1968) (P. O. Hubinont, F. Leroy, C. Robyn and P. Leleux, editors) (Basel, Munchen, New York: S. Karger)
267. Tachi, S., Tachi, C. and Lindner, H. R. (1970). Ultrastructural features of blastocyst attachment and trophoblastic invasion in the rat. *J. Reprod. Fert.*, **21** 37
268. Medawar, P. B. (1953). Some immunological and endocrinological problems raised by evolution of viviparity in vertebrates. *Symp. Soc. Exp. Biol., VII, Evolution*, 320 (Cambridge University Press)
269. Kirby, D. R. S. (1960). Development of the mouse eggs beneath the kidney capsule. *Nature* (*London*), **187,** 707
270. Kirby, D. R. S. (1963). The development of the mouse blastocysts transplanted to the scrotal and cryptorchid testis. *J. Anat.*, **97,** 119
271. Kirby, D. R. S. (1963). Development of the mouse blastocyst transplanted to the spleen. *J. Reprod. Fert.*, **5,** 1
272. Kirby, D. R. S. (1966). *Egg Implantation* (discussion p. 19) (G. E. W. Wolstenholme and M. O'Connor, editors) (London: Churchill)
273. Kirby, D. R. S. (1967). Ectopic autografts of blastocysts in mice maintained in delayed implantation. *J. Reprod. Fert.*, **14,** 515
274. Kirby, D. R. S. (1968). Immunological aspects of pregnancy. *Advances in Reproductive Physiology*, **3,** 33
275. Kirby, D. R. S. (1969). Immunology of implantation: *Proc. 1st Symposium of the Int. Coordination Committee for the Immunology of Reproduction* (Geneva: 1968): *Immunology and Reproduction*, 231 (R. G. Edwards, editor) (I.P.P.F.: London)
276. Kirby, D. R. S., Billington, W. D. and James, D. A. (1966). Transplantation of eggs to kidney and uterus of immunized mice. *Transplantation*, **4,** 713

277. Kirby, D. R. S., McWhirter, K. G., Teitelbaum, M. S. and Darlington, C. D. (1967). A possible immunological influence on sex ratio. *Lancet*, **ii,** 139
278. Kirby, D. R. S., Potts, D. M. and Wilson, I. B. (1967). On the orientation of the implanting blastocyst. *J. Embryol. Exp. Morph.*, **17,** 527
279. Breyere, E. J. and Barrett, M. K. (1960). Prolonged survival of skin homografts in parous female mice. *J. Nat. Cancer Inst.*, **23,** 1405
280. Anderson, J. M. (1970). The transplantation immunology of certain mammalian mothers and progeny. *Proc. Roy. Soc. B*, **176,** 115
281. Currie, G. A. (1969). The foetus as an allograft: The role of maternal unresponsiveness to paternally derived foetal antigens: *Foetal Autonomy*, 32 (G. E. W. Wolstenholme and M. O'Connor, editors) (London: Churchill)
282. Currie, G. A. (1970). Effect of interstrain pregnancy on the immune status of female mice sensitized to paternal antigens. *J. Reprod. Fert.*, **23,** 501

5
Hormonal Control of Oviductal Motility and Secretory Functions

E. M. COUTINHO
Federal University of Bahia, Brazil

5.1 INTRODUCTION

The Fallopian tubes or oviducts are convoluted muscular structures deriving embryonically from the Müllerian ducts, which connect the uterus and ovaries. Their main function in reproduction is to transport ova from the ovarian surface to the uterus, and to provide an appropriate environment for the gametes to meet and fertilisation to occur. To accomplish this the tube must adjust the contractility of its smooth musculature in such a way as to provide the proper amount of propulsion to the ova while controlling its permanence in the tube for a sufficiently long time to allow fertilisation. It must also control the activity of its secretory cells to assure secretion of the proper amount of a tubal fluid suitable to the early development of the zygote. Both the activity of the musculature and the ability of the oviduct to secrete tubal fluid are under endocrine control.

5.2 ANATOMICAL, STRUCTURAL, AND FUNCTIONAL CONSIDERATIONS

Four segments of the oviduct may be distinguished anatomically and functionally[1], the infundibulum with its fimbriae, the ampulla, the isthmus, and the uterotubal junction or junctura. The *infundibulum* has a very thin muscular wall, an epithelium formed of ciliated cells and its main function is collection of ova from the ovarian surface and its safe transport to the ampulla. The *ampulla* has well developed muscular layers and its epithelium has both ciliated and non-ciliated cells. The non-ciliated cells have secretory activity and contribute with their secretion to the formation of tubal fluid. In the ampulla, the ovum is detained for many hours or several days. It is here that fertilisation takes place. Early development of the zygote also occurs in the ampulla. The *isthmus* has thick muscular layers and its lumen is very narrow. The epithelium has non-ciliated cells but very few ciliated cells. At the isthmoampullar junction, a block to further transport of the ovum seems to be established at the time of ovulation. The block is apparently caused by activation of the isthmic musculature and may last several days. This ability to control the interval of permanence of the ovum in the tube through activation of its musculature seems to be the most important function of this portion of the oviduct. The isthmus ends at the *uterotubal junction* or *junctura*. This portion of the tube has the most complex musculature in view of the fusion of muscle fibres from the uterus with those of the tubes. The thickness and the arrangement of the muscular layers vary with the species. In man, the anthochthonous musculature of the uterus which forms the inner layer of the uterine wall consists of four bundles[2]. This arrangement enables a constrictor effect to be placed upon the larger part of the intramural tube and leaves this portion of the oviduct under myometrial control.

The musculature of the oviduct consists of longitudinal and circular muscle layers. In man, there are three layers: an inner longitudinal, an intermediary circular and an outer longitudinal layer. The muscular layers are covered by a serosal layer composed of mesothelium continuous with that of the peritonium and connective tissue. This layer is well vascularised. Smooth muscle fibres are found subperitonially and around the blood vessels. Muscle cells are also believed to exist in the mucosa which is the innermost layer of the oviductal wall. Innervation of the oviduct is predominantly adrenergic[3]. The adrenergic nerves supply both the musculature and the blood vessels. The amount of the musculature innervation in the ampulla and the isthmus is moderate but at the ampullary-isthmic junction the innervation is very dense[4].

5.3 HORMONAL INFLUENCES ON TUBAL MOTILITY

5.3.1 Methods

Several techniques have been used to investigate the motility of the Fallopian tube. *In vitro* methods are widely used and consist mainly in recording the contractions of strips or segments of tubal muscle with an optic lever or a force displacement transducer[5,6]. These techniques are used mostly in pharmacological studies. A combination of recording techniques may be used to evaluate separately but simultaneously, the activity of the circular and longitudinal muscles. One technique currently used (Ueda and Coutinho, unpublished) consists in recording tension of segments of the oviduct with a force displacement transducer, while a balloon tipped catheter inserted into the tubal lumen and connected to a pressure transducer is used to record changes in intratubal pressure resulting from contraction of circular muscle (Figure 5.1 (a and b)).

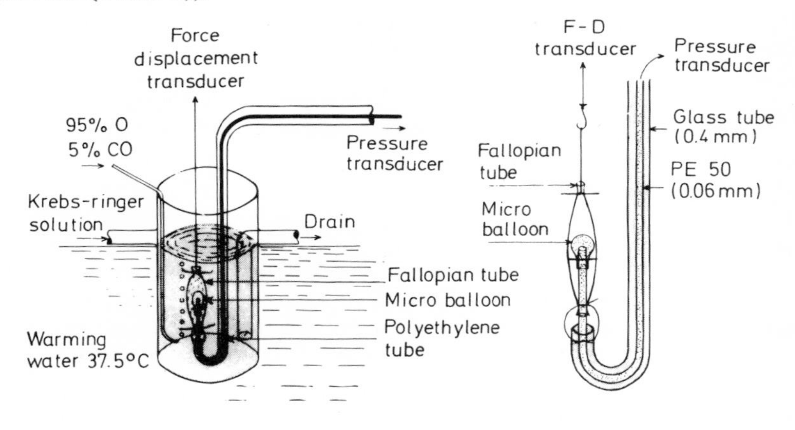

Figure 5.1a Technique for simultaneous recording of longitudinal and circular muscle contractions *in vitro*

A variety of techniques have been used to study the motility of the tube *in vivo*. Direct observation of tubal movements during laparotomy, laparoscopy or culdoscopy have been used to relate anatomical changes with physiological functions, but these techniques provide for only curtailed visual observation[2]. The movements of radio-opaque substances injected into the tubal lumen through the uterus may be studied by fluoroscopy, serial x-ray and cinematography[7]. Transuterine tubal insufflation of gas has probably been the most widely used technique to evaluate the muscular activity of the human oviduct[8]. This technique has been especially useful

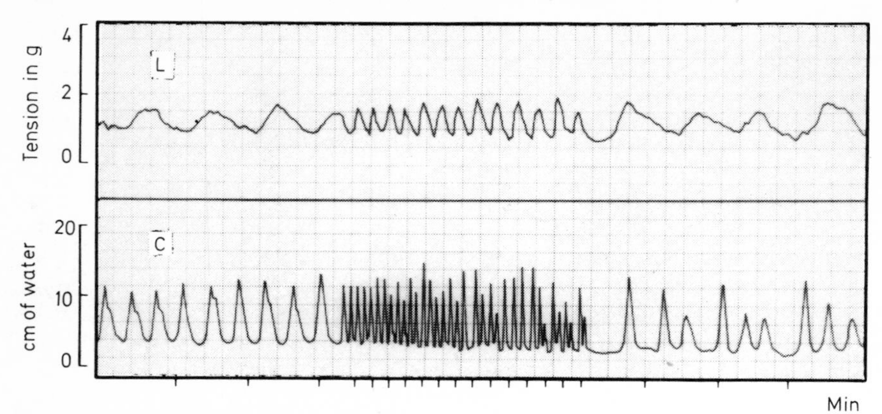

Figure 5.1b Simultaneous recording of changes in intratubal pressure resulting from the activity of the circular muscle (C) and changes in the longitudinal muscle tension (L), in the human tube *in vitro*. Technique illustrated in Figure 5.1a. Note that the contractions of the circular muscle are more frequent than those of the longitudinal muscle

in studies of the uterotubal junction[2]. When gas is injected into the uterus, intrauterine pressure rises because the uterotubal junction is closed. As the pressure increases, the junction is forced open, and once gas enters the oviduct the intrauterine pressure drops. The pressure necessary to force the junction open varies according to the species and the stage of the reproductive cycle[2]. Fluctuations in intrauterine pressure resulting from the passage of the gas from the uterus and through the tube into the peritonial cavity may be recorded. Tracings obtained with this technique provide information regarding the level of the uterotubal opening pressure and the patency of the tubes.

Direct recording of the muscular activity of the various portions of the Fallopian tube have been carried out *in vivo* in both animals and in women. Ichijo[5] recorded isotonic and isometric contractions of the rabbit oviduct during laparotomy with an optic lever and a force displacement transducer. Greenwald[6] recorded changes in intratubal pressure in the same animal with a pressure transducer also during laparotomy. Extratubal recording has been used in a limited way in the rabbit by Maistrello[9] who developed a special type of transducer for this purpose. Chronic recording of tubal motility, which is probably the most informative of the available techniques, has been used more recently in rabbits[10,11] and in monkeys[12].

In the human, direct recording of tubal activity has been carried out during laparotomy[13] and also chronically[14–19]. Sica-Blanco *et al.*[17,18] have used an open end catheter inserted into the tubal lumen to record variations in intratubal pressure. This technique, which is similar to that used by Neri *et al.*[12] to record tubal activity of the Rhesus monkey, has the disadvantage of the need for frequent flushing of the catheter to avoid obstructions. In view of the small calibre of the tube and its marked sensitivity to intraluminal volume changes, flushing with saline or water may become an important source of artifacts. In the techniques introduced by Maia and Coutinho in 1968[14], recording of intratubal pressure is carried out with a closed system. A polyvinyl catheter whose end is fenestrated and sheathed by a thin rubber diaphragm is inserted during laparotomy inside the tube. This end of the tube is placed in the portion of the oviduct whose pressure is to be recorded. The free end of the catheter is brought to the outside through an incision in the abdominal wall and held in place with adhesive tape. Changes in pressure in different portions of the tube may be recorded by using two or more catheters placed at the appropriate sites (see Ref. 16 and Figure 5.2). Recording may be carried out for two or more menstrual

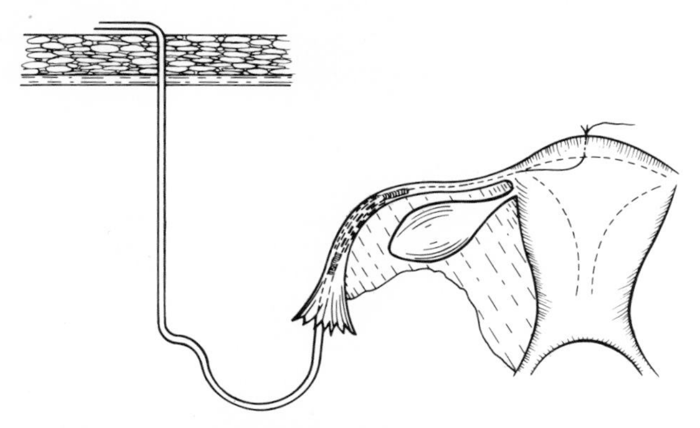

Figure 5.2 Diagrammatic representation of method for recording human tubal activity by converting the luminal portion of an indwelling catheter into a pressure receptor (From Maia and Coutinho[14], by courtesy of The C. V. Mosby Co.)

cycles without discomfort to the patients. The most stringent criticism to this technique is the possibility that the intratubal catheter may induce abnormal contractility patterns by mechanical stimulation of the tubal musculature. Because this possibility cannot be disregarded one should rely only on comparative studies where the position of the catheter is unchanged throughout the study period and when the patient is used as its own control.

5.3.2 Patterns of tubal motility

Tubal movements result not only from the muscular activity of the various layers of the smooth musculature of the tube itself but also from the activity of associated structures. Contractions of the mesosalpinx and other membranes, and even uterine activity may contribute to the motility of the tube. However, the rhythmic changes in intraluminal pressure ocurring in any given portion of the tube result probably from activity of tubal musculature. Recording of intratubal pressure from the ampulla or isthmus in women[14] and the Rhesus monkey[12] reveal a pattern of activity characterised by small contractions of 5 to 10 mmHg interspersed at regular intervals by outbursts of increased activity when intratubal pressure exceeds 20 mmHg. Periods of quiescence lasting several minutes may be observed following an outburst.

The phenomenon has been recorded also in the ampullary and isthmic portions of the rabbit oviduct and seems to be the most typical feature of tubal activity[11].

5.3.3 The influence of the menstrual cycle, pregnancy, and the puerperium on the patterns of tubal motility

In women, outbursts of increased activity occur with greatest frequency at the time of menstruation and during the early proliferative phase of the cycle. The contractions become more frequent during ovulation but although outbursts still occur, they are not followed by long lasting periods of quiescence, as they are during the early proliferative phase of the cycle.

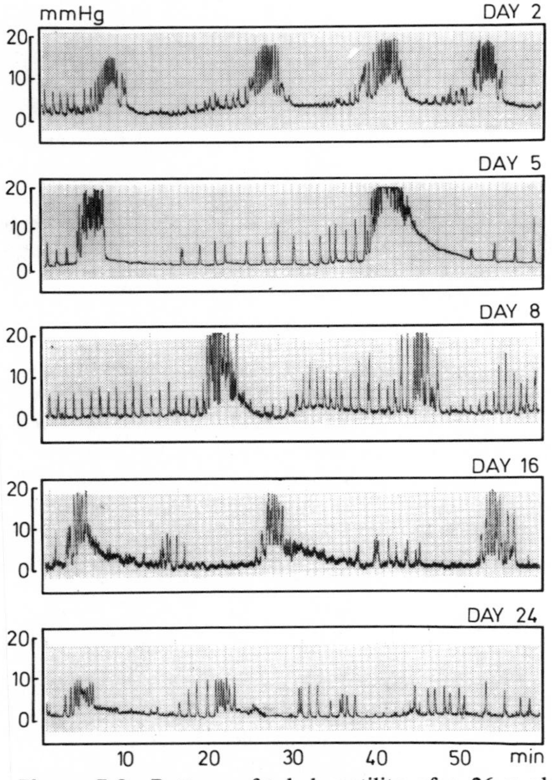

Figure 5.3 Patterns of tubal motility of a 26 y old woman during the menstrual cycle. Recordings were taken every other day for several hours but only representative one-hour tracings are shown. Note the outbursts of increased activity occurring at intervals of 15–20 min and lasting 5–10 min. Note also that the interval between the outbursts of increased activity was shorter during menstruation (From Maia and Coutinho[14] by courtesy of The C. V. Mosby Co.)

After ovulation, tubal activity diminishes markedly. The contractions become less frequent, and outbursts decrease in their intensity and duration (see Ref. 20 and Figure 5.3).

Although overall activity of the tube is reduced during the luteal phase, complete quiescence never occurs. The relative refractoriness of tubal musculature to the inhibitory effects of the corpus luteum hormones, is in marked contrast to the responsiveness of the myometrium and may be necessary to assure ovum transport during this period of uterine quiescence[21]. This inability of the ovarian hormones to supress tubal motility is emphasised during pregnancy. Recording of intra-ampullar pressure during the second trimester of gestation in women, reveals a pattern of activity similar to that recorded during the late luteal, or very early proliferative phase of the menstrual cycle (Vieira-Lopes and Coutinho, unpublished observations). Tubal activity of two patients recorded during the third trimester indicated

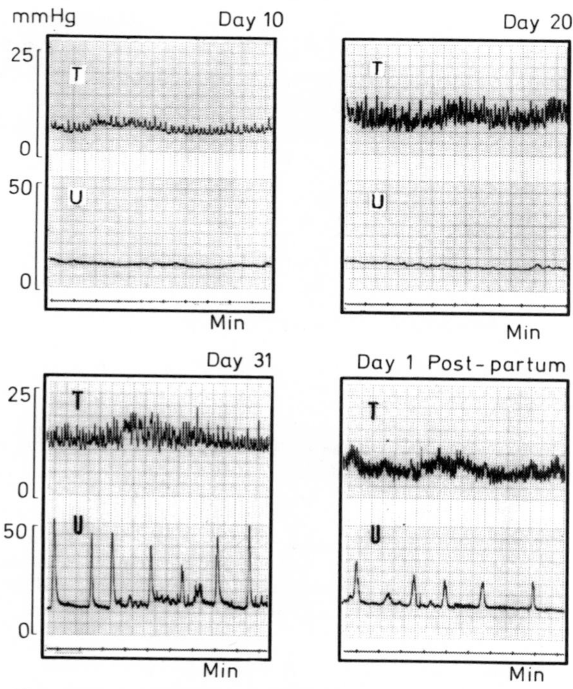

Figure 5.4 Motility of the Fallopian tube and the uterus of the rabbit during pregnancy. Note that on Day 10 both the frequency and the amplitude of the tubal contractions are reduced but activity is not abolished as it is in the uterus. During the third week of pregnancy (Day 20) tubal activity increases but uterine activity remains supressed. During delivery (Day 31) tubal activity remains at the same high level of late pregnancy. Only at this stage does uterine activity develop markedly. At the post-partum period both uterine and tubal motility return to the non-pregnant oestrus level (From Mattos and Coutinho[11], by courtesy of the Endocrine Society.)

that during this part of gestation tubal musculature is more active than during the early luteal phase of the menstrual cycle. In the rabbit, except for a slight decrease in frequency, no change was observed in oviductal motility during the first week of pregnancy. During the second week, the amplitude of the contractions decreased but no further decrease in frequency was observed. After the second week, the activity increases progressively and by the beginning of the fourth week, the pattern of motility is similar to the non-pregnant pattern (see Ref. 11 and Figure 5.4).

No consistent changes in tubal motility were observed from late pregnancy to the puerperium in the rabbit[11]. In the human, the early puerperal tube displays a pattern of activity similar to that recorded during the late luteal phase of the cycle or during menstruation. In lactating women, tubal musculature is markedly activated during suckling. In view of the marked responsiveness of the human oviduct to the stimulatory effect of oxytocin[20,21], it has been assumed that activation results from endogenous oxytocin released by the neural reflex. It should be noted however, that activation of tubal motility by suckling occurs several weeks after delivery apparently only after cyclic ovarian function is re-established (Coutinho and Vieira-Lopes, unpublished).

Stimulation of tubal motility following mechanical stimulation of the female breast occurs even in the absence of lactation (Figure 5.5). The

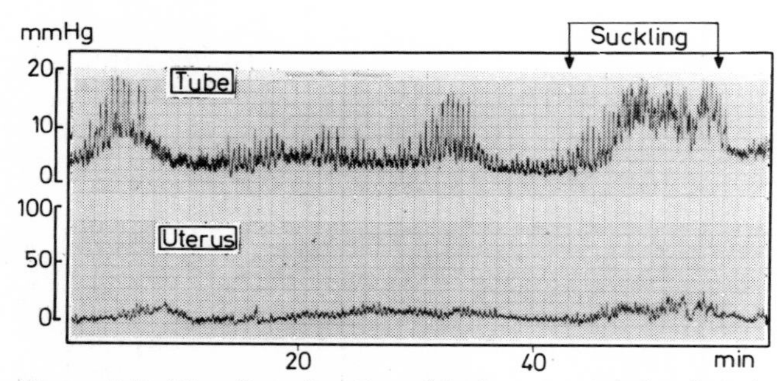

Figure 5.5 The effect of suction of the breast on tubal and uterine motility. Simultaneous recording of intrauterine and intratubal pressures on Day 16 of the menstrual cycle. Note the marked stimulation of tubal activity during suckling (From Coutinho[22], by courtesy of the Williams and Wilkins Co.)

response is more easily elicited during the early proliferative phase of the cycle[21]. This is in agreement with the observation that the response to exogenous oxytocin diminishes during the luteal phase[20].

5.3.4 Correlations between uterine and oviductal motility during the menstrual cycle and pregnancy

Except for the changes in pattern recorded after ovulation when tubal motility is progressively depressed, the tube is much less influenced by ovarian activity during the menstrual cycle than the uterus[22,23]. Like the uterus, however, the tube displays three distinct patterns of motility which may be correlated with phases of the ovarian cycle. Both the tube and the uterus reach two peaks of increased activity during the cycle. The first, during menstruation, when ovarian secretion of oestrogen should be lowest, and the second, during ovulation or just preceding ovulation when the ovarian oestrogen secretion should be the highest. The patterns of activity at these two stages of the cycle are however, quite different for both the uterus and the tube. During menstruation, the uterus has a low tonus and displays frequent contractions of high amplitude, reaching over 100 mmHg or higher. The tube has frequent outbursts of increased activity with contractions of high amplitude followed by periods of quiescence. During ovulation or just preceding ovulation when the reproductive tract is supposedly under oestrogen domination, uterine tonus increases considerably and uterine contractions become very frequent but of extremely short amplitude. During this phase, the uterus seems unable to contract or relax completely[24]. A similar picture of intense activity is described for the oviduct at this stage of the cycle. The tubal contractions are also very frequent and of small amplitude. Periods of quiescence are not observed during this phase. The third pattern of activity which may be correlated with a specific stage of the ovarian cycle, is that observed during the luteal phase when

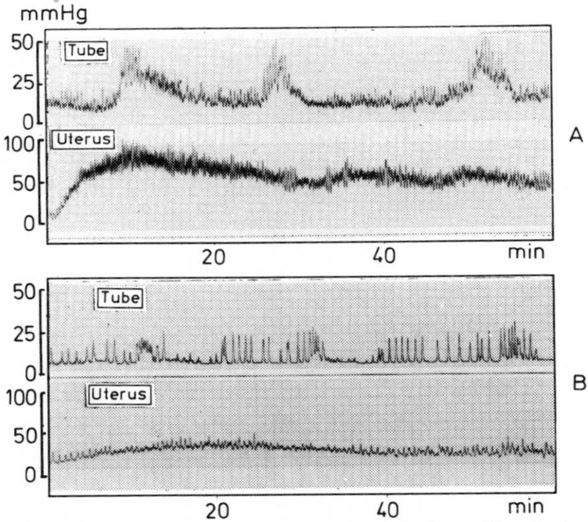

Figure 5.6 Patterns of uterine and tubal motility during the preovulatory (Day 12 (A)) and postovulatory (Day 23 (B)) phase. Simultaneous recording in the same patient (From Coutinho[23], by courtesy of the Nobel Foundation.)

both uterine and oviductal motility are depressed. It should be noted that the uterus is much more responsive to the inhibitory effect of the corpus luteum hormone, as a moderate degree of tubal activity always remains during the luteal phase (Figure 5.6).

Because the tube is very active during periods of low ovarian activity such as menstruation and the puerperium, it is conceivable that tubal motility is activated by withdrawal of the ovarian hormones. That ovarian withdrawal may activate myometrial motility has already been shown in rabbits[25]. Following ovariectomy in pseudopregnant, pregnant, and also oestrous animals, uterine activity increases progressively and reaches a peak 4–6 days after the operation (Figure 5.7). A similar increase in tubal

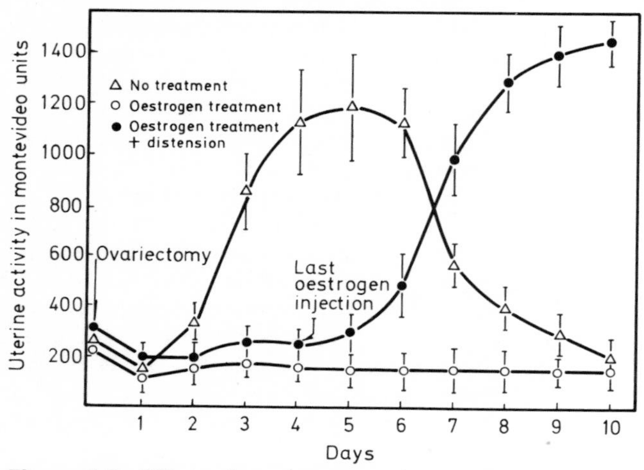

Figure 5.7 Effects of ovariectomy and oestrogen treatment on the activity of the rabbit uterus *in vivo*. Each curve represents means and SD for six animals. Except for the group of animals in which progressive distension was produced (solid circles), the rabbits represented had a constant uterine volume between 0.8 and 1.0 ml. The Montevideo Unit is calculated by multiplying the frequency of the contractions by their amplitude and is expressed in mmHg in 10 min[13]. This type of representation is restricted to the quantitative changes (From Coutinho and Mattos (1968). *Endocrinology*, **83,** 422 by courtesy of the Endocrine Society.)

activity was detected in oestrous rabbits, following ovariectomy, although changes were less marked than those observed in the uterus[11].

The hypothesis that oestrogen withdrawal plays an important role in the regulation of tubal activity is defended by Boling[26,27] who has repeatedly pointed out, that, in many species including man, there is a decrease in oestrogens immediately prior to ovulation. This drop in oestrogen rather than oestrogen itself would therefore be the cause for activation of tubal motility during ovulation. That oestrogen is not essential to sustain the spontaneous motility of the smooth musculature of the oviduct is indicated by the observation that the tubes of ovariectomised animals remained active long after castrate atrophy of the uterus reduced myometrial activity to a minimum[6,11]. However, it should be pointed out that withdrawal of oestrogen in the human may allow the release of tubo-stimulating substances from the pituitary, such as oxytocin, which would further activate the oviduct. A similar mechanism has been proposed to explain the exaggerated motility of the uterus during menstruation. In the latter case, vasopressin rather than oxytocin would be the activating substance[28–30].

Whatever the mechanism, activation by oestrogen or oestrogen+progesterone withdrawal represents a condition in which the muscle is set free from the controlling influences of those hormones and behaves therefore as an undisturbed spontaneously contracting smooth muscle. The patterns of tubal activity under oestrogen domination on the other hand, suggest that oestrogen prevents the tube from contracting and relaxing and from responding normally to physiological and pharmacological stimuli. By increasing the excitability of the muscle cell membrane, the muscle becomes over-reactive and a minimal degree of mechanical stimulation—such as that resulting from an increase in tubal fluid volume—will maintain the musculature constantly activated. The muscle never relaxes completely because before it does another contraction begins. In the narrower parts of the tube such an effect will produce a stricture and impede the passage of ova and even of fluid. In the rat and certain other rodents, the ampulla is distended by the fluid retained therein at the time of fertilisation[1]. A similar constrictor effect is induced by oestrogen on the uterine cervix of the rat during the day before pro-oestrus, causing fluid accumulation also in the uterine lumen[31]. Relaxation of the cervices occurring late in oestrus and allowing escape of fluid into the vagina, seems to be caused by the acute pro-oestrus increase in progesterone secretion from both the luteal and non-luteal elements of the ovary[32]. The constrictor effect of oestrogen and the relaxing action of progesterone on both the tubal isthmus and uterine cervix can be reproduced experimentally[32–34]. The isthmic block can be prolonged

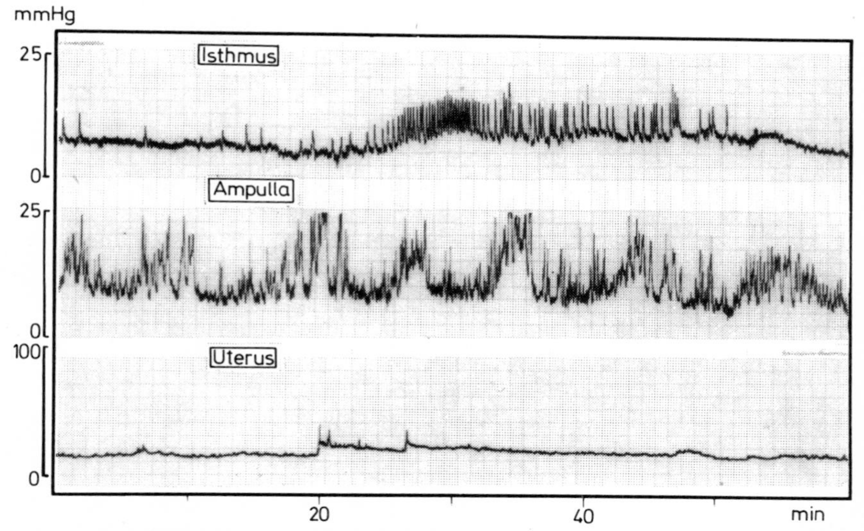

Figure 5.8 Simultaneous recording of intratubal pressure and intrauterine pressure during early pregnancy. 36 y old patient who underwent surgery for tubal ligation 4 days after having intercourse unprotected. Receptors were inserted into the isthmus and the ampulla. Recordings of both tubal and uterine activity were carried out every other day until its was found that the patient had become pregnant before the operation. The figure illustrates tubal and uterine activity at the end of the first week after the missing period. Note the long intervals of quiescence of the tubal isthmus and the activity of the ampulla. Note also the complete quiescence of the pregnant uterus

by oestrogen injections causing 'tube locking of ova'[35,36] or prevented by progesterone administration[37].

That the progestational hormones from the corpus luteum or extra luteal sources counteract the effects of oestrogen is also indicated by the change in pattern of tubal motility which takes place after ovulation. Relaxation becomes possible and short periods of quiescence reappear. Although a drop in oestrogen secretion may facilitate the release of the isthmic block, it is very likely that it is only when the progestational influence predominates that the free passage of the tubal contents towards the uterus becomes possible. During early pregnancy when luteal dominance is complete, recording of the isthmic portion of the oviduct reveals periods of activity followed by long intervals of quiescence, suggestive of complete and lasting relaxation. Simultaneous recording of the ampullar portion of the tube and the uterus shows a continuously active ampulla and a completely quiescent uterus (Figure 5.8). Daily administration of the progestin megestrol acetate as a contraceptive induces similar changes on tubal activity and prevents the

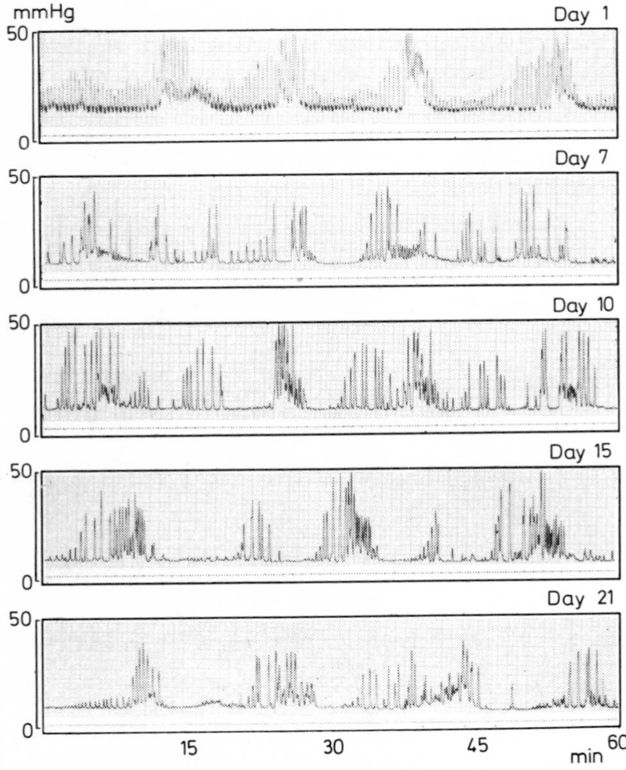

Figure 5.9 The effect of megestrol acetate on tubal activity. Recordings were carried out in a patient receiving 0.350 μg of megestrol acetate daily starting on Day 1. Note that there is almost no change in pattern throughout the cycle. No marked increase of contraction rate at mid-cycle suggestive of oestrogen domination is observed

changes in pattern which occur normally during the cycle (Figure 5.9). This ability of the progestins to prevent oestrogen induced isthmic closure during the ovulatory period may be the mechanism of its antifertility effect (see below).

5.3.5 The relationship between adrenergic activation and endocrine control of tubal motility

Although autonomic innervation is not essential to the function of the oviduct, there is ample evidence indicating that tubal motility is under adrenergic influence. The ampulla and the isthmic portions of the oviduct are well endowed with adrenergic innervation, and the isthmo-ampullar junction is richly supplied by adrenergic nerve endings[34,38]. Norepinephrine content of the oviduct varies with the density of the adrenergic innervation which appears to be concentrated in the smooth muscle coats and around the blood vessels.

The high degree of sensitivity of the tubes to both epinephrine and norepinephrine and the similarity of their response to the amines, with the outbursts of activity occurring spontaneously, led to the suggestion that the latter result from local release of small quantities of norepinephrine[39,40]. There is evidence that norepinephrine released by nerve endings combines with cell receptors and after exerting its effects is reincorporated by the neuron[41,42]. The rapid removal of the amine by nerves prevents diffusion to the neighbouring area and into the systemic circulation thus restricting the effect of the amine to the muscle layers associated with the corresponding nerve endings. There is a second type of uptake process for catecholamines (uptake II) which, in contrast with the intraneuronal uptake I, has a low affinity but high capacity for norepinephrine[42]. There are marked differences in affinities of various catecholamines to uptake I and uptake II and in the inhibition of both uptake processes by drugs or catecholamine metabolites. In a series of experiments carried out recently in chronically and repeatedly recorded rabbits, we have been able to suppress tubal activity partially or completely with the blocking agents phentolamine and phenoxbenzamine. This phenomenon is illustrated in Figure 5.10. This finding is in agreement with the observation that the effect of hypogastric nerve stimulation on the tube is blocked by phentolamine[43]. Simultaneous recordings of tubal and uterine motility in these animals show that during phentolamine blockade of tubal motility, myometrial activity is unaltered or may even be enhanced. When phentolamine is used, not only the outbursts, but all tubal activity can be reduced or abolished. With phenoxobenzamine however, repression of outbursts may be the only effect (Figure 5.10). In the human, attempts to supress tubal motility with therapeutic doses of phentolamine were unsuccessful. In fact, intravenous injections of phentolamine in women stimulate tubal motility causing outbursts of activity very similar to those which occur spontaneously[22]. This paradoxical response may result from potentiation of the effect of endogenous norepinephrine caused by blockade of the neuronal uptake[44]. This phenomenon has been observed in several other sympathetically innervated tissues and is called supersensitivity[45]. Supersensitivity due to the elimination of neuronal uptake is called 'presynaptic', whereas supersensitivity that is induced by changes in the responsiveness of the effector cells is named 'postsynaptic'. The degree of potentiation of the functional response to sympathomimetic amines uptake blockade in a certain tissue is determined by the affinity of the specific amine to the uptake sites and its affinity and intrinsic activity to the receptor sites. In tissues with a dense sympathetic innervation, like the tubal isthmus, it is to be expected that uptake blockade will produce a more marked potentiation than in poorly innervated tissues, where other mechanisms of inactivation are predominant. It is probably of interest to note that various steroids including 17-β-oestradiol and cholesterol are the most potent inhibitors of uptake

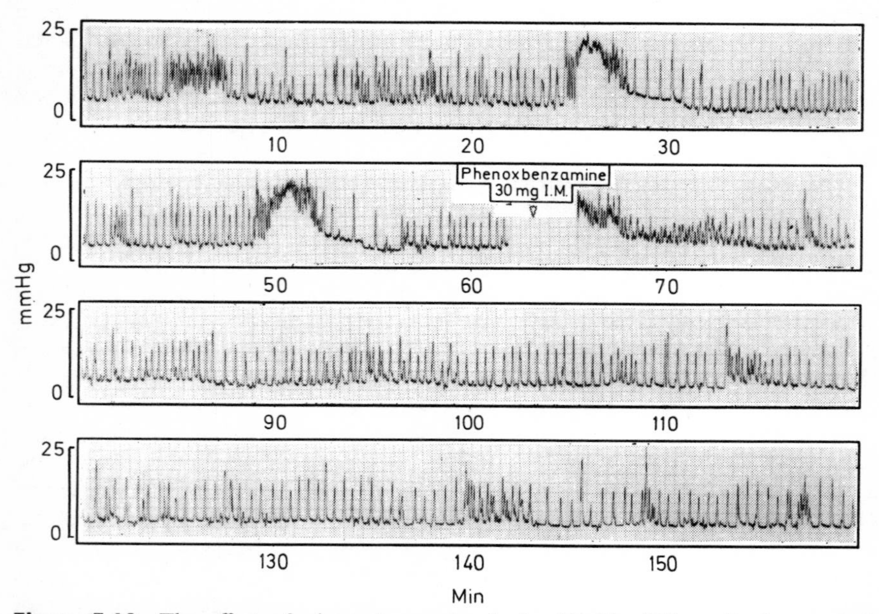

Figure 5.10 The effect of phenoxbenzamine hydrochloride (Dibenzyline) on tubal motility of the rabbit. The upper tracings show outbursts of increased activity occurring at 20 min intervals. Note that following the injection of Dibenzyline the phenomenon is suppressed (From Coutinho[22] by courtesy of the Williams Wilkins Co.)

II[46]. The inhibition of uptake II by the various steroid hormones occurs at concentrations which exceed those normally found in the plasma. However, in tissues like the oviduct and the uterus, which bind and accumulate oestrogens[45], inhibition of uptake II by these hormones probably occurs. It has been pointed out[46] that in tissues in which uptake I predominates, the blockade of uptake II by steroids or drugs will be of minor importance. On the other hand, in tissues in which for anatomical reasons the sympathetic nerve terminals are dissociated from the receptor bearing structures, as in the vascular bed or in tissues in which uptake was eliminated, uptake II will prevail and any interference with this mechanism now dominating will produce changes in tissue reactivity towards sympathomimetic amines.

Activation of tubal motility by oestrogens may be a result of supersensitivity caused by blocking of norepinephrine uptake. This is suggested by the greater sensitivity of the tube to the catecholamine during this phase[10,39,40], and by the demonstration that pharmacological blockade of norepinephrine uptake by moderate doses of phentolamine will exaggerate the phenomenon.

5.3.6 The influence of the ovarian steroids on the pharmacological responsiveness of the oviduct

The response of the oviduct to tubo-stimulating substances varies considerably during the menstrual cycle. In general, the response to various pharmacological agents is greater during the proliferative phase than the luteal phase, but it is greatest during menstruation. This applies to oxytocin[19], adrenergic compounds[39] and also to ergot derivatives such as ergonovine which are powerful tubal stimulants[23]. The response to these compounds is markedly diminished during pregnancy and progestin treatment. On the other hand, under luteal dominance the oviduct becomes more responsive to the inhibitory effects of β-adrenergic activators and α-adrenergic blockers.

The response of the tube to the prostaglandins E_2 and $F_{2\alpha}$ appear to be unaltered during the cycle[47]. However, the relaxing effect of prostaglandin E_2 seems to last longer during the luteal phase. Because prostaglandin E_2 acts as an α-adrenergic blocker, its inhibitory effects are more dramatic in those portions of the tube which are more densely innervated. Although the relaxing effects of prostaglandin E_2 are apparently associated with an increase in cyclic AMP attempts to prolong its relaxing action by the administration of aminophylline[48] were not successful (Coutinho, unpublished observations).

5.3.7 The control of gamete transport by oviductal musculature

The evidence presently available indicates that although ciliary movements may contribute to ovum pick-up at the surface of the ovary and its transport into the ampulla, it is probably through muscular activity that egg transport from the ampulla through the isthmus and into the uterus is accomplished. The extent of communication between the fimbriae and the ovary varies with the stage of the menstrual cycle. During ovulation they are brought in closer contact and it is conceivable that an increase in the muscular activity of the fimbriated end of the tube may 'massage' the ovary thus facilitating follicle rupture[26]. In view of the activating effect of oxytocin on the tubes[19,24] it has been proposed that reflex released oxytocin, resulting from stimulation of the breasts by manipulation or suction during sexual foreplay, may contribute to precipitate ovulation by activating the tube[22]. After ovulation the egg is propelled rapidly through the ampulla, but no further progress is possible in view of the block at the isthmo-ampullar junction. The oestrogen induced isthmic block, resulting probably from adrenergic activation through a process of 'supersensitivity', lasts a few days increasing thereby the chances of fertilisation. During this period of oestrogen domination, passage through the isthmus may be possible only through adrenergic inactivation or blockade[49]. α-receptor adrenergic blockers or β-receptor activators may release the isthmic block. It may be through such a mechanism that prostaglandin E_2 from semen absorbed through the vagina facilitates sperm penetration into the tube[47]. Block of α-adrenergic receptors maintain the whole tube quiescent obviating expulsion of the egg. The effect of α-adrenergic blockers or β-receptor activators is short-lived, lasting only a few minutes. When their effect is over, the musculature is reactivated and the isthmic block re-established. This will allow fertilisation and early development of the zygote to occur in the tube. However, as the levels of progesterone rise, oestrogen domination of the tube diminishes and the isthmic block is released. Passage becomes possible and the peristaltic contractions depressed but not suppressed by progesterone allow for a gentle propulsion of the egg through the isthmus and into the uterus.

Excesses of either oestrogens or progestins interfere with ovum transport causing ectopic implantations or infertility. This interference with the transport process or with the mechanism of ovum retention within the tube may contribute to the antifertility effect of both post-coital oestrogens and microdose progestins[50,51]. A prolonged and firmer isthmic block would certainly result from excess oestrogen, whereas microdose continuous progestins would prevent oestrogen-induced isthmic closure and allow the rapid passage of the unfertilised egg into the uterus shortly after ovulation. The increased incidence of ectopic implantations, observed in patients bearing subcutaneous capsules of progestins, may be explained by excessive inhibition of the propulsion mechanism rather than interference with isthmic passage. It should be noted that continuous progestin domination may, in addition to its effect on the musculature, depress tubal fluid production (see below) and render the fluid more concentrated. It may also depress cilia regeneration, depriving the tube of an efficient accessory mechanism of ovum transport.

5.4 HORMONAL INFLUENCES ON OVIDUCTAL SECRETION

Tubal fluid may be considered a mixture of secretions and transudates originating in the oviduct itself, the uterus and the peritoneum. At ovulation follicular fluid may also contribute to the mixture. To study tubal secretions it becomes necessary therefore, to prevent the possibility of inflow of fluid from the uterus and the peritoneum. This is usually done by allowing accumulation of fluid between ligatures. Both the fluid volume and composition are influenced by the ovarian hormones.

5.4.1. Fluid volume

In oestrous rabbits fluid volume averages 1.2 ml/day per oviduct. Following ovariectomy, tubal fluid volume drops to 0.2 ml/day per oviduct. Oestrogen treatment restores fluid production to oestrous levels. During pregnancy in the rabbit, there is progressive reduction in fluid volume from the oestrous level which reaches 0.3 ml/day per oviduct at the beginning of the third week[52]. Administration of progesterone in combination with an oestrogen to oestrogen primed castrates, reduces the secretion rate by 50%[53]. A similar ovarian dependent fluctuation in tubal fluid volume is observed in other species. In the sheep oviduct fluid volume at oestrus reaches 1.4 ml/day or more, whereas during the luteal phase it averages only 0.4 ml/day[54,55]. In the cow[55] and the monkey[56] maximum fluid volume is also reached at oestrus. The increase in fluid volume induced by oestrus or oestrogen treatment has a considerable diluting effect. Osmolarity of tubal fluid which ranges from 302 to 310 m-osmol in the oestrous rabbit[57] increases significantly following ovariectomy to reach an average of 388 m-osmol[53]. In oestrogen-primed castrates treated by progesterone+oestrogen, osmolarity also increases significantly over the oestrogen treated levels to reach values of 371 m-osmol or higher[53].

5.4.2 Fluid electrolytes

Despite the general diluting effect of oestrogen and the concentrating effects of oestrogen withdrawal and progesterone, the concentration of some constituents of tubal fluid is unchanged or increased under oestrogen influence. No significant changes in chloride and potassium concentration occurs in tubal fluid of rabbits between oestrus and pseudopregnancy[55]. Sodium concentration is significantly lowered in the rabbit in the first 3 days of pregnancy[58]. Progesterone treatment of oestrogen treated castrates lowered chloride concentration, while hormonal withdrawal significantly decreased the concentration of potassium. Hormonal withdrawal causes a significant increase in magnesium levels. Oestrogen treatment lowers magnesium and calcium levels, whereas progesterone treatment increases calcium but not magnesium concentration.

The effects of the ovarian hormones on the concentration of tubal fluid constituents is not the same for the various species studied. In the cow for example, calcium is highest at oestrus[59]. In ewes, magnesium levels are lowest at oestrus, whereas sodium and potassium are lower at metoestrus than dioestrus[60]. Fluctuations in bicarbonate secretion induced by the ovarian hormones, cause important changes in tubal fluid pH. In the rabbit and sheep, the highest pH is reached during oestrus[61,62]. Bicarbonate ion seems to be an important facilitating factor for corona cell dispersion in the rabbit. At a concentration of 66 mequiv/l it will give complete denution of the egg in only 2 h and without mechanical agitation. Acetazolamide, a carbonic anhydrase inhibitor, inhibits corona cell dispersion *in vivo*[63]. Bicarbonate is also a stimulant of glycolysis and respiration of the spermatozoon[64]. According to Hamner and Williams[65], washed human spermatozoa respire at a rate of 13.5 $\mu l/10^8$ sperm/h in the presence of bicarbonate and only 1.4 $\mu l/10^8$ sperm/h without bicarbonate.

5.4.3 Organic compounds

Both pyruvate and lactate concentrations are influenced by hormonal changes. Pyruvate concentration increases considerably after ovulation[66] but the changes in lactate are the most dramatic. Lactate levels at oestrus in the rabbit which average 290 μmol lactate/g increase very rapidly after ovulation to reach an average of 430 μmol lactate/g 8 days into pregnancy[67]. This is important since lactate seems to be the most important energy source for the developing zygote. Whitten[68] reported that mouse ova at the two-cell stage failed to develop in an artificial medium except some late two-cell zygotes which continued cleavage after the addition of lactate. As development proceeds the late two-cell stage can metabolise in addition to lactate, pyruvate, phosphoenol pyruvate and oxaloacetate. With further maturation other energy sources are usable so that in the eight-cell zygote, cleavage may be supported by malate, ketoglutarate and glucose[69,70].

Several other sugars which are present in small amounts in oviductal fluid seem to have their concentrations influenced by the ovarian steroids. Glucose concentration in the rabbit oviductal fluid is significantly increased after ovulation. Oligosaccharide levels are increased by oestrogen and decreased by oestrogen withdrawal or progesterone[57]. Inositol concentration is not influenced by oestrogen but is doubled by progesterone treatment[71]. The protein patterns of human and preovulatory primate tubal fluid appear quite similar to those of blood serum, suggesting their origin by transudation[72,73]. However, some proteins which are not found in the blood serum have been detected in tubal fluid. Mastroianni *et al.*[73] have described a protein in the tubal fluid of the Rhesus monkey which appears only after ovulation. Moghissi[74] has reported a β-glycoprotein in human tubal fluid which is absent from blood serum and immunologically identical with an ovarian tissue antigen. Moghissi[74] has also reported that compared to normal serum, the concentrations of amino acids of human tubal fluid were higher during the proliferative phase and lower during the luteal phase.

Post-partum tubal secretion contained the least amount of amino acids. There appeared to be also a steady decline in the amount of individual amino acids from proliferative to luteal phase and post-partum period. Feigelson and Key[75] using acrylamide gel disc electrophoresis Ouchterlony double-immunodiffusion and acrylamide gel immunophoresis have been able to show recently that although several protein components are common to oviductal fluid and serum of the rabbit, they are present in different proportions in these two fluids. They have also found that many serum proteins have no counterparts in oviductal fluid and that conversely several proteins are unique to oviductal fluid being undetectable in serum.

Acknowledgement

Unpublished studies referred to in this chapter were supported by a grant from the Ford Foundation.

References

1. Nilsson, O. and Reinius, S. (1969). Light and Electron Microscopic Structure of the Oviduct. *The Mammalian Oviduct* (E. S. E. Hafez and R. Blandau editors) (Chicago: The University of Chicago Press)
2. Hafez, E. S. E. and Black, D. L. (1969). The Mammalian Utero-Tubal Junction. *The Mammalian Oviduct* (E. S. E. Hafez and R. Blandau editors) (Chicago: The University of Chicago Press)

3. Brundin, J. and Wirsen, C. (1964). Adrenergic nerve terminals in the human Fallopian tube examined by fluorescence microscopy. *Acta Physiol. Scand.*, **61,** 505
4. Sjoberg, N. C. (1968). Increase in the transmitter content of adrenergic nerves in the reproductive tract of female rabbits after estrogen treatment. *Acta Endocrinol.*, **57,** 405
5. Ichijo, M. (1960). Studies on the motile function of the Fallopian tube. Report 1. Analytic Studies on the motile function of the Fallopian tube. *Tohoku J. Exp. Med.*, **72,** 211
6. Greenwald, G. S. (1963). *In vivo* recording of intraluminal pressure changes in the rabbit oviduct. *Fert. Steril.*, **14,** 666
7. Marshall, R. and May, J. W. (1960). Cine-(serial)photography of microtome sections. *Med. Biol. Illus.*, **10,** 267
8. Rubin, I. C. (1947). *Utero-tubal Insufflation* (St. Louis: Mosby)
9. Maistrello, I. (1971). Extraluminal recording of oviductal contractions in the unanesthetized rabbit. *J. Appl. Physiol.*, **31,** 768
10. Coutinho, E. M., Mattos, C. E. R. and da Silva, A. R. (1971). The effect of ovarian hormones on the adrenergic stimulation of the rabbit Fallopian tube. *Fert. Steril.*, **22,** 311
11. Mattos, C. E. R. and Coutinho, E. M. (1971). Effects of the ovarian hormones on tubal motility of the rabbit. *Endocrinology*, **89,** 912
12. Neri, A., Marcus, S. L. and Fuchs, F. (1972). Motility of the oviduct in the Rhesus monkey. *Obstet. Gynecol.*, **39,** 205
13. Gomez-Rogers, C., Ibarra-Polo, A. A., Garcia-Huidobro, M., Moran, A., Guiloff, E. and Millan, C. (1967). Physiology of the human Fallopian tube *in vivo* and *in vitro* studies. *Proc. VIII I.P.P.F. Congr.*, (R. K. B. Hankinson, R. L. Kleinman, and P. Eckstein editors) (London: I.P.P.F.)
14. Maia, H. and Coutinho, E. M. (1968). A new technique for recording human tubal activity *in vivo. Amer. J. Obstet. Gynecol.*, **102,** 1043
15. Coutinho, E. M. and Maia, H. (1969). Asynchronism between tubal and uterine activity in women. *J. Reprod. Fert.*, **15,** 591
16. Maia, H. and Coutinho, E. M. (1970). Peristalsis and antiperistalsis of the human Fallopian tube during the menstrual cycle. *Biol. Reprod.*, **2,** 305
17. Sica-Blanco, Y., Rozada, H., Remedio, M. R., Hendricks, C. H. and Alvarez, H. (1970). Human tubal motility. *Amer. J. Obstet. Gynecol.*, **106,** 79
18. Rozada, H., Sica-Blanco, Y., Cibils, L., Remedio, M. R. and Gil, B. E. (1971). Isthmic and ampullar contractility of the human oviduct *in vivo. Amer. J. Obstet. Gynecol.*, **111,** 91
19. Cibils, L. A., Sica-Blanco, Y., Remedio, M. R., Rozada, H. and Gil, B. E. (1971). The effect of sympathomimetic drugs upon the human oviduct *in vivo. Amer. J. Obstet. Gynecol.*, **110,** 481
20. Coutinho, E. M. and Maia, H. (1970). The influence of the ovarian steroids on the response of the human Fallopian tube to neurohypophyseal hormones *in vivo. Amer. J. Obstet. Gynecol.*, **108,** 194
21. Countinho, E. M. (1973). Hormonal Control of Tubal Musculature. *Regulation of Mammalian Reproduction* (S. Segal, P. Corfman, R. Crozier and P. Concliffe, editors) (Springfield: C. Thomas)
22. Coutinho, E. M. (1971). Physiologic and pharmacologic studies of the human oviduct. *Fert. Steril.*, **22,** 807
23. Coutinho, E. M. (1971). Tubal and uterine motility. *Nobel Symp.* **15**, *Control of human fertility* (E. Diczfalusy and U. Borell editors) (Stockholm: Almgvist and Wiksell and New York: John Wiley and Sons)
24. Coutinho, E. M. (1968). The effects of vasopressin and oxytocin on the genital tract of women. *Proc. VI World Congr. of Fert. Steril.* The Israel Academy of Sciences and Humanities (New York: Gordon and Breach)
25. Coutinho, E. M. and Mattos, C. E. R. (1968). Effects of estrogen on the motility of the non-atrophic estrogen deficient rabbit uterus. *Endocrinology*, **83,** 422
26. Boling, J. L. (1969). Endocrinology of Oviductal Musculature. *The Mammalian Oviduct* (E. S. E. Hafez and R. J. Blandau, editors) (Chicago: The University of Chicago Press)

27. Boling, J. L. (1971). Endocrinology of the oviduct. *Pathways to Conception* (A. I. Sherman editor) (Springfield: C. Thomas)
28. Coutinho, E. M. (1968). Uterine activity in non-pregnant women. *Proc. VIII I.P.P.F. World Conf., Santiago, Chile, 1967* (R. K. B. Hankinson, R. L. Kleinman and P. Eckstein editors) (London: I.P.P.F.)
29. Coutinho, E. M. (1969). Hormonal effects on the non-pregnant human uterus. *Progress in Endocrinology* (C. Gual editor) (Amsterdam: Excerpta Medica Int. Congr. Series No. 184. Excerpta Medica Foundation)
30. Coutinho, E. M. (1969b). Oxytocic and antiduretic effects of nausea in women. *Amer. J. Obstet. Gynecol.*, **105**, 127
31. Blandau, R. (1945). On the factors involved in sperm transport through the cervix uteri of the albino rat. *Amer. J. Anat.*, **77**, 263
32. Armstrong, D. T. (1968). Gonadotropins, ovarian metabolism and steroid biosynthesis. *Rect. Prog. Hormone Res.*, **24**, 255
33. Harper, M. J. K. (1966). Hormonal control of transport of eggs in cumulus through the ampulla of the rabbit oviduct. *Endocrinology*, **78**, 568
34. Kennedy, T. G. and Armstrong, D. T. (1972). Extra-ovarian action of prolactin in the regulation of uterine lumen fluid accumulation in rats. *Endocrinology*, **90**, 1503
35. Burdick, H. O. and Pincus, G. (1935). The effect of oestrin injections upon the developing ova of mice and rabbits. *Amer. J. Physiol.*, **111**, 201
36. Greenwald, G. S. (1961). A study of the transport of ova through the rabbit oviduct. *Fert. Steril.*, **12**, 80
37. Chang, M. C. (1966). Effects of oral administration of medroxyprogesterone acetate and ethinyl estradiol on the transportation and development of rabbit eggs. *Endocrinology*, **79**, 939
38. Nakanishi, H., Wansbrough, H. and Wood, C. (1967). Postganglionic sympathetic nerve innervating human Fallopian tube. *Amer. J. Physiol.*, **213**, 613
39. Coutinho, E. M., Maia, H. and Adeodato, J. (1970). Response of the human Fallopian tube to adrenergic stimulation. *Fert. Steril.*, **21**, 590
40. Erb, H. and Wenner, R. (1971). Influence of hormonal and neural substances on the motility of the Fallopian tube. *Excerpta Medica Foundation, Int. Congr. Series 234a, Abstracts of the VII World Congr. on Fert. Steril., Tokyo, 1971. Abstr. 203*
41. Axelrod, J. (1971). Noradrenaline: Fate and control of its biosynthesis. *Science*, **173**, 598
42. Iversen, L. L. (1968). Adrenergic neuro transmission. *Ciba Foundation Study Group* 33 (G. E. W. Wolstenholme and M. O'Connor editors) (London: Churchill)
43. Brundin, J. (1969). Pharmacology of the oviduct. *The Mammalian Oviduct* (E. S. E. Hafez and R. J. Blandau editors) (Chicago: The University of Chicago Press)
44. Hertling, G. and Suko, J. (1972). Influence of neuronal and extraneuronal uptake on disposition, metabolism and potency of catecholamines. *Perspectives in Neuro Pharmacology* (H. Snyder editor) (New York: Oxford University Press)
45. Kim, Y. J., Coats, J. B. and Flickinger, G. L. (1967). Incorporation and fate of estradiol-17-β-6-7H^3 in rabbit oviducts. *Proc. Soc. Exptl. Biol. Med.*, **126**, 918
46. Salt, P. J. and Iversen, L. L. (1972). Inhibition of the extraneuronal uptake of catecholamine in the isolated rat heart by cholesterol. *Nature (London)*, **238**, 91
47. Coutinho, E. M. and Maia, H. (1971). The contractile response of the human uterus, Fallopian tubes and ovary to prostaglandins *in vivo*. *Fert. Steril.*, **22**, 539
48. Coutinho, E. M. (1971). Inhibition of uterine motility by aminophylline. *Amer. J. Obstet. Gynecol.*, **110**, 726
49. Pauerstein, C. J. and Woodruff, J. D. (1970). Estrogen induced tubal arrest of ovum: antagonism by α-adrenergic blockade. *Obstet. Gynecol.*, **35**, 671
50. Croxatto, H., Diaz, S., Vera, R., Etchart, M. and Atria, P. (1969). Fertility control on women with a progestogen released in micro-quantities from subcutaneous capsules. *Amer. J. Obstet. Gynecol.*, **105**, 1135
51. Coutinho, E. M., Mattos, C. E. R., da Silva, A. R., Adeodato, J., Silva, M. C. and Tatum, H. (1970). Long term contraception by subcutaneous silastic capsules containing megestrol acetate. *Contraception*, **2**, 313
52. Bishop, D. W. (1956). Active secretion in the rabbit oviduct. *Amer. J. Physiol.*, **187**, 347

53. Hamner, E. and Fox, S. B. (1969). Biochemistry of oviductal secretions. *The Mammalian Oviduct* (E. S. E. Hafez and R. Blandau editors) (Chicago: The University of Chicago Press)
54. Black, D. L., Duby, R. T. and Spilman, C. H. (1968). Apparatus for the continuous collection of sheept oviduct fluid. *J. Reprod. Fert.*, **6,** 257
55. Edgerton, L. A., Martin, C. E., Troutt, H. F. and Foley, C. W. (1966). Collection of fluid from the uterus and oviducts. *J. Anim. Sci.*, **25,** 1265
56. Mastroianni, L. Jr., Shah, U. and Abdul-Karim, R. (1961). Prolonged volumetric collection of oviduct fluid in the Rhesus monkey. *Fert. Steril.*, **12,** 417
57. Stambaugh, R., Noreiga, C. and Mastroianni, L. Jr (1969). Bicarbonate ion; corona cell dispersing factor of the rabbit tubal fluid. *J. Reprod. Fert.*, **18,** 51
58. Holmdahl, T. H. and Mastroianni, L. Jr (1965). Continuous collection of rabbit oviduct secretions at low temperature. *Fert. Steril.*, **16,** 587
59. Olds, D. and Van Demark, N. L. (1957). Composition of luminal fluids in bovine female genitalia. *Fert. Steril.*, **8,** 345
60. Restall, B. J. and Wales, R. G. (1966). The Fallopian tube of the sheep. III The chemical composition of the fluid from the Fallopian tube. *Austr. J. Biol. Sci.*, **19,** 687
61. Bishop, D. W. (1957). Metabolic conditions within the oviduct of the rabbit. *Int. J. Fert.*, **2,** 11
62. Hadek, R. (1953). Alteration of pH of sheep's oviduct. *Nature (London)*, **171,** 976
63. Noriega, C. and Mastroianni, L. Jr (1969). Effect of carbonic anhydrase inhibitor on tubal ova. *Fert. Steril.*, **20,** 799
64. Jones, E. E. and Salisbury, G. W. (1962). The action of carbon dioxide as a reversible inhibitor of mammalian spermatozoan respiration. *Fed. Proc.*, **21,** 86
65. Hamner, C. E. and Williams, W. L. (1964). Effect of bicarbonate on the respiration of spermatozoa. *Fed. Proc.*, **23,** 430
66. Mastroianni, L. Jr and Brackett, B. G. (1968). Environmental conditions within the Fallopian tube. *Progress in Infertility* (S. J. Behrmen and R. Kistner editors) (Boston: Little Brown & Co)
67. Mounib, M. S. and Chang, M. C. (1964). Metabolism of endometrium and Fallopian tube in the rabbit. *Fed. Proc.*, **23,** 361
68. Whitten, W. K. (1957). Culture of tubal ova. *Nature (London)*, **179,** 1081
69. Brinster, R. L. (1965). Studies of the development of mouse embryos *in vitro. Preimplantation stages of pregnancy* (G. E. W. Wolstenholme and M. O'Connor editors) (Boston: Little Brown and Co)
70. Brinster, R. L. and Thomson, J. (1966). Development of eight cell mouse embryos *in vitro. Exp. Cell Res.*, **42,** 308
71. Gregoire, A. T., Gougsakdi, D. and Rakoff, A. I. (1962). The presence of inositol in genital tract secretions of the female rabbit. *Fert. Steril.*, **13,** 432
72. Moghissi, K. S. (1970). Human Fallopian tube fluid. L Protein composition. *Fert. Steril.*, **21,** 821
73. Mastroianni, L. Jr, Urzua, M., Avalos, M. and Stambaugh, R. (1969). Some observations of Fallopian tube fluid in the monkey. *Amer. J. Obstet. Gynecol.*, **103,** 703
74. Moghissi, K. S. (1971). Proteins and amino acids of human tubal fluid. *Excerpta Medica, Int. Congr. Series 234a* (R. Kleinman and V. R. Pickles editors) (Amsterdam: Excerpta Medica Foundation)
75. Feigelson, M. and Kay, E. (1972). Protein patterns of rabbit oviductal fluid. *Biology of Reproduction*, **6,** 244

6
Blood Concentration and Interplay of Pituitary and Gonadal Hormones governing the Reproductive Cycle in Female Mammals

TAMOTSU MIYAKE
Shionogi Research Laboratory, Osaka, Japan

6.1 INTRODUCTION

Profound knowledge of the mechanism regulating menstrual cycle and determination of the precise time of ovulation are of great importance in dealing with problems of infertility, family planning and in the diagnosis and treatment of menstrual disorders. However, considerable variations in the length of the menstrual cycle of different individuals and variability in the length of different cycles of the same individual make it impossible to predict the exact date of ovulation in each human female, although it is generally accepted that the most common length of the menstrual cycle is 28 days, the average duration of menstrual flow is 5 days and the ovulation usually occurs at the mid-point of the cycle. As menstruation is a visible sign of the periodic activity of ovaries and uterus in non-pregnant women, the first day of haemorrhage has been customarily taken as day 1 of the menstrual cycle. The ovulation divides the cycle into two parts, follicular (preovulatory) phase and luteal (postovulatory) phase. The biphasic nature of the basal body temperature, which shifts from relatively low (preovulatory phase) to relatively high (postovulatory phase) at approximately mid-cycle, has been used as the most practical available index of ovulation[1]. The increase in urinary pregnanediol excretion during the luteal phase of the cycle has been used as an obvious sign of an active corpus luteum established after ovulation[2]. In addition to the above criteria, simultaneous determination of the circulating blood levels of pituitary gonadotropins and ovarian steroid hormones is now possible because of the recent advent of extremely sensitive, precise and specific micromethods, e.g. radioimmunoassays for gonadotropins[3] and protein-binding radioassays or radioimmunoassays for steroid hormones[4], so that the regulatory mechanism of the ovulatory or menstrual cycle can be discussed on the basis of the cyclic fluctuations of the participating hormones in the blood and their alteration by administration of hormones or removal of endogenous hormones.

Because of the limitation of experlmental approaches in human studies, the use of non-human primates, particularly rhesus monkey (*Macaca mulata*), has been accelerated in the study of reproductive physiology, and the resemblance between the human and rhesus monkey has been shown in respect of the duration of the menstrual cycle, the timing of ovulation and the cyclic fluctuation patterns of pituitary and ovarian hormones in the circulating blood[5]. On the other hand, rats as well as primates are spontaneous ovulators and still the most popular laboratory animals in the study of reproduction. Among the enormous amount of information accumulated over many years, some selected data have been organised to set up a model for the regulation of ovulation in the rat[6]. It may be worthwhile, therefore, to compare the recent important observations in the human, the monkey and the rat for better understanding of the mechanism controlling the menstrual cycle and ovulation and also for pointing out the problems to be solved in the future.

6.2 HUMAN MENSTRUAL CYCLE

6.2.1 Circulating LH and FSH levels

According to the recent radioimmunoassay data of Abraham *et al.*[7], who made simultaneous determinations of LH, FSH, oestrogen and progestins on the same blood samples, the pattern of plasma levels of gonadotropins during the menstrual cycle is characterised by a consistent and apparent mid-cycle peak on the same day for either LH or FSH, the peak of LH being more remarkable (about five fold increase) than the peak of FSH (about two fold increase). In the normal ovulatory cycle, the luteal phase length (post-LH peak duration) appears to be in the range 12–16 days and the follicular phase length (pre-LH peak duration) seems to be in the range 11–21 days, indicating less variability in the length of the luteal phase than in the duration of the follicular phase. The plasma LH levels during the follicular phase (before the peak) are somewhat higher and relatively constant, and those during the luteal phase (after the peak) are relatively low and gradually decrease towards the menstrual period. The FSH pattern before the mid-cycle peak shows an early follicular rise (during the menstrual period) followed by a gradual decline until the abrupt rise at mid-cycle. After the mid-cycle peak, the FSH level gradually decreases and starts to increase a few days before the onset of the menses. The coincidence of the LH and FSH peaks at mid-cycle and the roughly parallel fluctuations of both hormones suggest that LH and FSH may be released simultaneously because of the pituitary response to the hypothalamic LH/FSH-releasing hormone which has been identified as a decapeptide by Schally *et al.*[8]. In relation to this, there is evidence that barely detectable amounts of LH-releasing hormone activity are found in the plasma of non-ovulatory women and increased LH-releasing hormone activity appears in the peripheral plasma during mid-cycle[9].

According to the recent study of Midgley and Jaffe[10] who analysed LH and FSH serum concentrations on consecutive hourly samples obtained during the menstrual cycle, LH levels are not maintained in a steady fashion but by frequent surges, which are most vigorous during the ascending and descending phases of the main peak. FSH levels also seem to fluctuate with a similar frequency but with a much smaller magnitude. This indicates the periodic secretion of LH from the anterior pituitary. It is not certain, however, whether it is the result of periodic response of the pituitary to the hypothalamic hormone or a consequence of the steady response of the pituitary to the pulsatile secretion of LH-releasing hormone from the hypothalamus.

6.2.2 Circulating oestrogen and progestin levels

Plasma oestradiol levels show the characteristic biphasic pattern of the ovulatory menstrual cycle with a relatively constant low level during the early follicular phase, rapid rise during the late follicular phase to reach a maximum one or two days prior to the LH peak, a subsequent fall and a second gradual rise and fall during the luteal phase[7]. The lowest level of plasma oestradiol is around 50 pg ml^{-1}, the peak level is in the range from 130–400 pg ml^{-1} and the plateau in the slow luteal rise is *c.* 120 pg ml^{-1}. In most cases, follicular rise of plasma oestradiol begins 4–5 days prior to the LH peak[7].

In contrast to oestrogen, plasma progesterone levels never increase before the onset of the mid-cycle LH surge[7]. During the follicular phase, plasma progesterone is maintained at a low level of 0.5 ng ml^{-1} or less increasing to 1–2 ng ml^{-1} during the LH peak, this small rise lasting for 2 days and being followed by a marked increase during the next 3–4 days of post-LH peak to reach a plateau of 10–20 ng ml^{-1}, which persists for about 5 days and decreases thereafter to reach levels below 1 ng ml^{-1} on the first day of menses[7].

Plasma levels of 17*a*-hydroxyprogesterone (17-OHP) also fluctuate during the menstrual cycle, and the pattern is different from that of progesterone and rather resembles that of oestrogen[7]. The consistent low levels of 17-OHP at *c*. 0.2–0.5 ng ml^{-1} during the follicular phase are followed by a significant rise at mid-cycle to reach a peak of *c*. 2 ng ml^{-1} coincident with the timing of the LH peak. From 2–4 days after the LH peak, 17-OHP decreases by a half, rises again during the luteal phase to a level of *c*. 2 ng ml^{-1} and decreases at the end of the cycle[7].

6.2.3 Control of gonadotropin levels by oestrogen and progestin

The concept that the earlier rise of plasma oestrogen may trigger LH release has been supported by early investigations that ovulation can be induced by the administration of oestrogens in many anovulatory women[11].

Ethynyloestradiol given orally in the early follicular phase of the cycle causes a prompt decrease in both FSH and LH levels during the treatment and a subsequent rebound increase in LH (but not FSH) approximately 36 h after cessation of the treatment[12]. When administered in mid-follicular phase, as ethynyloestradiol given orally, as oestradiol benzoate injected subcutaneously or as oestradiol infused intravenously, there is an initial suppression followed by an acute surge of both gonadotropins, which occurs 12–24 h after the termination of treatment[13]. Acute LH surge induced by exogenous oestrogen, however, appears to be incapable of inducing ovulation, for there is no concomitant increase in progesterone secretion and basal body temperature as seen in the spontaneous LH surge. Thus, exogenous oestrogen given before ovulation postpones spontaneous LH surge to produce a prolonged follicular phase, which may be due to the suppression by oestrogen of tonic secretion of gonadotropins, particularly of FSH[12,13]. Ethynyloestradiol given after ovulation, however, produces only a minor effect, namely shortening of the period for increased progesterone secretion within the normal limits[14]. It is likely, therefore, that the increased level of plasma oestrogen is essential for triggering the gonadotropin release, which is able to link with ovulation under the normal condition of follicular maturation.

The temporal relationship between plasma LH, progesterone and 17-OHP suggests that progesterone reflects only corpus luteum function, whereas 17-OHP serves as an index of both follicular maturation and corpus luteum function[15]. It is unlikely that 17-OHP or progesterone may play a role in triggering or facilitating gonadotropin surge during the normal menstrual cycle, since 17-OHP and progesterone in the peripheral blood never rise before the LH surge. Progesterone and synthetic progestins administered to normal menstrual women, act as ovulation inhibitors and reduce the urinary pregnanediol excretion during the luteal phase[16]. Further, synthetic progestins, when given after ovulation, have been found to produce a decrease in both progesterone and oestradiol synthesis in the human corpus luteum[17]. It should be noted, however, that a rapid rise in plasma LH can be elicited by a single intramuscular injection of progesterone following daily oral administration of ethynyloestradiol in the patients with primary ovarian insufficiency, castrated women and postmenopausal women[18].

The triggering effect of oestrogen and the biphasic action of progesterone, inhibitory and stimulatory on LH secretion, and the temporal relationship between plasma oestradiol, LH and progesterone suggest that the earlier rise in plasma oestrogen triggers the onset of plasma LH rise which promptly stimulates progesterone synthesis and secretion, the increased progesterone output during the preovulatory stage, in turn, regulating the height and duration of the LH surge, and progesterone secreted after ovulation definitely suppressing the secretion of LH to the basal level during the luteal phase. Obviously, the site of action of gonadal steroids in the regulation of gonadotropin secretion has not been determined in humans.

6.2.4 Control of ovarian steroid secretion by gonadotropins

Stimulatory effects of exogenous gonadotropins on the ovarian steroid secretion in women can be observed during gonadotropin therapy in patients with secondary amenorrhea or anovulatory cycles[11,19,21,22]. After the induction of ovulation by exogenous gonadotropins, the life-span and function of the corpus luteum appear to be largely independent of further gonadotropin stimulation, but the continued low levels of LH seem to be necessary for the maintenance of normal luteal function in hypophysectomised women[11]. It is strongly suggested that LH is the principal luteotropic factor in women[11], but prolactin appears not to be[20].

6.3 MONKEY MENSTRUAL CYCLE

6.3.1 Circulating LH, oestrogens and progesterone levels

The time courses of oestradiol, LH and progesterone concentrations in the peripheral blood during the menstrual cycle of the rhesus monkey are quite similar to those in the human female. Hotchkiss *et al.*[23] determined the mean daily oestradiol and LH levels in the same serum samples during 10 menstrual cycles of eight rhesus monkeys, and found that

(a) a rapid rise in oestradiol levels is noted about 4 days prior to the LH peak when LH levels are still kept lowest,

(b) the oestrogen increase is accelerated further as the LH release initiates,

(c) the oestrogen peak coincides with the LH peak, mostly observed between day 9 and 13,

(d) 24 h after the peak, oestradiol levels fall abruptly and reach their lowest point 48 h after the peak, while the LH levels decrease more slowly, and

(e) during the luteal phase of the cycle, oestradiol levels rise again to the levels observed during the early follicular phase, while the serum LH levels are low.

Kirton *et al.*[24] determined LH and progesterone in peripheral serum samples obtained daily through four menstrual cycles in four rhesus monkeys and made laparoscopic examinations to determine the incidence and time of ovulation. Their results indicated that serum progesterone increases with the LH surge, levels off or decreases slightly near the time of ovulation and then increases markedly with the functional development of the corpus luteum; ovulation usually occurs within 30 h of the LH peak. Daily determinations of plasma progesterone levels and serial laparotomies for the detection of ovulation were made at the periovulatory stage by Johansson *et al.*[25] who demonstrated that the plasma progesterone concentration begins to increase on the day prior to ovulation, confirming the preovulatory increase in progesterone secretion in the rhesus monkey. The base-line levels and the peak values of LH and oestradiol during the menstrual cycle in the rhesus monkey are almost the same as those in the human, but the usual coincidence of the oestrogen peak with the LH peak, and the rarity of the sustained oestrogen peak during the luteal phase, are characteristic of the rhesus monkey. In the human, the oestrogen peak usually precedes the LH peak, and its rapid fall is followed by a sustained increase associated with the marked increase in progesterone levels during the luteal phase[7].

6.3.2 Control of LH levels by oestrogen and progesterone

The similarity in the length of the menstrual cycle, the patterns of the mid-cycle LH peak and the ovarian steroid secretions between human and rhesus monkey may suggest the same cause and effect relationship between pituitary LH and gonadal steroids in rhesus monkey as in humans. Spies and Niswender[26] reported that the LH peak, which normally occurs between day 10 and 15 in the cycle, can be blocked by 0.5 mg per day of progesterone injected subcutaneously between days 2 to 10, 8 to 16 or 8 to 22 of the cycle, and that the base-line levels of serum LH do not change for several days after the cessation of progesterone treatment, suggesting no occurrence of rebound surge of LH. Further, oestradiol injected at increasing dosages of 100, 200 and 300 μg for 3 days after the progesterone treatment induced a marked LH peak followed by ovulation in three monkeys and a minor LH rise without ovulation in another two out of the nine monkeys tried. Yamaji *et al.*[27] demonstrated that a single injection of oestradiol benzoate, of an amount capable of mimicking the spontaneous elevation of circulating oestrogen levels for at least 12 h, can produce an advanced LH surge occurring 24–48 h after the injection during days 2 and 3 of the cycle. In most of the cases so treated, delayed spontaneous surges of oestrogen and LH are observed between day 20 and 25, and this is followed by a marked increase in progesterone levels indicative of ovulation. Such an oestrogen-induced LH surge at an earlier follicular phase can be prevented by progesterone administered simultaneously, and the injection of oestradiol benzoate is not able to induce an LH surge during the early luteal phase of the menstrual cycle, except when corpora lutea have been excised[28]. It is obvious, therefore, that oestrogen plays a role of triggering LH release whereas progesterone acts to suppress LH secretion in rhesus monkeys as well as in humans.

The pulsatile LH discharges found by hourly sampling of the blood in human menstrual cycles[10] have not been observed in the normal cycles of rhesus monkeys. According to Knobil *et al.*[28], however, this can be seen in chronically ovariectomised monkeys. The 'circhoral' oscillation of LH level is promptly suppressed by a single intravenous injection of oestradiol-17β (0.1 μg kg^{-1}) but not by progesterone administration, and its abrupt re-appearence following several hours of suppression is noticed with higher amplitude and lower frequency when the exogenously increased oestrogen concentration in the blood falls down to the base-line level[29]. The interruption of pulsatile discharge of LH can be brought about by single injections of α-adrenergic blocking agents (e.g. phentolamine, phenoxybenzamine) or of neuroleptic drugs (e.g. chlorpromazine, haloperidol), and the resumption of LH oscillation is induced in a similar fashion, which is indistinguishable from that seen following oestrogen blockade, and further the β-adrenergic blocker, propranolol, and the deep anaesthetic dose of pentobarbital produce no significant effect in this regard[30]. From these observations, Knobil *et al.*[28] concluded that tonic LH secretion, as reflected in the circulating levels of this hormone during the follicular phase of the cycle, can be explained by the negative feed-back action of oestrogen alone, and this negative feed-back loop probably involves a dopaminergic and/or an α-adrenergic component. Further, they[28] proposed that the oestrogen peak coincident with the LH surge seen in oestrogen-treated intact females, as well as in normally cycling monkeys, is a consequence rather than the cause of increased LH secretion. The trigger for LH release must be the smaller and earlier rise in the blood oestrogen levels associated with the decline in LH levels which usually precedes the surge, since the exogenously increased plasma oestrogen levels fall progressively during the period of increased LH secretion in ovariectomised monkeys.

6.3.3 Control of luteal function by gonadotropin

According to Macdonald and Greep[31], exogenous LH given during the luteal phase of the menstrual cycle causes a dose-dependent increase in progesterone concentration in peripheral blood, but has no effect on progesterone levels during the follicular phase. Moudgal *et al.*[32] reported that neutralisation of monkey LH with the antiserum to HCG during the critical period of LH surge leads to blockade of both ovulation and corpus luteum formation. They[33] further demonstrated that a large amount of LH specific antibody, when injected for four consecutive days (day 15 to 18 inclusive) into normally cycling monkeys (*Macaca fascicularis*), produces a shortening of the luteal phase and a decrease in progesterone levels. Exogenous FSH also increases progesterone levels during the luteal phase, but it appears to act indirectly through the release of endogenous LH or by affecting the receptor site for LH[31]. Prolactin has no effect in this respect during all stages of the cycle[31]. It is emphasised, therefore, that the preovulatory surge of LH is essential for ovulation and luteinisation, and the reduced levels of LH during the luteal phase are necessary for the maintenance of luteal function.

6.4 RAT OESTROUS CYCLE

6.4.1 Circulating LH, FSH and prolactin levels

The recent development of radioimmunoassay (RIA) for gonadotropins and competitive protein-binding radio-assay for steroid hormones has permitted the determination of pituitary and gonadal hormone concentrations in blood samples or samples collected repeatedly from the same individual, so that more detailed patterns of blood LH, FSH, prolactin, oestrogens and progestins are now capable of being delineated even for small laboratory animals such as rats. Interlaboratory determinations[34] (on the same samples) of both blood LH and pituitary LH obtained by ovarian ascorbic acid depletion (OAAD) assay and those obtained by RIA showed a considerable identity in the estimates, although it was noticed for blood LH that OAAD assay tended to measure something else which RIA could not detect. However, a similar comparison between Steelman–Pohley assay (SPA) and RIA for FSH determination has not yet been made, as SPA is not sensitive enough to detect the FSH concentration in rat circulating blood[34]. Danne and Parlow[35] have made simultaneous RIA determinations of serum LH and FSH during the oestrous cycle of 4-day cycling rats and demonstrated that the serum LH level only increased on the late afternoon of pro-oestrus (3 p.m. to 6 p.m.) and, to a lesser extent, on the evening of pro-oestrus. The serum FSH concentration, largely undetectable during dioestrus, also began to rise on the late afternoon of pro-oestrus, but did not attain its maximum value until 8 to 9 p.m. pro-oestrus. The maximal elevation of serum FSH persisted until 4 to 5 p.m. of oestrus and slowly declined thereafter. Such a disagreement in LH and FSH patterns is also found in the data of Linkie and Niswender[36].

The great similarity in all the data showing a dramatic LH surge on the afternoon of pro-oestrus *c.* 10–12 h before ovulation provides direct support for the earlier discovery of Everett *et al.*[37] who indirectly determined the 'critical period' for ovulating hormone release by the use of timed hypophysectomy and pharmacological blocking agents. Considering the importance of the role of LH in ovulation[38], the pro-oestrous LH surge is no doubt an essential step for the induction of ovulation. FSH surge occurring synchronously with the LH surge, however, does not seem to be essential for the expected ovulation, because the role of FSH in the acute process of ovulation is still obscure[39]. Danne and Parlow[35] revealed that, in contrast with the sharp prominency and precipitousness of LH surge, the pro-oestrous surge of FSH was relatively slow and protracted through the oestrous and metoestrous stages of the cycle, and that pentobarbital, highly effective in inhibiting the pro-oestrous surge of LH, was only partially effective in inhibiting the pro-oestrous surge of FSH. More recently, Cobbs and Schwartz[40] have reported that the ovulatory effects of exogenous FSH or

LH, both immunologically purified, were blocked by previous treatment of antiserum to LH and not blocked by antiserum to FSH in pentobarbitalised rats, suggesting that, in the absence of LH, ovulation does not occur in the rat and LH alone is adequate for producing ovulation. These facts support the idea[35] that LH may stimulate the differentiation of a set of well-developed follicles culminating in ovulation and that FSH may stimulate the next crop of follicles scheduled for ovulation. Nevertheless, the possibility exists that the concurrent interplay between LH and FSH during the period of pro-oestrous surge may operate in the accomplishment of ovulation.

Prolactin surge in the blood of cycling rats occurs only on the afternoon of pro-oestrus[36,41]. This seems to synchronise with the apparent decrease in prolactin content in the pituitary noticed on the afternoon of pro-oestrus[42]. Either phenomenon can be blocked by pentobarbital given at pro-oestrus before the spontaneous occurrence of prolactin surge[42], thus confirming the existence of a neural mechanism controlling prolactin release. It is generally accepted that prolactin released by coital stimuli acts as a luteotropic factor to maintain newly formed corpora lutea in a functional state and to bring on pseudopregnancy. The role of prolactin surge occurring a few hours earlier, and lasting longer than LH surge[36,43], without any mechanical stimulus, is not known. It may be suggested, however, that

(a) prolactin may participate in the progesterone synthesis by regulating cholesterol dynamics in the progesterone-producing cells in cooperation with LH[44,45], and

(b) prolactin may act as a luteolytic factor on the previous crop of corpora lutea formed during each cycle, since ergocornine, an inhibitor of prolactin release, given for 1–3 oestrous cycles, produced singificant increases in weight of the ovaries and in the number of corpora lutea[46].

The neural mechanism controlling prolactin surge at pro-oestrus may be quite different from that controlling prolactin secretion triggered by coital stimulus, the pattern of which shows a daily nocturnal surge during the period of pseudopregnancy[47].

6.4.2 Oestrogen secretion

Hori *et al.*[48] first demonstrated a cyclic fluctuation of oestrogen concentration in ovarian venous blood during the rat oestrous cycle, which was characterised by the following three points,

(a) oestrogen secretion starts to increase at least 24 h before the onset of ovulatory release of LH,

(b) oestrogen secretion attains a maximum value about 6 h before the onset of LH release and

(c) a drastic fall of oestrogen secretion occurs during the period of LH surge.

Such an oestrogen surge, confirmed by Yoshinaga *et al.*[49] and also by Shaikh[50], demonstrates that oestradiol is the main component in the oestrogen surge. Brown-Grant *et al.*[51] determined LH (by RIA) and oestradiol concentrations (by competitive protein-binding assay) in peripheral blood plasma during the oestrous cycle of 4-day cyclic rats, and again confirmed the oestrogen surge preceding LH surge.

It is likely that the highly stimulated oestrogen secretion under the base-line levels of blood FSH and LH is sustained by anterior pituitary function which may continuously secrete both LH and FSH to some extent. In this context, Hori *et al.*[48,52] demonstrated that a highly elevated ovarian oestrogen secretion at pro-oestrus was minimised within 3 h after hypophysectomy, but was restored in 1 h either by the homogenate of rat anterior pituitaries or by a minute amount of FSH in combination with LH given intravenously, either of which alone was unable to produce a sufficient restoration even at greater dosages. Furthermore, there is other evidence that oestrogen secretion on the day before pro-oestrus can be prevented by antiserum to LH, but not by antiserum to FSH[39], thus revealing the necessity of LH secretion for oestrogen secretion to take place. Ovarian responsiveness of oestrogen secretion to gonadotropins seems to be coupled with follicular development[53], but the relationship between folliculogenesis and oestrogen secretion remains to be elucidated.

A rapid fall of elevated oestrogen secretion during the time of LH release in this species may be due to the deficiency of pituitary factor(s) to stimulate oestrogen secretion or to a decrease in ovarian responsiveness to these pituitary factor(s). The former possibility is unlikely because of the high levels of blood LH and FSH and because the pituitary extract obtained either before or after LH release has enough activity to stimulate ovarian oestrogen secretion[52]. The latter possibility is supported by the fact that neither the rat pituitary extract nor LH when given after the completion of LH release was at all capable of elevating oestrogen secretion. It was further demonstrated that either LH or FSH given several hours before the expected LH release terminated ovarian secretion of oestrogen earlier by a time interval corresponding to that between LH injection and expected LH surge, and the minimal effective doses necessary to terminate oestrogen secretion were almost equivalent to those for induction of ovulation[52,54].

During the short period of time when LH surge, drastic fall of oestrogen secretion and rapid rise of progesterone secretion are synchronously taking place, the morphologic changes characterised as 'preovulatory swelling'[55] may be seen. This point was chronologically elucidated by Hori *et al.*[54] who observed a slight swelling of theca interna with hyperemia and histochemically decreased distribution of glucose-6-phosphate dehydrogenase activity, a distension of intercellular space in the granulosa layer and the disorientated appearance of the nuclei of basal granulosa cells in mature follicles within 6 h after the occurrence of LH release. Such histological changes were correspondingly advanced by exogenous LH given earlier than the spontaneous LH surge. Thus it could be postulated that under the exposure of LH surge ripe follicles start to differentiate towards ovulation, during which process unfavourable conditions for oestrogen synthesis may be brought about.

6.4.3 Progestin secretion

By the use of gas–liquid chromatography following thin layer chromatography, Uchida *et al.*[56] determined both progesterone and 20α-OHP in the ovarian venous blood during the 4-day oestrous cycle, taking particular account of the critical time of LH release, which was determined by Kobayashi *et al.*[57] on the same laboratory rats under the same breeding conditions. Their data clearly demonstrated that more than a tenfold increase in progesterone secretion rate occurs in a short period of time from the lowest level (0.26 $\mu g\ h^{-1}\ ovary^{-1}$) to the highest level (3.31 $\mu g\ h^{-1}\ ovary^{-1}$) on the afternoon of pro-oestrus, and the elevation almost synchronises with LH surge but never occurs before LH release. The maximally-elevated progesterone secretion, which continues for at least 4 h, decreases rapidly in the morning and gradually in the afternoon through the day of oestrus, then increases again to make a small peak on the afternoon of dioestrus–1. After the small rise on the afternoon of dioestrus–1, it returns to the lowest level on the afternoon of dioestrus–2, which lasts until the time of LH release. Such a cyclic pattern of progesterone level in the ovarian venous blood in relation to LH surge was further confirmed by Barraclough *et al.*[58] who used the competitive protein-binding method and also by Piacsek *et al.*[59] who used thin layer chromatography followed by gas–liquid chromatography.

Drastic preovulatory rise of progesterone secretion can be blocked by hypophysectomy[56] or by pentobarbital injection[58], suggesting that LH release is the cause of increased progesterone output. The increase in progesterone concentrations in ovarian venous blood seems to be a reflection of increased progesterone synthesis rather than progesterone release, since progesterone concentrations in the ovarian tissues do not fluctuate much during the oestrous cycle[56]. More direct evidence on this point has been given by Chatterton, Macdonald and Greep[60], who demonstrated LH stimulation of *in vitro* progestin synthesis in ovaries obtained from adult rats throughout the oestrous cycle, and this steroidogenic effect of LH was confirmed *in vivo* by Yoshinaga *et al.*[61] and Uchida *et al.*[62,63]. In comparison with LH, a huge amount (more than 50 μg NIH–FSH–S_1) is needed for FSH to cause a significant stimulation of ovarian progesterone secretion. A similar stimulation was produced by a minute amount (0.5 μg NIH–LH–S_{12} per rat) of LH[63]. Thus, the particular gonadotropin responsible for pre-ovulatory progesterone output is very likely to be LH, although substantial data has not been available to exclude the possibility of LH–FSH synergism in ovarian progesterone secretion.

Blood prolactin levels rise on the afternoon of pro-oestrus[41] and seem to start earlier than that of LH release[36]. Although exogenous prolactin when tested in hypophysectomised rats, has no acute stimulatory effect on the ovarian progesterone output[61,63], the possible participation of prolactin in the preovulatory increase of progesterone output may be explained by the concept of Armstrong[44] and Greep[64]. Briefly summarising their views;

(a) prolactin increases the store of cholesterol, particularly of the esterified form, in the steroid-producing cells of the ovary by the induction and maintenance of cholesterol synthetase that promotes the synthesis of cholesterol esters from circulating cholesterol and free fatty acids,

(b) cholesterol esters are pooled in the intracellular lipid droplets,

(c) prolactin induces, but does not activate, the cholesterol esterase which hydrolyses cholesterol ester to produce free cholesterol that enters the steroid biosynthetic pathway starting with the oxidation of cholesterol to 20α-hydroxycholesterol, and

(d) LH not only activates the mitochondrial enzyme to promote side-chain cleavage of cholesterol but also activates cholesterol esterase to accelerate the hydrolysis of stored cholesterol ester.

It is emphasised, therefore, that both prolactin and LH participate in the progesterone biosynthesis in the rat luteal cells by affecting the enzymes involved in cholesterol turnover, and that the maintenance of an adequate store of cholesterol ester may be a prerequisite to sustained biosynthesis of progesterone.

In contrast with pituitary dependency of the preovulatory rise of progesterone secretion, the increase in progestin secretion on the afternoon of dioestrus–1 is independent of the pituitary function at this stage[56], and probably resulted from the intrinsic function of the newly formed corpora lutea, because it occurs after the blood LH has fallen to the base-line level. Its appearance cannot be blocked by hypophysectomy[57] or by pentobarbital injection, using the same amount as that which blocks the preovulatory surge of LH[58]. A possible autonomy in the corpus luteum function at this stage may be supported by the data of Chatterton *et al.*[60] who demonstrated *in vitro* that the rat ovarian tissue isolated at any stage of the cycle is able to synthesise a considerable amount of progesterone and 20α-hydroxypregn-4-en-3-one (20α-OHP) in the medium without LH, and the incorporation of acetate-1-^{14}C into progesterone and 20α-OHP is markedly higher during the luteal phase but it is not affected by the addition of LH. It was further demonstrated that the contribution of corpora lutea of metoestrous ovary to the total synthesis of progestins is much greater than that of the remaining tissue, but the response of ovarian progestin synthesis to LH is not superior at this stage. A gradual withdrawal of post-ovulatory progesterone surge may be explained by the exhaustion of the cholesterol stores in the luteinised tissue and by a possible participation of prostaglandin $F_{2\alpha}$ having luteolytic activity, the mechanism of which will be discussed by Behrman (Chapter 10) in this volume.

According to the investigators, the patterns of 20α-hydroxypregn-4-en-3-one (20α-OHP) secretion during the oestrous cycle of the rat are very different in spite of a similarity in the patterns of progesterone secretion. It has been agreed, however, that the rate of 20α-OHP secretion during the oestrous cycle of the rat is usually higher and variable as compared with that of progesterone and 20α-OHP does not necessarily fluctuate in parallel with progesterone throughout the oestrous cycle[56]. Piacsek *et al.*[59] recently demonstrated that 20α-OHP secretion rose earlier than progesterone secretion either before or simultaneously with LH release, and reached a peak 4–6 h before progesterone secretion attained its maximum level, then fell to its lowest level when progesterone secretion was still at its highest. However, they[59] agreed with previous investigators that the pattern of 20α-OHP during the luteal stage was parallel with that of progesterone showing a small rise at dioestrus–1. These findings suggest different sources for the secretion of progesterone and 20α-OHP during the oestrous cycle. Granulosa cells of follicles may be the cellular origin of progesterone secretion at the preovulatory stage, and newly formed corpora lutea for progesterone at the luteal phase[55]. In this context, Leavitt *et al.*[65] demonstrated, by *in vitro* experiments using hamster ovary, that follicles and interstitium contribute approximately equal amounts of progesterone during the preovulatory period of the cycle. It is reasonable to consider aged corpora lutea as the source of 20α-OHP secretion, since 20α-hydroxysteroid dehydrogenase activity responsible for conversion of progesterone to 20α-OHP is rich in involuting corpora lutea but poor in freshly formed corpora lutea[66].

6.4.4 Control of gonadotropin levels by oestrogen

Since oestrogen surge preceeds LH surge, it is likely that oestrogen initiates ovulatory release of LH during the oestrous cycle. In rats showing a 5-day oestrous cycle, oestradiol benzoate given at the mid-dioestrous stage advances ovulation by about 24 h[67]. Also in 4-day cyclic rats, a single injection of oestradiol on the morning of dioestrus–1, at the critical period between 10 a.m. and 12 noon, is able to advance ovulation by 24 h in about 50% of cases[68]. The oestrogen dependency of ovulatory LH surge was further proved by the following evidence. Removal of oestrogen surge by ovariectomy on the morning of dioestrus–2 abolishes LH release, and oestrogen replacement (subcutaneous injection of oestradiol) immediately after ovariectomy restores the LH release at the regular time nearly to the normal level[69]. Chemical oestrogen antagonists[70–72], or antiserum to oestradiol[73] given on the day before pro-oestrus are able to block expected ovulation as well as vaginal cornification.

Kobayashi *et al.*[71,74] demonstrated that the integrated amount of oestrogen secreted during dioestrus through pro-oestrus until the onset of LH release is not required for LH release, but oestrogen secreted *c.* 10–12 h before LH release is enough to cause LH release. It is presumed, therefore, that oestrogen activation of a neural mechanism controlling the ovulatory surge of pituitary gonadotropin starts *c.* 24 h before ovulating hormone release and is complete within 12 h. Further increase in oestrogen secretion beyond that time, the pattern of which is almost parallel with the increase in pituitary LH content[74], may be responsible for the biosynthesis and storage of gonadotropins in the anterior pituitary[76] and the increase in pituitary responsiveness to the hypothalamic-releasing hormone[77,78].

Since FSH surge almost synchronises with LH surge during the oestrous cycle, and a single hypothalamic releasing hormone, LH–RH, is able to release both LH and FSH simultaneously from the pituitary[8], it is likely that oestrogen initiates both LH and FSH surge by the same mechanism.

Spontaneous prolactin surge occurring on the afternoon of pro-oestrus almost in parallel with LH surge can be blocked by pentobarbital, suggesting a neural mechanism controlling the prolactin release[42]. Not only LH surge but also prolactin surge is suppressed by antiserum to oestradiol given on the morning of dioestrus–2, and the suppression can be removed by the simultaneous administration of diethylstilboestrol[43]. Moreover, antiserum to LH given on the morning of dioestrus–2 also blocks the prolactin surge as the result of the suppression of oestrogen secretion which, otherwise, is expected to increase[79]. It is suggested, therefore, that the neural mechanism controlling prolactin release may also be activated by oestrogen which starts to increase during the afternoon of dioestrus–2. Although the mechanism of oestrogen dependency of prolactin surge at pro-oestrus has not been clearly elucidated, the following possibilities may be considered.

(a) Oestrogen may suppress a prolactin-inhibiting centre in the rat hypothalamus, the location of which appears to be in the thalamo–hypothalamic border[80].

(b) Oestrogen may suppress the function of 'hypophysiotropic' cells secreting prolactin-inhibiting factor in the hypothalamus[81].

(c) Oestrogen may act directly on the pituitary to decrease its sensitivity at pro-oestrus to a tonic negative stimulus, which seems to be mediated by prolactin-inhibiting factor.

(d) Since the possible existence of a specific prolactin-releasing factor (PRF) has been suggested[82], oestrogen may act on either a hypothalamic or pituitary level to increase the secretion and/or action of PRF.

6.4.5 Control of gonadotropin levels by progestin

Ever since Everett[67] first found that progesterone given on the day of dioestrus–3 accelerated ovulation and vaginal cornification by *ca.* 24 h in the normal 5-day cyclic rats, many investigators have confirmed the facilitatory effect of progesterone on spontaneous ovulation and ovulating hormone release[80]. In normal 4-day cyclic rats, progesterone given on the morning of pro-oestrus, but not on the morning of dioestrus–2, advances ovulatory release of LH by a few hours[86,87]. The optimal time necessary for progesterone injection to cause advancement of LH release is *c.* 6 h before the normal onset of LH release[86], and this corresponds to the time at which oestrogen output is at its maximum level[48]. Under the above conditions the timing of ovulation is also correspondingly advanced by the advance release of LH[86]. The facilitation by exogenous progesterone of LH release can be blocked by pentobarbital, indicating that progesterone may affect the neural mechanism controlling LH release[86,87].

Although there is no doubt that exogenous progesterone is able to facilitate LH release and ovulation under certain experimental conditions[88,89], it is still controversial whether endogenous progesterone or 20α-OHP in the circulating blood participates in the process of spontaneous LH release during the oestrous cycle. It may be that the increase in ovarian venous progesterone does not occur before LH release[56,58,59]. Ferin *et al.*[73] have shown by the use of the antiserum to progesterone that the preovulatory rise in progesterone output is not essential for subsequent ovulation. Barraclough *et al.*[58] demonstrated, however, that peripheral venous progesterone increased at the time of the LH peak and before the time of ovarian venous progesterone increase, and suggested the possibility of adrenal participation in triggering LH release. In fact, Resko[90] reported that in the ovariectomised rat the adrenal gland secretes progesterone, the circulating level of which is regulated by adrenocorticotropin, so that adrenal progesterone may have important influences on the reproductive processes. In this connection, Lynn *et al.*[91] postulated the possibility of adrenal participation in the timing of mating and LH release in cyclic rats. Piacsek *et al.*[59] found a remarkable rise in the rate of 20α-OHP secretion a few hours before the time of LH release. Uchida *et al.*[62] reported that unlike progesterone, exogenous 20α-OHP had no facilitatory effect on the preovulatory progesterone increase induced by LH release. Kobayashi *et al.*[92] also noted no ability of 20α-OHP to accelerate the delayed ovulation caused by progesterone given at dioestrus–2. Swerdloff *et al.*[94], however, demonstrated that both progesterone and 20α-OHP were capable of inducing a surge of LH in oestrogen-treated castrated female rats, and suggested that 20α-OHP acts by synergising with oestradiol in triggering the preovulatory LH peak and that both 20α-OHP and progesterone may augment the height of this response. At any rate, further substantiation may be necessary for the explanation of the role of 20α-OHP secreted at the preovulatory stage.

A rapid fall of blood LH occurs immediately after the rise in preovulatory progesterone, and remains at the base-line level until the next LH surge. This may be explained by the suppressive effect of progesterone on the pituitary gonadotropin secretion, which has commonly been accepted as a negative feedback action of the steriod. According to Kobayashi *et al.*[86,93], exogenous progesterone given at any time on any day except pro-oestrus prevents the expected ovulation partially or totally, and progesterone given on the day before pro-oestrus (dioestrus–2 in a 4-day rat) or until 2 a.m. of pro-oestrus produces a complete block of ovulation and vaginal cornification on the expected day of oestrus. However, progesterone given at 5 a.m. of pro-oestrus is only partially effective in preventing ovulation, and the steroid given at or after 9 a.m. of pro-oestrus facilitates both LH release and ovulation by a few hours. Since the oestrogen secreted until 2 a.m. of pro-oestrus is essential for the ovulatory discharge of LH in the 4-day rats used[74], the most effective period for progesterone causing ovulation block corresponds to the period for oestrogen secretion which is essential to complete the activation of the neural component having 24-h rhythmicity[83].

It is likely, therefore, that exogenous progesterone given before the completion of oestrogen activation of the neural mechanism suppresses the pituitary gonadotropin secretion (probably LH) responsible for oestrogen secretion, and thus prevents normal occurrence of the neural activation leading to LH–RH release. Thus, endogenous progesterone secreted during the luteal phase, which may reflect the life-span of the corpus luteum, may be considered to have a role in adjusting the onset of the increase in oestrogen secretion.

6.4.6 Active sites of ovarian steroids in the regulation of gonadotropin secretion

Recent autoradiographic studies on the localisation of oestrogen have revealed that ^{3}H-oestradiol is found concentrated in the nuclei of cells in all of the known target tissues for oestrogen, including anterior pituitary and brain, suggesting a dual feed-back regulation of gonadotropin secretion[95]. In fact, specific oestradiol-binding macromolecules have been found in both pituitary and hypothalamus[96,97], and the oestrogen activation of hypothalamus is induced in harmony with sexual maturation[98]. In contrast to oestrogen receptors, neither biochemical nor autoradiographic data showing the existence of progesterone receptors in the brain or pituitary is available at the present time.

It has been generally accepted that the preoptic–suprachiasmatic area of the hypothalamus plays a role as the focus for the neural mechanism that acts regularly, with a 24-h periodicity, through the hypophysiotropic area to elicit ovulatory discharge of gonadotropins from the pituitary, while the hypophysiotropic area is concerned with the basal tonic secretion of FSH and LH other than their ovulatory discharge[80,84,85]. The preoptic–suprachiasmatic area has been postulated as the positive feed-back site of oestrogen by the fact that oestrogen implants at this area evoke sex behavioural interests in the gonadectomised female rat[99] and also promote ovulation in the intact immature rat, whereas lesions in that area block the response[100,101]. On the other hand, the medial basal tuberal region and the hypophysiotropic area[80] have been thought of as the negative feed-back site of oestrogen, since female rats with lesions in the median eminence did not show a rise in plasma LH following ovariectomy[101,102], and oestrogen implants in this area inhibited pituitary–gonadal function[99].

Anterior pituitary has also been involved as a steroid feedback site, since there is evidence that intrapituitary implantation of oestrogen promotes gonadotropin secretion and progesterone implants suppress ovulation[103,104]. Furthermore, intrapituitary implantation of oestradiol benzoate on dioestrus–2 in the 5-day cyclic rats caused a full day advancement of ovulation, vaginal cornification and mating behaviour[105], whereas progesterone (at 1/10 the effective subcutaneous dose) implanted into the anterior pituitary or median eminence region on the day before pro-oestrus proved significantly active in interfering with ovulation[106]. In relation to this point, it should be noted that oestrogen is capable of augmenting the pituitary response (of LH release) to exogenous LH-releasing hormone[77], whereas progesterone reduces it[107], suggesting direct effects of oestrogen and progesterone on the pituitary.

Although the medial basal hypothalamus and the anterior pituitary appear to be the negative feedback sites of progesterone, the site(s) of the facilitatory action of progesterone on ovulatory LH release is not yet conclusively defined. For example, the medial preoptic area on the one hand[108–110] and the ventromedial arcuate region on the other[111] have been suggested as the sites of the acute facilitatory action of progesterone. Recent studies of Kobayashi *et al.*[86,112], using the techniques of intracranial steroid implantation and hypothalamic deafferentation by the Halász's knife[113], have led to the conclusion that the facilitation of ovulating hormone release by exogenous progesterone is due to the increase in the responsiveness of the medial basal hypothalamus to the ovulatory stimulus coming from the higher neural centre via the preoptic–suprachiasmatic region, rather than the direct stimulation of the hypothalamus to secrete hypothalamic-releasing hormone. In their experiment, intrapituitary implants of progesterone were not effective in facilitating ovulation. On the other hand, by determining the effect of ovarian steroids on hypothalamic thresholds for ovulation in

4-day cyclic rats, McDonald and Gilmore[114] suggested that both oestradiol and progesterone exert a positive feedback effect at or below the level of the median eminence and that negative feedback effects of progesterone are exerted on the preoptic area.

It is well established that the preoptic–suprachiasmatic area acts through the hypophysiotropic area of the hypothalamus to promote an ovulatory discharge of gonadotropins[84]. Recent studies of Velasco and Taleisnik[115–117] have suggested that the release of gonadotropins is under the modulatory influence of the amygdala and hippocampus, both of which have an input through the preoptic area directly into the medial basal hypothalamus, the influence of the amygdala being facilitatory while that of the hippocampus is inhibitory. Kalra and McCann[118] have recently reported that oestrogen receptors, which modify release of LH and/or FSH, appear to be located in the preoptic area, median eminence arcuate region, amygdala and hippocampus; progesterone-sensitive nervous elements, which stimulate release of LH and/or FSH, may be present in the preoptic area and amygdala.

6.5 CONCLUDING REMARKS

On the basis of blood concentrations and interplay of pituitary and gonadal hormones, the following view may be proposed for the explanation of hormonal control of menstrual cycle and ovulation in normal women. During the period of menstruation, circulating FSH and LH levels are relatively high. This is caused by the regression of corpus luteum resulting in the removal of progesterone suppression to the tonic secretion of gonadotropins. FSH and LH stimulate follicular growth. The oestrogen secretion coupling with follicular development is sustained by circulating gonadotropins, probably FSH and LH. When plasma oestrogens reach a certain threshold level, an acute release of gonadotropins from the pituitary occurs to produce synchronous surges of LH and FSH producing mid-cycle peaks respectively. During the gonadotropin surges, luteinisation of granulosa cells proceeds in the developed follicle[119]. This probably links with the preovulatory rise of blood progesterone levels, since high concentrations of progesterone have been demonstrated in the follicular fluid[120] as well as in the venous blood collected from the ovary bearing the mature follicle[121]. Ovulation occurs within 2 days after mid-cycle LH peak, and progesterone output increases rapidly and progressively in accordance with the formation of the corpus luteum, which has the autonomy to secrete progesterone and oestrogens. Under the influence of both steroids, uterine endometrium hypertrophies and shows the characteristic change of progestational proliferation. During the luteal phase, the length of which depends on the life-span of the corpus luteum both FSH and LH levels are suppressed by the negative feed-back action of progesterone. Regression of the corpus luteum accompanied by the fall in blood levels of ovarian steroids induces menstruation, and FSH and LH levels increase gradually as the result of progesterone withdrawal.

There is no doubt that oestrogen surge triggers LH surge in the human, monkey and the rat in the above cycle. LH surge is responsible for ovulation, but the significance of the FSH surge, normally synchronised with the LH surge in women, is still unknown. Induction of ovulation does not seem to require FSH either in the human[22] or in the rat[35,39,40].

The preovulatory rise in blood progesterone levels in the primate is not as marked as that in the rat, and progesterone appears not to be the steroid responsible for the gonadotropin surge, but it may have a role in preventing hyperstimulation or superovulation in women[22]. In the rat also, the preovulatory progesterone rise is not a cause but a consequence of the LH surge[56]. There is no doubt that exogenous progesterone given at the proper time is able to facilitate the LH surge by a few hours in the rat, but the physiological significance of endogenous preovulatory progesterone is not fully elucidated. A possibility is that adrenal progesterone participates in the process of LH surge in the rat[90,91].

The life-span of the corpus luteum in non-pregnant primates appears to be regulated mainly by the interplay between LH and progesterone, since reduced levels of LH appear to be required for the maintenance of functional corpus luteum in monkeys[33] and also in women[11]. The major roles of the ovarian steroids secreted during the luteal phase are the preparation of uterine endometrium for the implantation of the blastocyst and the suppression of pituitary gonadotropin secretion, by which maturation of follicles is inhibited. A possibility is suggested, however, that high intra-ovarian levels of steroids during the luteal phase participate in inhibiting follicular growth[11]. The corpus luteum, therefore, plays a key role in adjusting the onset of follicular development for the next cycle.

Timing of the oestrogen surge occurring under the base-line levels of gonadotropins may be regulated by the rate of follicular growth, since the ovarian responsiveness to gonadotropins for oestrogen secretion depends on the follicular maturation[53]. The regulatory mechanism for follicular growth coupled with oestrogen secretion is an important problem remaining to be clarified.

It has been suggested that hypothalamic control of the anterior pituitary in secretion of LH and FSH is mediated by a single hypothalamic hormone, LH–RH/FSH–RH[8]. In the rat, the ovulatory surge of LH is linked with the circadian rhythmicity of the hypothalamus, possibly the suprachiasmatic and preoptic regions, which are activated by oestrogen at a certain period of time[83–85]. It is known that pituitary responsiveness to exogenous hypothalamic LH-releasing hormone can be altered by the ovarian steroids, augmented by oestrogen[77] and suppressed by progesterone[107]. It is conceivable that oestrogen, accepted in the 'oestrogen receptor area' in the brain, stimulates the hypothalamic function to evoke LH–RH release into the hypophysial portal blood, and, at the same time, augments pituitary sensitivity to LH–RH and thus induces the LH surge. The reverse effects on both hypothalamus and pituitary may be provided by progesterone, the secretion of which is increased after the LH surge. Feed-back sites of steroids have been considerably substantiated in the rat, but not in the primate. In contrast to the situation in the rat, the 'hypothalamic clock' which determines the timing of ovulation does not seem to exist in the rhesus monkey[5,28]. There is a similarity, however, between rats and rhesus monkeys in that a full course of blood oestrogen rise is not required for hypothalamic stimulation, but only an initial and partial rise retaining for about half a day is a prerequisite to produce the LH ovulatory surge.

References

1. Hartman, C. G. (1962). *Science and the Safe Period* (Baltimore: Williams and Wilkins)
2. Klopper, A. I. (1900). Pregnandiol and pregnantriol, *Methods in Hormone Research*, Vol. 1, 140 (R. I. Dorfman editor) (New York: Academic Press)
3. Diczfalusy, E. (editor) (1969). Immunoassay of gonadotrophins., *Acta Endocrinol.*, **Suppl. 142,** 9
4. Diczfalusy, E. (editor) (1970). Steroid assay by protein binding, *Acta Endocrinol.*, **Suppl. 147,** 9
5. Knobil, E. (1971). Hormonal control of the menstrual cycle and ovulation in the rhesus monkey, *Acta Endocrinol.*, **Suppl. 166,** 137
6. Schwartz, N. B. (1969). A model for the regulation of ova ovulation in the rat, *Rec. Prog. Hormone Res.*, **25,** 1
7. Abraham, G. E., Odell, W. D., Swerdoloff, R. S. and Hopper, K. (1972). Simultaneous radioimmunoassay of plasma FSH, LH, progesterone and 17-hydroxyprogesterone, and estradiol-17β during the menstrual cycle, *J. Clin. Endocrinol. Metab.*, **34,** 312
8. Schally, A. V., Kastin, A. B. and Arimura, A. (1971). Hypothalamic follicle-stimulating hormone and luteinising hormone-releasing hormone: Structure, physiology and clinical studies, *Fert. Steril.*, **22,** 703
9. Malcacara, J. M., Seyler, L. E., Jr and Reichlin, S. (1972). Luteinizing hormone-releasing factor activity in peripheral blood from women during the midcycle luteinizing hormone ovulatory surge, *J. Clin. Endocrinyl.*, **34,** 271
10. Midgley, A. R. and Jaffe, R. B. (1971). Regulation of human gonadotropins: X. Episodic fluctuation of LH during the menstrual cycle, *J. Clin. Endocrinol.*, **33,** 962
11. Vande Wiele, R. L., Bogumil, J., Dyrenfurth, I., Ferin, M., Jewelewicz, R., Warren, M., Rizkallah, T. and Mikhail, G. (1970). Mechanisms regulating the menstrual cycle in women, *Rec. Prog. Hormone Res.*, **26,** 63
12. Tai, C. C. and Yen, S. S. C. (1971). The effect of ethinylestradiol administration during early follicular phase of the cycle on the gonadotropin levels and ovarian function, *J. Clin. Endocrinol.*, **33,** 917
13. Yen, S. S. C. and Tai, C. C. (1972). Acute gonadotropin release induced by exogenous estradiol during the mid-follicular phase of the menstrual cycle, *J. Clin. Endocrinol.*, **34,** 298
14. Johansson, E. D. B. and Gemzell, C. (1971). Plasma levels of progesterone during the luteal phase in normal women treated with synthetic oestrogens. (RS 2874, F 6103 and ethinyloestradiol), *Acta Endocrinol.*, **68,** 551
15. Strott, C. A., Yoshimi, T., Ross, G. T. and Lipsett, M. B. (1969). Ovarian physiology: Relationship between plasma LH and steroidogenesis by the follicle and corpus luteum; effect of HCG, *J. Clin. Endocrinol.*, **29,** 1157
16. Pincus, G. (1965). *Control of Fertility* (New York: Academic Press)
17. Johansson, E. D. B., Wide, L. and Gemzell, C. (1971). Luteinizing hormone (LH) and progesterone in plasma and LH and oestrogens in urine during 42 normal menstrual cycles, *Acta Endocrinol.*, **68,** 502
18. Leyendecker, G., Warldlaw, S. and Nocke, W. (1971). Steroid-induced positive feedback in the human female: New aspects on the control of ovulation, *Gonadotropins*, 720 (Saxena, Beling and Gandy, editors) (New York: Wiley-Interscience)
19. Bertrand, P. V., Cloeman, J. R., Crooke, A. C., Macnaughton, M. C. and Mills, I. H. (1972). Human ovarian response to gonadotrophins with different ratios of follicle-stimulating hormone: luteinizing hormone assessed by different parameters, *J. Endocrinol.*, **53,** 231

20. Niswender, G. D., Menon, K. M. J. and Jaffe, R. B. (1972). Regulation of the corpus luteum during the menstrual cycle and early pregnancy, *Fert. Steril.*, **23,** 432
21. Bettendorf, G., Lehman, F., Neale, Ch. and Breckwoldt, M. (1971). Plasma steroid pate pattern during gonadotropin stimulation, *Gonadotropins*, 749 (Saxena, Beling and Gandy, editors) (New York: Wiley-Interscience)
22. Ross, G. T., Cargille, C. M., Lipsett, M. B., Rayford, P. L., Marshall, J. R., Strott, C. A. and Rodbard, D. (1970). Pituitary and gonadal hormones in women during spontaneous and induced ovulatory cycles, *Rec. Prog. Hormone Res.*, **26,** 1
23. Hotchkiss, J., Atkinson, L. E. and Knobil, E. (1971). Time course of serum estrogens and luteinizing hormone (LH) concentrations during the menstrual cycle, *Endocrinology*, **89,** 177
24. Kirton, K. T., Niswender, G. G., Midgley, A. R., Jr, Jaffe, R. B. and Forbes, A. D. (1970). Serum luteinizing hormone and progesterone concentration during the menstrual cycle of the rhesus monkey, *J. Clin. Endocrinol.*, **30,** 105
25. Johansson, E. D. B., Neill, J. D. and Knobil, E. (1968). Periovulatory progesterone concentration in the peripheral plasma of the rhesus monkey with methodologic note on the detection of ovulation, *Endocrinology*, **82,** 143
26. Spies, H. G. and Niswender, G. G. (1972). Effect of progesterone and estradiol on LH release and ovulation in rhesus monkeys, *Endocrinology*, **90,** 257
27. Yamaji, T., Diershke, D. J., Hotchkiss, J., Bhattacharya, A. N., Surve, A. H. and Knobil, E. (1971). Estrogen induction of LH release in the rhesus monkey, *Endocrinology*, **89,** 1034
28. Knobil, E., Dierschke, D. J., Yamaji, Y., Karsch, F. J., Hotchkiss, J. and Weick, R. F. (1971). Role of estrogen in positive and negative feedback control of LH secretion during the menstrual cycle of the rhesus monkey, *Gonadotropins*, 72 (Saxena, Beling and Gandy, editors) (New York: Wiley-Interscience)
29. Yamaji, T., Dierschke, D. J., Bhattacharya, A. N. and Knobil, E. (1972). The negative feedback control by estradiol and progesterone of LH secretion in the ovariectomized rhesus monkey, *Endocrinology*, **90,** 771
30. Bhattacharya, A. N., Dierschke, D. J., Yamaji, T. and Knobil, E. (1972). The pharmacologic blockade of the circhoral mode of LH secretion in the ovariectomized rhesus monkey, *Endocrinology*, **90,** 778
31. Macdonald, G. J. and Greep, R. O. (1972). Ability of luteinizing hormone (LH) to acutely increase serum progesterone levels during the secretory phase of the rhesus menstrual cycle, *Fert. Steril.*, **23,** 466
32. Moudgal, N. R. Macdonald, G. J. and Greep, R. O. (1971). Effect of HCG antiserum on ovulation and corpus luteum formation in the monkey (*Macaca fascicularis*), *J. Clin. Endocrinol.*, **32,** 579
33. Moudgal, N. R., Macdonald, G. J. and Greep, R. O. (1972). Role of endogenous primate LH in maintaining corpus luteum function of the monkey, *J. Clin. Endocrinol.*, **35,** 113
34. Bogdanove, E. M., Schwartz, N. B. and Reichert, L. E., Jr. (1971). Comparisons of pituitary: Serum luteinizing hormone (LH) ratios in the castrated rat by radioimmunoassay and OAAD bioassay, *Endocrinology*, **88,** 644
35. Danne, T. A. and Parlow, A. F. (1971). Periovulatory patterns of rat serum follicle-stimulating hormone and luteinizing hormone during the normal estrous cycle: Effects of pentobarbital, *Endocrinology*, **88,** 653
36. Linkie, D. M. and Niswender, G. D. (1972). Serum levels of prolactin, luteinizing hormone and follicle-stimulating hormone during pregnancy in the rat, *Endocrinology*, **90,** 632
37. Everett, J. W., Sawyer, C. H. and Markee, J. E. (1949). A neurogenic timing factor in control of the ovulatory discharge of luteinizing hormone in the cyclic rat, *Endocrinology*, **44,** 234
38. Greep, R. O. (1961). Physiology of the anterior hypophysis in the regulation to reproduction, *Sex and Internal Secretions* Vol. 1, 240 (W. C. Young, editor) (Baltimore: Williams and Wilkins)
39. Schwartz, N. B. (1972). Reproduction, gonadal function and its regulation, *Ann. Rev. Physiol.*, **34,** 425

40. Cobbs, S. B. and Schwartz, N. B. (1972). The influence of immunologically purified LH and FSH on ovulation, *The 5th Annual Meeting of the Society for the Study of Reproduction* (East Lancing), Abstract p. 79
41. Kwa, H. G. and Verhofstad, F. (1967). Prolactin levels in the plasma of female rats, *J. Endocrinol.*, **39,** 455
42. Yokoyama, A., Tomogane, H. and Ota, K. (1971). Prolactin surge on the afternoon of proestrus in the rat and its blockade by pentobarbitone, *Experientia*, **27,** 578
43. Neill, J. D., Freeman, M. E. and Tillson, S. A. (1971). Control of the proestrus surge of prolactin and luteinizing hormone secretion by estrogens in the rat, *Endocrinology*, **89,** 1448
44. Armstrong, D. T. (1968). Gonadotropins, ovarian metabolism and steroid synthesis, *Rec. Prog. Hormone Res.*, **24,** 255
45. Behrman, H. R., Orczyk, G. P., Macdonald, G. J. and Greep, R. O. (1970). Prolactin induction of enzymes controlling luteal cholesterol ester turnover, *Endocrinology*, **87,** 1251
46. Wuttke, W. and Meites, J. (1971). Luteolytic role of prolactin during the estrous cycle of the rat, *Proc. Soc. Exp. Biol. Med.*, **137,** 988
47. Freeman, M. E. and Neill, J. D. (1972). The pattern of prolactin secretion during pseudopregnancy in the rat: A daily nocturnal surge, *Endocrinology*, **90,** 1292
48. Hori, T., Ide, M. and Miyake, T. (1968). Ovarian estrogen secretion during the estrous cycle and under the influence of exogenous gonadotropins in rats, *Endocrinol. Jap.*, **15,** 215
49. Yoshinaga, K., Hawkins, R. A. and Stocker, J. F. (1969). Estrogen secretion by the rat ovary *in vivo* during the estrous cycle and pregnancy, *Endocrinology*, **85,** 103
50. Shaikh, A. A. (1971). Estrone and estradiol levels in the ovarian venous blood from rats during the estrous cycle and pregnancy, *Biology of Reproduction*, **5,** 297
51. Brown-Grant, K., Exley, D. and Naftolin, F. (1970). Peripheral plasma oestradiol and luteinizing hormone concentrations during the oestrous cycle of the rat, *J. Endocrinol.*, **48,** 295
52. Hori, T., Ide, M. and Miyake, T. (1969). Pituitary regulation of preovulatory estrogen secretion in the rat, *Endocrinol., Jap.*, **16,** 351
53. Chatterton, R. T., Jr, Chatterton, A. J. and Greep, R. O. (1969). *In vitro* biosynthesis of estrone and estradiol-17β by cycling rat ovaries. Effect of luteinizing hormone, *Endocrinology*, **84,** 252
54. Hori, T., Ide, M., Kato, G. and Myiake, T. (1970). Relation between estrogen secretion and follicular morphology in the rat ovary under the influence of ovulating hormone or exogenous gonadotropins, *Endocrinol. Jap.*, **17,** 489
55. Young, W. C. (1961). The mammarian ovaries, *Sex and Internal Secretions*, Vol. 1, 449 (W. C. Young, editor) (Baltimore: Williams and Wilkins Co.)
56. Uchida, K., Kadowaki, M. and Miyake, T. (1969). Ovarian secretion of progesterone and 20α-hydroxypregn-4-en-3-one during rat estrous cycle in chronological relation to pituitary release of luteinizing hormone, *Endocrinol. Jap.*, **16,** 227
57. Kobayashi, K., Hara, K. and Miyake, T. (1968). Luteinizing hormone concentrations in pituitary and in blood plasma during the estrous cycle of the rat, *Endocrinol. Jap.*, **15,** 313
58. Barraclough, C. A., Collu, R., Massa, R. and Martini, L. (1971). Temporal interrelationships between plasma LH, ovarian secretion rates and peripheral plasma progestin concentrations in the rat: Effects of nembutal and exogenous gonadotropins, *Endocrinology*, **88,** 1437
59. Piacsek, B. E., Schneider, T. C. and Gay, V. L. (1971). Sequential study of luteinizing hormone (LH) and progestin secretion on the afternoon of proestrus in the rat, *Endocrinology*, **89,** 39
60. Chatterton, R. T., Jr, Macdonald, G. J. and Greep, R. O. (1968). Biosynthesis of progesterone and 20α-hydroxypregn-4-en-3-one by the rat ovary during the estrous cycle and early pregnancy, *Endocrinology*, **83,** 1
61. Yoshinaga, K., Grieves, S. A. and Short, R. V. (1967). Steroidogenic effects of luteinizing hormone and prolactin on the rat ovary *in vivo*, *J. Endocrinol.*, **38,** 423

62. Uchida, K., Kadowaki, M. and Miyake, T. (1969). Effect of exogenous progesterone on the preovulatory progesterone secretion in the rat, *Endocrinol. Jap.*, **16,** 485
63. Uchida, K., Kadowaki, M. and Miyake, T. (1969). Acute effects of various gonadotropins and other pituitary hormones on preovulatory ovarian progestin secretion in hypophysectomized rats, *Endocrinol. Jap.*, **16,** 239
64. Greep, R. O. (1971). Regulation of luteal cell function. *Hormonal Steroids*, 670 James and Martini, editors) (Amsterdam: Excerpta Medica)
65. Leavitt, W. W., Bosley, C. G. and Blaha, G. C. (1971). Source of ovarian preovulatory progesterone, *Nature (New Biology) (London)*, **234,** 283
66. Wiest, W. G., Kidwell, W. R. and Balogh, K., Jr (1968). Progesterone catabolism in the rat ovary: A regulatory mechanism for progestational potency during pregnancy, *Endocrinology*, **82,** 844
67. Everett, J. W. (1948). Progesterone and estrogen in the experimental control of ovulation time and other features of the estrous cycle in the rat, *Endocrinology*, **43,** 389
68. Kobayashi, F., Hara, K. and Miyake, T. (1971). Induction of delayed or advanced ovulation by estrogen in 4-day cyclic rat, *Endocrinol. Jap.*, **18,** 389
69. Miyake, T. (1969). Causal relationship between ovarian steroid secretion and pituitary luteinizing hormone release in the rat estrous cycle, *Endocrinol. Jap.*, **Suppl. 1,** 83
70. Shirley, B., Wolinsky, J. and Schwartz, N. B. (1968). Effects of a single injection of estrogen antagonist on the estrous cycle of the rat, *Endocrinology*, **82,** 959
71. Kobayashi, F., Hara, K. and Miyake, T. (1969). Causal relationship between luteinizing hormone release and oestrogen secretion in the rat, *Endocrinol. Jap.*, **16,** 261
72. Labhsetwar, A. P. (1970). Role of estrogens in ovulation: A study using the estrogen-antagonist I.C.I. 46,474, *Endocrinology*, **87,** 542
73. Ferin, M., Tempone, A., Zimmering, P. E. and Vande Wiele, R. L. (1969). Effect of antibodies to 17β-estradiol and progesterone on the estrous cycle of the rat, *Endocrinology*, **85,** 1070
74. Kobayashi, F., Hara, K. and Miyake, T. (1969). Further studies on the causal relationship between the secretion of estrogen and the release of luteinizing hormone in the rat, *Endocrinol. Jap.*, **16,** 501
75. Miyake, T. (1968). Interrelationship between the release of pituitary luteinizing hormone and the secretions of ovarian estrogen and progestin during estrus cycle of the rat, *Integrative Mechanism of Neuroendocrine System*, Vol 1, 139 (S. Itch, editor) (Sapporo, Japan: Hokkaido Univ. Med. Library Series)
76. Ieiri, T., Akikusa, Y. and Yamamoto, K. (1971). Synthesis and release of prolactin and growth hormone, *Endocrinology*, **89,** 1533
77. Arimura, A. and Schally, A. V. (1971). Augmentation of pituitary responsiveness to LH-releasing hormone (LH–RH) by oestrogen, *Proc. Soc. Exp. Biol. Med.*, **136,** 290
78. Debeljuk, L., Arimura, A. and Schally, A. V. (1972). Effect of estradiol and progesterone on the LH release induced by LH-releasing hormone (LH–RH) in intact diestrous rats and anestrous ewes, *Proc. Soc. Exp. Biol. Med.*, **139,** 774
79. Freeman, M. E., Reichert, L. E., Jr. and Neill, J. D. (1972). Regulation of the proestrus surge of prolactin by gonadotropin and estrogens in the rat, *Endocrinology*, **90,** 232
80. Flerko, B. (1966). Control of gonadotropin secretion in the female, *Neuroendocrinology*, Vol 1, 613 L. Martini and W. F. Ganong, editors) (New York: Academic Press)
81. Meites, J. (1966). Control of mammary growth and lactation, *Neuroendocrinology*, Vol 1, 669 (L. Martini and W. F. Ganong, editors) (New York: Academic Press)
82. Mshkinsky, J., Khazen, K. and Sulman, F. G. (1968). Prolactin-releasing activity of the hypothalamus in postpartum rats, *Endocrinology*, **82,** 611
83. Everett, J. W. (1961). The mammalian female reproductive cycle and its controlling mechanisms, *Sex and Internal Secretions*, Vol 1, 497 (W. C. Young, editor) (Baltimore: Williams and Wilkins)

84. Everett, J. W. (1964). Central neural control of reproductive functions of adenohypophysis, *Physiol. Rev.*, **44,** 373
85. Everett, J. W. (1969). Neuroendocrine aspects of mammalian reproduction, *Ann. Rev. Physiol.*, **31,** 383
86. Kobayashi, F., Hara, K. and Miyake, T. (1970). Facilitation of luteinizing hormone release by progesterone in proestrous rats, *Endocrinol. Jap.*, **17,** 149
87. Brown-Grant, K. and Naftolin, F. (1971). Facilitation of luteinizing hormone secretion in the female rat by progesterone, *J. Endocrinol.*, **53,** 37
88. Taleisnik, S., Calligaris, A. and Astrada, J. J. (1971). Feedback effect of gonadal steroids on the release of gonadotropins, *Hormonal Steroids*, 699 (V. H. T. James and L. Martini, editors) (Amsterdam: Excerpta Medica)
89. Sawyer, C. H. and Hilliard, J. (1971). Sites of feedback action of estrogen and progesterone, *Hormonal Steroids*, 716 (V. H. T. James and L. Martini, editors) (Amsterdam: Excerpta Medica)
90. Resko, J. A. (1969). Endocrine control of adrenal progesterone secretion in the ovariectomized rat, *Science*, **164,** 70
91. Nequin, L. G. and Schwartz, N. B. (1971). Adrenal participation in the timing of mating and LH release in the cyclic rat, *Endocrinology*, **88,** 325
92. Kobayashi, F., Hara, K. and Miyake, T. (1969). Inhibitory and facilitatory effects of steroids on the release of luteinizing hormone in the rat, *Endocrinol. Jap.*, **16,** 493
93. Kobayashi, F., Hara, K. and Miyake, T. (1969). Effects of steroids on the release of luteinizing hormone in the rat, *Endocrinol. Jap.*, **16,** 251
94. Swerdloff, R. S., Jacobs, H. S. and Odell, W. D. (1970). Synergistic role of progesterone in estrogen induction of LH and FSH surge, *Endocrinology*, **90,** 1529
95. Stumpf, W. E., Baerwaldt, C. and Sar, M. (1971). Autoradiographic cellular and subcellular localization of sexual steroids, *Basic Actions of Sex Steroids on Target Organs*, 3 (Basel: Karger)
96. Kato, J. (1970). Estrogen receptors in the hypothalamus and hypophysis in relation to reproduction. *Hormonal Steroids*, 764 (V. H. T. James and L. Martini, editors) (Amsterdam: Excerpta Medica)
97. Mowles, T. F., Ashkanazy, B., Mix, E., Jr and Sheppard, H. (1971). Hypothalamic and hypophyseal estradiol-binding complexes, *Endocrinology*, **89,** 484
98. Kato, J., Atsumi, Y. and Inaba, M. (1971). Development of estrogen receptors in the rat hypothalamus, *J. Biochem.*, **70,** 1051
99. Lisk, R. D. (1960). Estrogen-sensitive centers in the hypothalamus of the rat, *J. Exp. Zool.*, **145,** 197
100. Döcke, F. and Dörner, G. (1965). The mechanism of the induction of ovulation by oestrogens, *J. Endocrinol.*, **33,** 491
101. Smith, E. R. and Davidson, J. M. (1968). Role of estrogen in the cerebral control of puberty in female rat, *Endocrinology*, **82,** 100
102. Ramirez, V. D., Abrams, R. M. and McCann, S. M. (1964). Effect of estradiol implants in hypothalamo–hypophysial region of the rat on the secretion of luteinizing hormone, *Endocrinology*, **75,** 243
103. Dörner, G. and Döcke, F. (1967). Influence of intrahypothalamic and intra-hypophyseal implantation of estrogen or progesterone on gonadal gonadotrophin release, *Endocrinol. Exp.*, **1,** 65
104. Döcke, F. and Dörner, G. (1967). Mechanism of the gestagen effect on ovulation, *Acta Endocrinol.*, **Suppl. 119,** 163
105. Weick, R. F., Smith, E. R., Dominguez, R., Dhriwal, A. P. S. and Davidson, J. M. (1971). Mechanism of stimulatory feedback effect of estradiol benzoate on the pituitary, *Endocrinology*, **88,** 293
106. Labhsetwar, A. P. and Bainbridge, J. G. (1971). Inhibition of ovulation by intracranial implantation of progesterone in the 4-day cyclic rat, *J. Reprod. Fert.*, **27,** 445
107. Arimura, A. and Schally, A. V. (1970). Progesterone suppression of LH-releasing hormone-induced stimulation of LH release in rats, *Endocrinology*, **87,** 653
108. Barraclough, C. A., Yrarrazaval, S. and Hatton, R. (1964). A possible hypothalamic site of action of progesterone in the facilitation of ovulation in the rat, *Endocrinology*, **75,** 838

109. Zeilmaker, G. H. and Moll, J. (1967). Effects of progesterone on ovulation in the rat. *Acta Endocrinol.*, **55,** 378
110. Taleisnik, S., Velasco, M. E. and Astrada, J. J. (1970). Effect of hypothalamic deafferentation on the control of luteinizing hormone secretion, *J. Endocrinol.*, **46,** 1
111. Döcke, F. and Dörner, G. (1969). A possible mechanism by which progesterone facilitates ovulation in the rat, *Neuroendocrinology*, **4,** 139
112. Kobayashi, F. and Miyake, T. (1971). Acute effect of hypothalamic deafferentation on progesterone-induced ovulating hormone release in the rat, *Endocrinol. Jap.*, **18,** 395
113. Halász, B. and Pupp, L. (1965). Hormone secretion of the anterior pituitary gland after physical interruption of all nervous pathways to the hypophysiotropic area, *Endocrinology*, **77,** 553
114. McDonald, P. G. and Gilmore, D. P. (1971). The effect of ovarian steroids on hypothalamic thresholds for ovulation in the female rat, *J. Endocrinol.*, **49,** 421
115. Velasco, M. E. and Taleisnik, S. (1969). Release of gonadotropins induced by amygdaloid stimulation in the rat, *Endocrinology*, **84,** 132
116. Velasco, M. E. and Taleisnik, S. (1969). Effect of hippocampal stimulation on the release of gonadotropin, *Endocrinology*, **85,** 1154
117. Velasco, M. E. and Taleisnik, S. (1971). Effects of interruption of amygdaloid and hypocampal afferents to the medial hypothalamus on gonadotropin release, *J. Endocrinol.*, **51,** 41
118. Kalra, P. S. and McCann, S. M. (1972). Effects of CNS implants of ovarian steroids on gonadotropin release, *4th International Congress of Endocrinology* (International congress series No. 256), Abstract p. 118 (Amsterdam: Excerpta Medica)
119. Delforge, J. P., Thomas, K., Roux, F., Carneiro de Siqueira, J. and Ferin, J. (1972). The relationships between granulosa cells growth and luteinization and plasma luteinizing hormone discharge in human. I. A morphometric analysis, *Fert. Steril.*, **23,** 1
120. Zander, J. (1958). Steroids in the human ovary, *J. Biol. Chem.*, **232,** 117
121. Mikhail, G. (1970). Hormone secretion by the human ovaries, *Gynec. Invest.*, **1,** 5

7
The Hormones of the Placenta and their Role in the Onset of Labour

A. KLOPPER
University of Aberdeen

Felix qui potuit rerum cognoscere causas
Virgil. Georgics II, 490

7.1 INTRODUCTION

If success is a measure of happiness, then the search for the cause of labour has not been a happy one for it has not been successful. In one sense it has always been doomed to failure. There is probably no single ultimate starting point or even one chain of events involved, but many processes, starting at different points and flowing together in different ways, to produce the same endpoint—the onset of labour. This review will be concerned with only one aspect of this multifarious event, the nature of the biological signal that sets it off. This is boggy ground where it is proper to proceed with care. Nearly the only certainty is that throughout the range of the mammalia this signal arises from the products of conception and not from the maternal organism. At first glance it would seem probable that the maternal organism has some means of measuring the maturation of the life within the womb, and can, when it is capable of independent existence, expel it into the world. It is a relatively new concept that the mother is, as it were, a passive agent and that the foetus itself should decide when it is capable of independent existence and pull the trigger which will precipitate it into the world. Although, in terms of this philosophy, the ultimate cause lies with the maturation processes in the foetus, we shall not be concerned with these, but purely with the nature of the biological signal which the foetus sends to the mother, requiring it to be expelled.

There are many ways in which the foetus can signal to the mother; simply stretching the uterus by its increasing bulk, or irritating the cervix by pressing upon it come readily to mind. No doubt such mechanical factors and other elements do play a part in the onset of labour. Our present concern is only with endocrine aspects; that is with compounds elaborated by the foetus or placenta which, when transmitted to the mother, influence uterine contractility. It is an oversimplification to think of labour only in terms of uterine contraction. The cervix, for instance, undergoes great changes in its physical state, its chemical composition and its microscopic architecture during pregnancy. These changes are as essential to the successful outcome of the pregnancy as contractions of the myometrium. There is reason to think that the cervical changes too are influenced by endocrine factors, but in the human at least they are not amenable to experimental examination. It is a matter for regret that so little is known about endocrine effects on the cervix during gestation that they cannot be reviewed here. Clinicians and research workers alike have tended to concentrate on uterine contractility and in doing so may have missed a key section of the process which leads to labour.

7.2 THE FOETO-PLACENTAL UNIT

The concept that 'the foetus and placenta should be considered as an integrated endocrine unit, exhibiting a rather high degree of autonomy' has long been the central theme of the work done in Stockholm under the direction of Egon Diczfalusy[1]. As far as the steroid hormones are concerned the evidence is quite clear. They are made in a series of steps, some of which are carried out in the foetus, some in the placenta and some in both. The picture that emerges is of a to-and-fro traffic of steroids between foetus and placenta, the molecules being altered in each location, presumably at one stage or another serving functions in the foeto-placental unit, sometimes being metabolised and ultimately being transferred to the mother, either as inactive waste products or as active chemical messengers. In this respect a difficulty arises. Some steroids, like progesterone, are secreted in an active form and in relatively large amounts by the foeto-placental unit into the maternal circulation, where presumably they play a role, albeit one the nature of which is the subject of dispute. Others, like oestriol, are also secreted in comparatively large amounts but it has not proved possible to assign to them a role peculiar to a particular molecular structure. There has accordingly arisen the suggestion that some steroids, notably oestriol, are mere waste products, rapidly inactivated by sulphurylation in the foetus in order to protect itself against the steroid and secreted to the mother, not as a biological signal, but as metabolic garbage. It must be owned that arguments in favour of oestriol having a function in its own right are in part frankly teleological. It is hard to credit that a steroid made in large quantities and peculiar to human pregnancy should have no function at all.

7.3 THE HORMONES OF THE FOETO-PLACENTAL UNIT

A great many of the substances produced in the foeto-placental unit fit in any definition of a hormone. Many, like a thyrotrophic substance, are newly discovered and almost nothing is known about them. The existence of others, like a compound having follicle stimulating properties is surmised rather than demonstrated. The protein hormones, placental lactogen and chorionic gonadotrophin are purely placental in origin and the foetus is not involved in their biogenesis. They are part of the barrage directed at the mother and aimed at inducing in her the metabolic changes essential to the survival of the foetus. There is, however, very little to suggest that they are concerned with the onset of labour and they will be dealt with very lightly in this review. Among the great array of steroid hormones produced by the foeto-placental unit are many, such as corticoids, androgens and aldosterone, which are also not germane to the topic of labour. In considering the endocrinology of pregnancy attention will, therefore, be mainly directed towards the progestogens and the oestrogens which may be presumed to bear some part in the onset of labour.

7.3.1 The precursors of placental oestrogens and progestogens

The basic unit for the biogenesis of steroids is acetate, a 2-carbon fragment. These small molecules are joined together to form other compounds containing more and more carbon atoms. The process culminates in a steroid, cholesterol, which contains 27 carbon atoms. Cholesterol is the raw material from which all steroid hormones are made. Many of the intermediates between acetate and cholesterol such as mevalonate, squalene and lanosterol are known. The question arises whether cholesterol can be made by either the foetus or the placenta. The evidence on this point is quite clear. Telegdy *et al.*[2] found that perfusing a placenta with radioactive precursors such as acetate did not lead to the formation of labelled cholesterol, but that perfusing the entire foeto-placental unit did so. Later experiments[3] showed that liver preparations from midterm foetuses can convert large quantities of acetate to cholesterol. The foetus can make cholesterol but the placenta cannot.

There is good evidence that all the steroid hormones of the foeto-placental unit are made from cholesterol. The question is whether this raw material is provided by the foetal liver or is supplied from the high concentration of cholesterol in the maternal circulation. The Stockholm school were at first

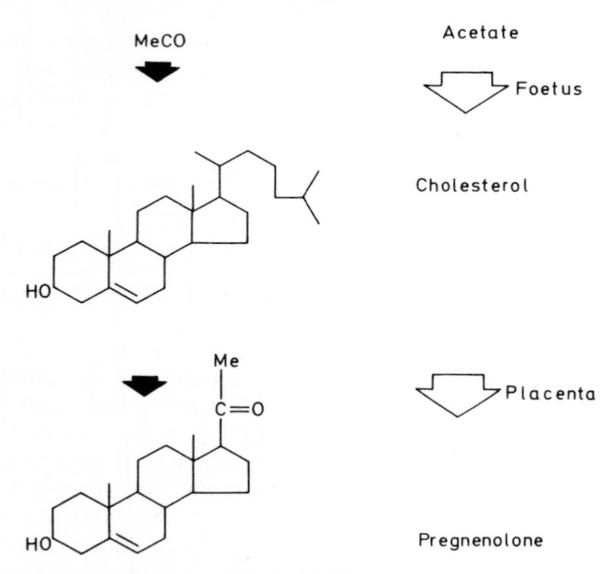

Figure 7.1 The initial biosynthetic steps common to all the steroids of the foeto-placental unit

inclined to favour the view that maternal cholesterol was the principal source of the steroid in the foeto-placental unit[4] but have lately stressed the quantitative importance of foetal biogenesis of cholesterol[2]. There is good evidence that the placenta can continue to produce steroid hormones after the foetus has been removed[5]. One is obliged to accept that at least the placenta can use maternal cholesterol if the need arises. This implies that the steps up to and including cholesterol are not rate-limiting; the control of hormone synthesis in the foeto-placental unit comes at a stage later than the formation of cholesterol.

Cholesterol has a side chain of six carbon atoms and the first step in the synthesis of all steroid hormones from cholesterol is the enzymatic removal of this side chain in several stages. This results in the formation of a 21-carbon compound, pregnenolone. When cholesterol is perfused through a term placenta, pregnenolone is formed but the foetus is unable to make significant amounts of pregnenolone from cholesterol[6]. Thus the interdependence of the foetus and placenta is found at an early stage in steroidogenesis; the foetus makes cholesterol and passes it to the placenta which converts it to pregnenolone.

Pregnenolone can be metabolised in different ways and it is at this point that pathways which lead to the oestrogens, to progesterone, the androgens and the corticosteroids begin to branch. The common pathway of all foeto-placental steroids to this point is illustrated in Figure 7.1.

7.3.2 Progesterone biosynthesis

When pregnenolone is injected into the foetal circulation, either at mid-pregnancy or in mature anencephalics, no progesterone is formed, provided that the placenta has been removed. The foetus can transform pregnenolone into other steroids but appears to be lacking in the 3β-hydroxysteroid dehydrogenase system for converting pregnenolone to progesterone or indeed for converting any β, γ-unsaturated steroid such as pregnenolone or dehydroepiandrosterone into the corresponding α, β-unsaturated 3-oxosteroids such as progesterone or androstenedione. On the other hand, placental homogenates or the intact term or pre-term placenta will readily convert pregenenolone into progesterone. Although under *in vivo* conditions the foetus shows no 3β-hydroxysteroid dehydrogenase activity it must be acknowledged that these perfusion experiments were often done under unphysiological conditions and that homogenates of foetal tissues may on occasion show such activity. The conclusion about the inability of the foetus to convert pregnenolone to progesterone rests on somewhat insecure ground.

The probable site of progesterone manufacture is the placenta. There is a high concentration of pregnenolone in the umbilical circulation and this may be the source of raw material for progesterone synthesis in the placenta. On the other hand, only a modest reduction in progesterone synthesis takes place when the cord circulation is cut off[7]. The placenta must therefore be able to draw on its own reserves of pregnenolone and cholesterol or to switch over to a much increased use of maternal precursors.

Conversion to progesterone is not the only metabolic pathway open to the placenta as far as pregnenolone is concerned. It can, for instance, produce 17α-hydroxypregnenolone or 17α-hydroxyprogesterone. But these are minor biosynthetic routes in the placenta; in the main it is progesterone that is produced. Once the progesterone is formed there are three possibilities as to its disposal. Firstly it can be further metabolised *in situ*. The available evidence is that very little such further metabolism takes place in the placenta and the progesterone in the peripheral venous blood of the mother, nine or ten times into the retroplacental maternal blood. There is no secure evidence about how much placental progesterone goes into the maternal circulation and what proportion goes to the foetus. There is a relatively very high concentration of progesterone in the peripheral venous blood of the mother nine or ten times as much as the next highest steroid[8]. In the retroplacental blood the progesterone concentration is even higher. Later the evidence that this point of highest progesterone concentration is also its site of action in the maternal organism will be considered. At this stage the question arises whether the

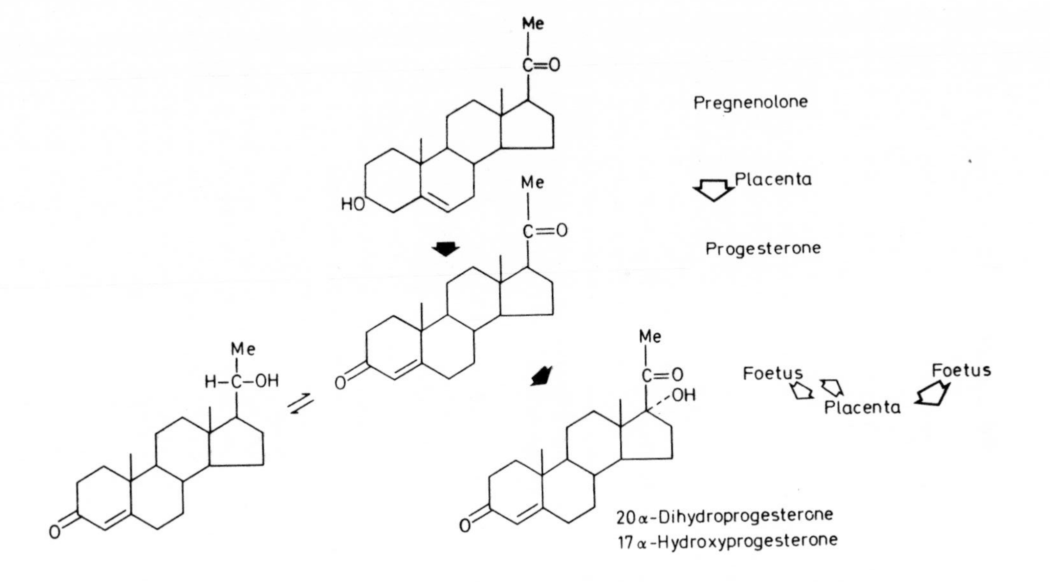

Figure 7.2 The main metabolic pathways of progesterone in the foeto-placental unit

placenta is able to vary the concentration of progesterone at the target organ by diverting more or less of the steroid to the foetal circulation. If this were so one would expect foetal progesterone concentration to rise or fall as the maternal levels fall or rise. There is very little direct evidence on this point but such findings as bear upon it point to control of progesterone concentration being exercised by alterations of placental production rather than changes in disposition. The facts that cholesterol supply to the placenta is not rate-limiting and that the biosynthetic steps from cholesterol to pregnenolone and from pregnenolone to progesterone take place in the placenta make it probable that the control over progesterone production and hence its presumed role in the onset of labour is situated in the placenta.

There is no doubt that a substantial proportion of the progesterone produced by the placenta is secreted into the foetal circulation. In the umbilical vein the concentration of progesterone is about 96 μg/100 ml of plasma while in the artery it is 54 μg/100 ml[9]. Thus, at term, the foetus metabolises about 40 μg of progesterone for every 100 ml of plasma that passes through it. This metabolism is mainly reductive, hydrogen being added to the keto groups at C-3 and C-20 to convert them to hydroxyls or to the unsaturated bond between C-4—C-5 to produce a fully saturated compound. Otherwise hydroxyls may be added at C-17, C-6, C-11, C-16 or C-21[10].

The main foetal metabolite of progesterone is the reduced form of the C-20 ketone, i.e. dihydroprogesterone. Two C-20 dihydroprogesterone isomers have been isolated, of which much more 20*a*- than 20*β*-dihydroprogesterone is produced. A curious interconversion of progesterone and 20*a*-dihydroprogesterone takes place in that the placenta readily converts the dihydroprogesterone back into progesterone. The placenta passes progesterone to the foetus, which converts it into 20*a*-dihydroprogesterone and passes this back to the placenta, which once more converts it to progesterone. The placenta is unable to do the same with 20*β*-dihydroprogesterone, which is not therefore involved in this closed circle. This interconversion appears to be a means by which progesterone is conserved in the foeto-placental unit. Some of the progesterone produced by the placenta is used in the foetus for the biogenesis of other steroids but the bulk ultimately passes out of the placenta to the maternal circulation. Judged in terms of its metabolism the progesterone produced in the placenta is directed at the mother, not the foetus. Whether this direction is in order to fulfil a rôle in parturition remains to be seen.

The main pathways of progesterone metabolism in the foeto-placental unit are shown in Figure 7.2.

7.3.3 Corticosteroid biosynthesis

Although there is evidence of the involvement of corticosteroids in the onset of labour in the sheep, very little is known about their role in human parturition and the placental metabolism of corticosteroids will not be examined in any detail.

Upon perfusion of the previable foetus with labelled progesterone radioactive cortisol and corticosterone can be isolated. Corticosterone, which does not have a 17α-hydroxyl group, is probably made directly from progesterone by hydroxylation at C-11 and C-21 in the foetal adrenal. Cortisol, which has a hydroxyl group at C-17, is made via 17α-hydroxyprogesterone.

An interesting facet of corticosteroid metabolism has recently come to light[11]. The foetus is able to convert cortisol, which is biologically active, to cortisone, which is biologically less active, but cannot do the reverse reaction. The placenta, on the other hand, is able to do the interconversion freely in both directions. The foetus is able further to protect itself against potent corticosteroids by reducing both cortisol and cortisone to their corresponding tetrahydro and hexahydro metabolites.

The placenta supplies to the foetal adrenal large amounts of corticosteroid substrate in the form of progesterone. There is no significant difference between the cord progesterone concentration of foetuses delivered by elective Caesarean section before the onset of labour and those delivered after labour, nor any other evidence to suggest that progesterone levels in the foetal circulation change consistently before labour. That being so, it is unlikely that progesterone supply, the only element of corticosteroid synthesis under placental control, is a rate-limiting step in corticosteroid biosynthesis. If corticosteroids play a part in the onset of labour this is a role under control of the foetal adrenal, not the placenta. This fits other evidence, such as the fact that premature labour can be precipitated in sheep by stimulating the foetal adrenal by the injection of ACTH into the foetal circulation, and that in the human when the foetal pituitary–adrenal axis is deficient, as in anencephaly, the onset of labour is delayed.

7.3.4 Oestrogen biosynthesis

Until 20 years or so ago it was accepted that the placenta was the source of the large amounts of oestrogen found in maternal urine during pregnancy. Oestrogens occur in high concentration in the placenta, and this organ will synthesise them from a variety of other steroids. In this sense, and in the sense that part of the biogenesis of each oestrogen molecule in the maternal organism has taken place in the placenta, it is still true that the placenta is the main source of oestrogen in pregnancy. This belief was tempered by the experimental work of Cassmer[7]. He found that when the foetal circulation was cut off *in vivo* the placental capacity to synthesise oestriol declined greatly, although it remained *in situ* and alive. At least as far as this particular oestrogen was concerned, the foetus was essential to its biogenesis in the foeto-placental unit.

If the foetus is involved in oestrogen biosynthesis, the question arises as to which foetal organs are responsible and in what manner. Much of the evidence points to the foetal adrenal. When fragments of placenta were transplanted into spayed mice, oestrogen was not produced as judged by vaginal smears. Nor did any oestrogenic activity result when portions of foetal adrenal were so transplanted. But when foetal adrenal and placental tissue were transplanted together, oestrous smears resulted[12]. Thus the combined action of foetal adrenals and the placenta is necessary for oestrogen synthesis. Later experiments have shown that other foetal tissues, notably the liver, are also involved in oestrogen biosynthesis.

The various routes by which oestrogens can be made in the foeto-placental unit are known, although the quantitative importance of particular pathways is open to dispute. It has long been known that there are high concentrations of dehydroepiandrosterone in the foetal circulation[13]. Incubation and perfusion experiments show that it is formed in the foetus, mainly in the adrenals. The foetus does not possess the desmolase for removing the 2-carbon side chain of C-21 steroids like progesterone with a Δ^4-3-ketone structure. When, however, pregnenolone or 17α-hydroxypregnenolone having a Δ^5-3β-hydroxy structure are used as precursors, the foetal adrenal readily turns them into dehydroepiandrosterone.

The concentration of dehydroepiandrosterone in the umbilical artery plasma is about 30 μg/100 ml higher than in the umbilical vein and it is probable that this difference represents abstraction of dehydroepiandrosterone for oestrogen synthesis by the placenta. The steps between dehydroepiandrosterone and oestrogen—hydroxylation at C-19 and subsequent aromatisation of ring A—are carried out very rapidly and efficiently by the placenta.

There are also high concentrations of dehydroepiandrosterone in the maternal circulation and the possibility arises that the placenta is also able to utilise these for oestrogen synthesis. Both in mothers bearing an anencephalic foetus where foetal adrenal steroid biosynthesis is much reduced, and in hydatidiform mole, where no foetus is present at all, oestrogen production, although lowered, still occurs. Under pathological circumstances at least, it is possible for the placenta to produce oestrogens from maternal dehydroepiandrosterone although probably *in vivo* the foetal adrenal is the main source of this oestrogen precursor.

So far oestrogen biosynthesis in general has been considered. More than 20 different oestrogens have been isolated from the urine of pregnant women and it is not feasible here to examine each compound individually. Some may be involved in the onset of labour and it is of interest to consider the biogenesis of these particular oestrogens briefly. The first of these is oestriol, an oestrogen characterised by having a hydroxyl group attached to C-16. The enzyme system capable of inserting a hydroxyl group at C-16 exists in the foetus but not in the placenta. Hydroxylation at C-16 is therefore a step peculiar to the foetus, taking place mainly but not exclusively in the foetal liver. The human foetus has a very powerful capacity for this step and the concentration of oestriol, relative to other oestrogens in the foetus is very high. The reason for the biosynthesis of large amounts of oestriol by the foetus is obscure; indeed within the foetus it occurs largely as an inert conjugate, oestriol-3-sulphate. Moreover it is peculiar to human, or at least primate, pregnancy. Why the human foetus should 16-hydroxylate a large proportion of the oestrogen it makes is mysterious as all other mammals appear to be able to carry through pregnancy without making this particular steroid.

The foetus has the capacity to 16-hydroxylate steroids much earlier in the biosynthetic sequence than the final oestrogen molecule. It can, for instance, insert a C-16 hydroxyl in such early C-21 molecules as pregnenolone or in

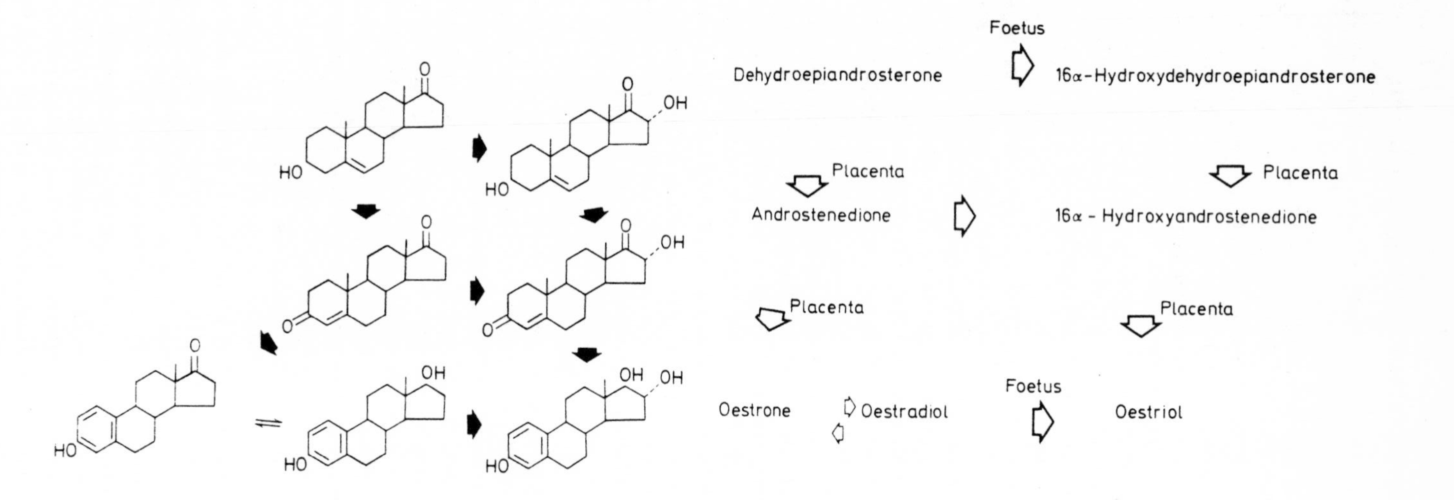

Figure 7.3 The biosynthetic pathways of oestrogens in the foeto-placental unit

later C-19 compounds such as dehydroepiandrosterone. Wherever in the biosynthetic sequence the 16-hydroxylation is done there is a strong possibility that this is a rate-limiting step in the formation of oestriol. Whether this oestrogen has any particular biological significance in human pregnancy is a matter of controversy. It has been argued that 16-hydroxylation is merely a mechanism for the disposal of potent oestrogens within the foeto-placental unit—a kind of metabolic garbage. If so, it is odd that this step should be built in so early in the biosynthetic sequence that molecules should be earmarked for disposal long before the endproduct is reached. On the other hand it must be owned that all efforts to demonstrate a role in human pregnancy peculiar to this particular oestrogen, especially in relation to the onset of pregnancy, have been fruitless. The lesson, perhaps, is that teleology is a poor basis for scientific hypothesis.

The other two major oestrogens, oestrone and oestradiol-17β, are much more widespread. Although other oestrogens, e.g. oestradiol-17α, may occur in relatively high concentration in some species, oestrone or oestradiol-17β is the major oestrogen in most mammalian pregnancies. The placenta features prominently in their biosynthesis, being able rapidly to convert dehydroepiandrosterone to oestrone. In molar pregnancy, where there is no foetal source of precursor dehydroepiandrosterone at all, oestrone production is much the same as in normal pregnancy[14]. It is possible therefore that a substantial proportion of the precursor for placental synthesis of oestrone and oestradiol comes from the maternal circulation.

Some of the oestriol in the urine of pregnant women comes from the foeto-placental unit, where it is made by a route not involving either oestrone or oestradiol. But some of the oestriol in urine is also the result of metabolism of oestradiol in the maternal organism. For the last 20 years assays of oestriol have been done almost entirely on urine, a fluid in which the oestriol content is greatly in excess over all other oestrogens. Not surprisingly therefore attention has focused on the biosynthesis and possible biological role of oestriol. In plasma, on the other hand, when only the presumably functionally active unconjugated compounds are considered, oestradiol is present in higher concentration than oestriol[8]. If oestradiol rather than oestriol is the functionally active oestrogen in human pregnancy, more stress is laid on the placental role. It has already been argued that precursor supply to the point of pregnenolone is unlikely to be a rate-limiting step in the foeto-placental synthesis of oestrogens. It is possible that an increase or reduction in the rate of the next step—removal of 2 carbon atoms to go from a C-21 to a C-19 compound—could increase or decrease the production of oestrogen. But this step can proceed by two routes, via progesterone in the placenta or via pregnenolone in the foetal adrenal, so it is possible that a change in one pathway could be compensated for by a corresponding change in the other. Two other steps likely to be rate-limiting remain. One, 16-hydroxylation, is entirely under foetal control, but leads only to oestriol. The other, aromatisation of ring A, is entirely under placental control, and leads to both oestradiol and oestriol. If foeto-placental oestrogen production is involved in the onset of labour the foetus is likely to exercise its effect through oestriol but the placenta can do so through either oestradiol or oestriol.

The main pathways of oestrogen biosynthesis are shown in Figure 7.3.

7.3.5 The protein hormones of the foeto-placental unit

The two protein hormones of the foeto-placental unit, human placental lactogen and chorionic gonadotrophin, are produced entirely in the placenta and appear to be directed mainly at the maternal organism. They are not therefore involved in the unitary concept of foetus and placenta. It is not proposed to consider the biogenesis of either hormone in any detail but the remote connections of both with parturition will be briefly explored.

7.3.5.1 Placental lactogen

This hormone (a brave but brief attempt to enforce the use of the term human chorionic somatomammotrophin has been defeated by the limitations of the spoken word) occurs in increasing concentration in plasma during the course of pregnancy. The levels reach a plateau of 5.6 ± 0.6 (SD) $\mu g\ ml^{-1}$ at 35 weeks gestation but do not fall after 40 weeks gestation[15]. There is no suggestion that any consistent changes in placental lactogen levels occur in association with the onset of labour although it has been found that they tend to rise when labour is fully established[16]. It is supposed that this is a mechanical effect resulting from increased uterine blood flow rather than a reflection of a physiological activity. Although the physiological activity of placental lactogen is directed at the mother, not the foetus, its role is with carbohydrate metabolism, not uterine contractility.

7.3.5.2 Chorionic gonadotrophin

The pattern of changing levels of chorionic gonadotrophin in plasma during pregnancy is quite unlike that of placental lactogen and it is likely that its biogenesis is unrelated and controlled by different factors. Serum gonadotropin reaches its peak concentration early in pregnancy, at 60–80 days. A small secondary peak can be discerned in late pregnancy[17]. It is seldom remarked upon but the timing is such as to suggest a distant possibility of an association with the onset of labour.

The level of chorionic gonadotrophin is higher in the umbilical vein than in the arteries[18]. It is possible that this foetal extraction of chorionic gonadotrophin reflects the fact that it is performing some function in the foetus, and various speculations concerning the possible effect of chorionic gonadotrophin on oestrogen metabolism have been brought forward. At first it was supposed that an increase in chorionic gonadotrophin would bring about a decline in oestrogen production[19], but later work raised an opposite possibility[20]. It was shown that chorionic gonadotrophin, when given to the newborn, stimulated the production of dehydroepiandrosterone and hence the supply of precursor for oestrogen synthesis. Not only is the foetal role in oestrogen synthesis affected by chorionic gonadotrophin, but it has also been shown to affect the placental activity in this respect by increasing the aromatisation of neutral steroids[21]. Curiously, although chorionic gonadotrophin clearly increases ovarian progesterone synthesis, intravenous infusion of the hormone in late pregnancy does not cause any rise in placental progesterone production[22].

Although it is by no means established that chorionic gonadotrophin levels affect oestrogen synthesis or that the increased oestrogen levels of late pregnancy have anything to do with the onset of labour, the time sequence of a late pregnancy rise in chorionic gonadotrophin, its effect on dehydroepiandrosterone synthesis, the rise in oestriol in the last weeks of pregnancy and the ultimate onset of labour is intriguing. It is a speculation worthy of experimental examination.

7.4 GENERAL CONSIDERATIONS CONCERNING THE ONSET OF LABOUR

The onset of labour is an event influenced by many factors. It is false reasoning to suppose that endocrine changes are the only, or indeed necessarily the dominating, factor in the process. There is, however, reason to suppose that they play some part in the onset of every labour. Before going on to consider what might be the role of particular hormones it is as well to review the nature of the processes in connection with labour which might be subject to endocrine control.

7.4.1 The control of myometrial contraction

The uterus, like other muscles, owes its contractile properties to two sets of protein filaments within the individual cell. The synthesis of both, actin and myosin, is under endocrine control. During pregnancy the human uterus increases from 75 g to a weight at term of 880 g. Both hyperplasia and hypertrophy are involved in this increase. Part of the hypertrophy is caused by an intracellular increase in actomyosin. This accumulation of contractile protein appears to be caused by oestrogen, for, in animals, oöphorectomy decreases the actomyosin content, and treatment of such a castrated animal with oestrogens restores the uterine actomyosin[23].

The energy for uterine contraction is derived in the first instance from the energy-rich phosphate bonds of adenosine triphosphate (ATP). In the non-pregnant state uterine muscle contains 0.6 μmol ATP and 1.2 μmol creatine phosphate (CP)/g of wet weight. In the term uterus the concentration of these energy sources is about doubled. The second source of uterine energy is derived by the Lohmann reaction involving the transformation of adenosine diphosphate (ADP) into the triphosphate, thus: $CP + ADP \rightleftharpoons creatine + ATP$. Creatine itself has no effect on actomyosin and is therefore not available as a primary energy source. The available reserves of ATP and of CP in the uterine muscle at term can supply energy for about 50 contractions. Thereafter further ATP has to be synthesised by the glycolysis of muscle glycogen to lactic acid.

A major determinant of muscle contraction is the membrane potential of the myometrial cell. This is to a large extent determined by the intracellular concentration of potassium. Outward diffusion of K^+ is balanced by inward movement of Na^+ and Cl^-. Membrane potential is therefore determined by the permeability of the membrane to these electrolytes and by their concentrations in the cell environment, a phenomenon under endocrine control[24].

Endocrine regulatory influences on myometrial contractility can be exercised at a number of points—on the contractile machinery, on the energy supply, on the membrane function, on the ionic environment and on the propagation of the contractile impulse[25].

Myometrial activity does not start suddenly with the onset of labour. The human uterus is quiescent until 20 weeks gestation. Then spontaneous activity starts and increases markedly during the last 8–10 weeks of pregnancy. At this time a second process, effacement and dilatation of the cervix, also starts. Although it may at first glance appear that this is merely the result of myometrial activity, the two processes may, to some extent, be independent, for they do not always occur in step. Taken together, the rate of increase in uterine contraction and of cervical dilatation during the last 8 weeks determine the time of onset of labour. Women with quiescent uteri and closed cervices go beyond 40 weeks before the commencement of labour, those with active uteri and open cervices tend to deliver early[26]. Csapo, in examining the interaction between mechanical factors such as the stretching of uterine muscle caused by the increasing volume of the uterine content, and endocrine factors, also concluded that critical changes began in the last 6 weeks of pregnancy[27]. There is much to support the view that the genesis of the endocrine changes which influence the onset of labour is to be sought a month or two before the onset of the event.

7.4.2 The role of the cervix in the onset of labour

Three kinds of changes in the cervix have some bearing on the onset of labour. They are the changes which take place in the composition of the cervix, in its mechanical properties and in its microscopic architecture. Only 8% of the cervix is made up of muscle fibres. The rest, for the most part, consists of collagen fibres set in a matrix of mucopolysaccharide ground substance. During pregnancy the collagen content of the cervix declines. It is unlikely that this of itself can account for the functional changes in the cervix which precede labour. More likely the basis of these functional changes lies in the mucopolysaccharides and in the physico-chemical relationship of collagen fibrils to ground substance. Some inkling of the great changes in the cervix wrought by pregnancy can be got by considering the difference between surgical dilatation of the cervix in the non-pregnant state and cervical dilatation in labour. In the first instance, the cervix can only be forced open with a thin steel rod, often tearing its fibres, while in the second case it stretches easily over the head of a baby.

In the rat the changes in the chemical composition of the cervix go *pari passu* with changes in its physical properties. During pregnancy there is a great increase in the distensibility of the cervix. It is likely that this change is related to the ease with which collagen fibrils can slip upon one another, as though there is a change in the molecular links between one fibril and another and in the viscosity of the ground substance binding them together[28]. At the same time, the water content of the cervix increases so that the collagen fibres are spread further apart, and the collagen concentration in terms of wet weight actually decreases while the total collagen content increases[28]. Although the rat cervix shows these changes to some degree in early pregnancy, the dramatic alterations take place in the 24 h before labour. No comparable data on cervical changes in women are available but it is a reasonable guess that in the human the time scale of these events is different, and the rapid changes of the last 24 h in the rat are spread over a month or 6 weeks in the human. Some information about the changes in the microscopic build of the human cervix during pregnancy is available. In the non-pregnant woman the collagen is a dense tightly woven felt. During pregnancy there is some dissociation of collagen bundles into fibrillar constituents[29]. There is a loosening of the collagen with an increase of interfibrillar ground substance. The ground substance, which increases greatly in mass, does so mainly by hydration and so influences the solubility of the collagen.

Studies on the part which placental hormones play in bringing about these changes are as yet in their infancy. A few scattered findings show interesting pointers along the way. It has been shown that in rats, oestrogen affects collagen metabolism, stimulating both synthesis and breakdown, particularly in the uterus[30]. It may be surmised that this results from a stimulatory effect on collagenase, the enzyme which hydrolyses collagen. Under the influence of oestrogen large amounts of short-lived acid mucopolysaccharides are produced in the uterus[31]. Progesterone blocks this oestrogen-induced production of mucopolysaccharides. It must be owned, however, that in these experiments the chief effect of oestrogen was on the ground substance of the endometrium, the mucopolysaccharides of the cervix being relatively resistant to oestrogen.

7.4.3 The foetal role in the onset of labour

The concept that labour starts as the result of a biological signal arising from the foetus has gained ground fast in recent years. As first clearly enunciated, it was mainly based on experimental findings in sheep[32], and this animal has continued to be the most convenient model to study. It was found that if the foetal lamb *in utero* had its pituitary destroyed or adrenals removed, the ewe did not go into labour at term. This suggested that labour was initiated by the activation of the pituitary–adrenal axis, a hypothesis much strengthened by the demonstration that infusion of cortisol or corticotrophin into the foetal lamb would lead to premature labour[33]. It remained to show that under normal circumstances the onset of labour in the ewe is preceded by a rise in foetal corticosteroids. This was demonstrated soon after[34]. Indeed the mechanism whereby the pituitary of the foetal lamb[35] and goat[36] initiates labour has been worked out in convincing detail.

The deductions from the findings on sheep cannot be applied without alteration to the onset of labour in the human. Of course the experimental situations cannot be repeated, but some pathological conditions have a bearing on foetal pituitary–adrenal function in the human foetus. Thus, in the anencephalic foetus, adrenal hypoplasia occurs and this malformation, if hydramnios does not develop, is associated with prolonged gestation. In general, the smaller the size of the adrenals, the greater the prolongation of the pregnancy beyond term[37]. Although it is sometimes very difficult to induce labour, particularly if the adrenal hypoplasia is severe, such patients do go into spontaneous labour eventually and, when mechanical factors such as uterine distension are dominant, may go into premature labour. Even in other primates, the evidence that the foetal adrenal gives the endocrine signal for the onset of labour is not clear. When foetal adrenalectomy was performed in 8 monkeys, 6 were delivered at or near term and only 2 went postmature[38]. Turnbull's[26] conclusion that 'whatever mechanisms control parturition, they are likely to be operative from within the uterus' is probably true but it is unlikely that the human foetal adrenal operates in the same manner as that of the sheep.

7.5 THE ROLE OF PLACENTAL HORMONES IN THE ONSET OF LABOUR

7.5.1 Progesterone

For more than 60 years the activity of progesterone has been known both in terms of secretory change in the endometrium and as an uncertain ingredient in the mix essential for the maintenance of pregnancy. First it got a local habitation and a name. Halban did as much as any man for the first when he wrote, 'Die aktiven Schwangerschaftssabstanzen sind ein Effekt der Placenta'[39] and Butenandt gave it a name in the sense of a chemical formula. Its putative role in pregnancy and the onset of labour will in our times always be associated with the name of Arpad Csapo. He built the scattered data about the nature of myometrial contraction and the physiological activity of progesterone into a coherent theory[40]. The central tenet of this theory was that progesterone was the agent which throughout gestation restrained myometrial contraction. As a corollary it followed that the onset of labour was due to the diminution of the influence of progesterone. This thesis was supported by many shrewd experimental observations on the physiological actions of progesterone—that its action on the energy potential or the ionic environment of the muscle cell and above all on the propagation of the contractile impulse was such as to dampen down uterine activity. In many mammalian species a simplistic formulation of this theory will still explain experimental observations with regard to the onset of labour. Thus in rabbits the administration of progesterone will postpone the onset of labour, while the most recent observations on plasma progesterone using sophisticated protein binding assay techniques have confirmed that in sheep[41] and goats[42] the onset of labour is preceded by a marked fall in progesterone levels.

When this theory is applied to the onset of labour in the human, discrepancies appear. Although the Csapo school have continued to maintain that a fall in the concentration of progesterone in the maternal plasma is associated with the onset of labour[43], other recent work has failed to demonstrate a consistent fall in maternal progesterone levels[8,44]. A consensus of opinion against a general decline in peripheral plasma progesterone levels has led Csapo to emphasise the importance of a localised fall in progesterone in the retroplacental blood and adjacent uterine wall. There is indeed substantial support for the view that as pregnancy progresses the myometrial concentration declines relative to the peripheral plasma level. In the uterus itself, progesterone is present in highest concentration near the placenta in early pregnancy, but in later pregnancy becomes much more homogeneously distributed throughout the uterus[45]. These findings cannot be construed as establishing the case for a localised fall in progesterone concentration as a major factor in the onset of labour. The hypothesis of progesterone withdrawal as a basis for the onset of labour in the human is beset with difficulties. It is probably not a central event but may well be one link in a chain of cause and effect.

It is improbable that the onset of labour is preceded by a fall in the progesterone concentration in the maternal circulation. It has not proved possible to establish clearly whether or not there is a decline of progesterone concentration in the intervillous space or in the myometrium overlying the placenta. It should not be forgotten that there is convincing evidence that the signal for the onset of labour arises from the foetus. If so, presumably the key changes take place in the foetal circulation. For obvious reasons experimental observations are difficult here but some information might be gained by comparing the plasma progesterone of foetuses whose mothers were in labour at Caesarean section with that of those in whom the mother was not in labour. Assays done in labour are, of course, open to many objections, not least that the alterations of uterine blood flow induced by labour can greatly change local progesterone concentration. Although a withdrawal of progesterone block can no longer be accepted as the primary change leading to labour, there is still reason to suppose that it is one element in the complex of factors associated with the onset of labour. It may be that the importance of declining progesterone levels lies not so much in the direct effect this may have on the contraction of muscle cells but on oxytocic agents, notably the sensitivity of the myometrium to pitocin or on the release of prostaglandin from decidua.

7.5.2 The role of oestrogens in the onset of labour

To date 22 different oestrogens have been identified in urine from pregnant women. It is very likely that each of these different types of molecule are also present in the circulation and may exert an effect on the uterus. If a role in the onset of labour is unique to one particular molecular species of oestrogen, it could be that the right oestrogen has never been examined, for, among the natural oestrogens, oestriol and oestradiol-17β are the only compounds which have been considered experimentally. There is some justification for the concentration of attention on these two oestrogens. Oestradiol-17β is the most potent natural oestrogen in terms of uterotrophic effect, and receptor proteins in the uterus which bind this oestrogen and not others have been described[46]. On the other hand oestriol is characteristic of human pregnancy, being produced in much larger quantity by women than by other mammals. It is also a form of oestrogen for the biosynthesis of which the foetus is essential, as the placenta does not possess the necessary 16-hydroxylase. Perhaps too much stress had been laid on the high levels of oestriol as compared with other oestrogens during pregnancy. This impression is derived largely from urinary assays where it is true that the oestriol content is greatly in excess of that of any other oestrogen. But in plasma, and therefore presumably near the point of oestrogen action, the concentration of unconjugated oestradiol-17β exceeds that of unconjugated oestriol[8], or approaches it closely[47].

There are two ways by which the putative role of oestrogens in labour might be tested. One is to examine the changes of oestrogen level which take place before and during spontaneous labour, and the other is to see what effect changing the levels by administration of oestrogens has on labour. Both these approaches have been explored.

7.5.2.1 Changes in oestrogen level associated with the onset of labour

The urinary output of oestriol rises very steeply during the last few weeks of pregnancy. It has been surmised that this represents a new element in oestriol biosynthesis, specifically foetal in origin[48]. There is some evidence that, as early as 36 weeks, the urinary oestriol output gives an indication of when a woman is likely to go into labour.

Urinary oestriol excretion is an unsatisfactory way of examining oestrogen changes at the myometrial level. It is an endpoint metabolically distant from the site of action. The oestriol output from day to day in the same normal subject is very variable and subtle changes in blood levels may easily be obscured. Urine collections over short periods of time, which might be presumed to demonstrate rapid changes in oestrogen production, give such variable results as to be useless. Not surprisingly, no firm conclusions can be drawn from the urinary findings. Changes in oestrogen level associated with the onset of labour have to be examined afresh now that suitable plasma assay techniques are available. In animals such as goats and sheep a very sharp rise of oestrogen concentration in the maternal circulation takes place in the last day or two before the onset of labour[35,36,41,42]. When assays are done on the foetal circulation, these changes are much more marked. It is probable that the primary change in oestrogen level is in the foetal circulation and that assays on maternal plasma can, at best, be said to view the situation, 'as through a glass, darkly'. Nevertheless, maternal levels are the only available data for the human, and need to be critically considered. The early studies on plasma oestrogens were beset with methodological difficulties. The introduction of radioimmunoassay techniques has made the results more reliable but has not changed the conclusions drawn from earlier work. The increased sensitivity of radioimmunoassay methods has also made it possible to examine changes in unconjugated oestrogens; presumably by the biologically active oestrogen component. It was found that in human pregnancy neither unconjugated oestradiol-17β nor unconjugated oestriol showed any sudden rise in concentration before the onset of labour[8,47,49] similar to that found for other oestrogens in sheep and goats. Nor, when labour was artificially initiated, either by rupture of the membranes at term, or by infusion of prostaglandins to induce abortion, did the oestrogen levels give any indication of the likelihood of the uterus to respond to the stimulus, or show any consistent changes associated with the onset of labour[50,51]. It cannot be concluded from these experiments that changes in plasma oestrogen play no part in the onset of labour in human pregnancy. The oestrogen level in peripheral venous blood is determined by outflow factors such as metabolism, conjugation, transfer to extra plasma compartments and renal excretion, as well as by inflow from the foeto-placental unit. In the uterine veins the dominant factors determining oestrogen concentration are inflow from the foeto-placental unit and uterine blood flow. It is possible, therefore, that measurements made on uterine blood may give a more accurate representation of any changes of oestrogen level in the foetal circulation which may precede labour.

7.5.2.2 The effect of oestrogen administration on labour

Ever since oestrogens have been available for clinical use, attempts have been made to influence the onset of labour by the administration of oestrogen. The early experiments were poorly designed and lacked controls, so that meaningful experiments really date from the work of Pinto in the early 1960s[52]. He used a double blind design, giving an intravenous infusion of either oestradiol-17β or the solvent vehicle to equal numbers of women at term. He claimed that those given the oestrogen tended to go into spontaneous labour sooner, and to respond more readily to intravenous oxytocin. Other work from this group[53,54] has emphasised the fact that the action of oestradiol is not only directly upon myometrial contractility, but also in causing changes in the cervix and in creating an enhanced sensitivity to the oxytocic effect of pitocin. Since then opinion has been divided between those who found that oestrogen administration would facilitate the onset of labour[55] and those who found that it had no effect[56,57].

The Aberdeen school have made some observations on the action of oestriol on the pregnant uterus which are relevant to the present discussion. Having failed to find any effect with either oestriol or stilboestrol[57], they reasoned that the oestrogen might be metabolised and inactivated before reaching the myometrium, and that intrauterine administration might approach the physiological situation more closely. They accordingly tried the effect of the intra-amniotic administration of oestriol sulphate, using a double blind experimental design[58]. At first it appeared that patients pretreated with oestriol in this way did tend to go into labour sooner and to deliver in a shorter time than the controls when a stimulus for the onset of labour (rupture of the membranes) was applied. Subsequent work, however, failed to show any effect of oestriol on spontaneous uterine activity in midpregnancy or at term or in the response to a stimulus for the onset of labour[59,60]. It must be acknowledged that, at least as far as oestriol is concerned, this compound does not affect myometrial contractility in the short term, even when given by the route and in the form in which it normally makes its way to the myometrium.

The negative results of experiments designed to influence the onset of labour by the administration of oestrogen can be criticised in three respects. Firstly, they have taken into consideration only the effect on myometrial contractility, whereas the main effect of oestrogen may be on the cervix.

Secondly, they have been short-term experiments lasting 2 or 3 days at most, whereas the action of oestrogen may take longer to have an effect on the uterus. Thirdly, the studies of the Aberdeen school have involved only oestriol and it is possible that in the natural situation more potent oestrogens, such as oestradiol-17β, are concerned. The main criticism that can be levelled against those studies which have shown a positive effect of oestrogens on the onset of labour is concerned with the selection of cases. Many factors, other than oestrogens, bear upon the onset of labour. Unless a blind design and rigorous standardisation of both control and treatment groups is used, erroneous conclusions may be drawn. It is essential that the patients studied should be homogeneous as regards parity, age, maternal and foetal size and cervical state.

It is not possible at this stage to give a final verdict on the role of oestrogens in the onset of labour. There are too many findings which point to some involvement of oestrogens to be ignored. On the other hand, a rise of oestrogen level is not always a dominant or indeed an essential feature of the onset of labour. Possibly the part that oestrogen has to play is that of one link in a chain of endocrine events which may lead to labour. Probably other routes can by-pass the oestrogen control point and possibly the whole endocrine series of events play only a permissive role.

7.6 A THEORY ON THE INTEGRATED ACTION OF HORMONES IN THE ONSET OF LABOUR

This review of endocrine factors in labour has taken no account of the role played by such hormones as pitocin and prostaglandins because they are not placental hormones. There are, however, grounds for believing that they are involved in labour and any theory of integrated endocrine action has to take them into account. Such a theory, based on the work of Liggins[35] and of Thorburn[42], can now be put forward. The primary event in the endocrine role is probably an activation of the foetal pituitary–adrenal axis resulting in a rise of foetal corticosteroid secretion. This leads to an increased oestrogen production by the foeto-placental unit. Oestrogens cause a release of prostaglandins from the uterine decidua and an increased sensitivity of the myometrium. Progesterone impedes prostaglandin release and raises the muscle threshold to oxytocin. Prostaglandins in turn lower the threshold to oxytocin so that uterine contractions result at a level of pituitary secretion which previously had no effect. Prostaglandins may also affect the uterine muscle directly. In species where the progesterone supply in pregnancy is derived from the ovary, prostaglandins cause a fall in progesterone production and it is possible that they may also do so in those species where progesterone is produced by the placenta.

This outline is a general hypothesis about the endocrine events associated with the onset of labour. It has to be modified for particular species, notably *homo sapiens*. Its weakness is that it depends on a time sequence of events which have not been clearly demonstrated for any species: events which have their genesis in the foetus and cannot be examined in human pregnancy. Also this hypothesis depends on four phenomena which have not been convincingly shown in the human—a rise in foetal corticosterone levels, an increased production of oestrogen and a decrease in progesterone production by the foeto-placental unit and the release of prostaglandins by the endometrium. Although a sequence of acute events along the lines suggested may occur in the last few days of pregnancy, it is probable that the origin of human labour is to be found further back in pregnancy, when, at 34–36 weeks, oestrogens rise sharply, progesterone concentration levels off, the cervix starts to take up and dilate and the myometrium becomes active.

References

1. Diczfalusy, E. (1962). Endocrinology of the foetus. *Acta Obstet. Gynecol. Scand.*, **41,** Suppl. 1, 45
2. Telegdy, G., Weeks, J. W., Lerner, U., Stakemann, G. and Diczfalusy, E. (1970). Acetate and cholesterol metabolism in the human foeto-placental unit at midgestation. *Acta Endocrinol.*, **63,** 91
3. Telegdy, G., Robins, M. and Diczfalusy, E. (1972). A study of sterol and steroid synthesis from sodium acetate by human foetal liver preparations. *J. Steroid Biochem.*, **3,** 693
4. Diczfalusy, E. (1968). *The Foeto-placental Unit*, 65 (A. Pecile and C. Finzi, editors) (Amsterdam: Excerpta Medica Foundation)
5. Kim, M. H., Borth, R., McCleary, P. H., Woolever, C. A. and Young, P. C. (1971). Sex hormone secretion of the placenta left in situ after ovarian pregnancy. *Amer. J. Obstet. Gynecol.*, **110,** 658
6. Jaffe, R. B. and Peterson, E. P. (1966). *In vivo* steroid biogenesis and metabolism in the human term placenta. *Steroids*, **8,** 695
7. Cassmer, O. (1959). Hormone production of the isolated human placenta. *Acta Endocrinol.*, Suppl. 45
8. Tulchinsky, D., Hobel, C. J., Yeager, B. S. and Marshall, J. R. (1972). Plasma estrone, estradiol, progesterone and 17-hydroxyprogesterone in human pregnancy. *Amer. J. Obstet. Gynecol.*, **112,** 1095
9. Scommegna, A., Bard, L. and Bieniarz, J. (1972). Progesterone and pregnenolone sulfate in pregnancy plasma. *Amer. J. Obstet. Gynecol.*, **113,** 60
10. YoungLai, E. V. and Solomon, S. (1969). *Foetus and Placenta*, 249 (A. Klopper and E. Diczfalusy, editors) (Oxford: Blackwells)
11. Pasqualini, J. R., Nguyen, E. L., Uhrich, F., Wiqvist, N. and Diczfalusy, E. (1970). Cortisol and cortisone metabolism in the human foeto-placental unit at midgestation. *J. Steroid Biochem.*, **1,** 209
12. Frandsen, V. A. and Stakemann, G. (1963). The site of production of oestrogenic hormones in pregnancy. *Acta Endocrinol.*, **43,** 184
13. Simmer, H. H., Easterling, W. E., Pion, R. J. and Dignam, W. J. (1964). Neutral C19-steroids and steroid sulfates in human pregnancy. *Steroids*, **4,** 125
14. Siiteri, P. K. and MacDonald, P. C. (1966). Placental estrogen biosynthesis during human pregnancy. *J. Clin. Endocrinol.*, **26,** 751
15. Letchworth, A. T., Boardman, R. J., Bristow, C., Landon, J. and Chard, T. (1971). A rapid semi-automated method for the measurement of human chorionic somatomammotropin. The normal range in the third trimester and its relation to foetal weight. *J. Obstet. Gynaec. Brit. Cwlth.*, **78,** 542
16. Pavlou, C., Chard, T. and Letchworth, A. T. (1972). Circulating levels of human chorionic somatomammotrophin in late pregnancy. Disappearance from the circulation after delivery, variation during labour and circadian variation. *J. Obstet. Gynaec. Brit. Cwlth.*, **79,** 629
17. Brody, S. (1969). *Foetus and Placenta*, 365 (A. Klopper and E. Diczfalusy, editors) (Oxford: Blackwells)
18. Lauritzen, Ch. and Lehmann, W. D. (1967). Levels of chorionic gonadotrophin in the newborn infant and their relation to adrenal dehydroepiandrosterone. *J. Endocrinol.*, **39,** 173

19. Smith, G. V. and Smith, O. W. (1948). Internal secretions and toxaemia of late pregnancy. *Physiol. Rev.*, **28,** 1
20. Lauritzen, C., Shackleton, C. H. and Mitchell, F. L. (1969). The effect of exogenous human chorionic gonadotrophin on steroid excretion in the newborn. *Acta Endocrinol.*, **61,** 83
21. Cedard, L., Varangot, J. and Yannotti, S. (1964). Influence des gonadotrophines chroniques sur le metabolisme des steroides dans les placentas humains perfuses in vitro. *C.R. Acad. Sci. (Paris)*, **258,** 3769
22. Runnebaum, B., Holzmann, K., Bierwirth, v. Münstermann and Zander, J. (1972). Effect of HCG on plasma progesterone during the luteal phase of the menstrual cycle and during pregnancy. *Acta Endocrinol.*, **69,** 739
23. Csapo, A. (1948). Actomyosin content of the uterus. *Nature (London)*, **162,** 218
24. Daniel, E. and Daniel, B. (1957). Effects of ovarian hormones on the content and distribution of cation in intact and extracted rabbit and cat uterus. *Canad. J. Biochem.*, **35,** 1205
25. Pulkkinen, M. O. (1970). Regulation of uterine contractility. *Acta Obstet. Gynecol. Scand.*, **49,** Suppl. 1
26. Turnbull, A. C. (1971). Myometrial contractility in pregnancy and its regulation. *Proc. Roy. Soc. Med.*, **64,** 1015
27. Csapo, A. (1970). The diagnostic significance of the intrauterine pressure. *Obstet. Gynecol. Survey*, **25,** 515
28. Harkness, M. L. and Harkness, R. D. (1959). Changes in the physical properties of the uterine cervix of the rat during pregnancy. *J. Physiol.*, **148,** 524
29. Buckingham, J. C., Seldon, R. and Danforth, D. N. (1962). Connective tissue changes in the cervix during pregnancy and labor. *Ann. N.Y. Acad. Sci.*, **97,** 733
30. Kao, Kung-Ying Tang, Hitt, W. E. and McGavack, T. H. (1965). The effect of oestradiol benzoate upon collagen synthesis by sponge biopsy connective tissue. *Proc. Soc. Exp. Biol. Med.*, **119,** 364
31. Zachariae, F. (1958). Autoradiographic and histochemical studies of sulphomucopolysaccharides in the rabbit uterus oviducts and vagina; variations under hormonal influence. *Acta Endocrinol.*, **29,** 118
32. Liggins, G. C., Kennedy, P. C. and Holm, L. W. (1967). Failure of initiation of parturition after electrocoagulation of the pituitary of the foetal lamb. *Amer. J. Obstet. Gynecol.*, **98,** 1080
33. Liggins, G. C. (1968). Premature parturition after infusion of corticotrophin or cortisol into foetal lambs. *J. Endocrinol.*, **42,** 323
34. Bassett, J. M. and Thorburn, G. D. (1969). Foetal plasma corticosteorids and the initiation of parturition in the sheep. *J. Endocrinol.*, **44,** 285
35. Liggins, G. C. (1973). *Endocrine Factors in Labour* (A. Klopper and J. Gardner, editors) (Bristol: Cambridge University Press)
36. Currie, W. B., Wong, M. S. F., Cox, R. I. and Thorburn, G. D. (1973). *Endocrine Factors in Labour* (A. Klopper and J. Gardner, editors) (Bristol: Cambridge University Press)
37. Anderson, A. M., Laurence, K. M. and Turnbull, A. C. (1969). The relationship in anencephaly between the size of the adrenal cortex and the length of gestation. *J. Obstet. Gynaecol. Brit. Cwlth.*, **76,** 196
38. Mueller-Heubach, E., Myers, R. E. and Adamsons, K. (1972). Effects of adrenalectomy on pregnancy length in the rhesus monkey. *Amer. J. Obstet. Gynecol.*, **112,** 221
39. Halban, J. (1905). *Arch. Gynak.*, **12,** 496. Cited in *Oestrogenen beim Menschen*, Lauritzen Ch. and Diczfalusy, E., 499 (Berlin: Springer-Verlag)
40. Csapo, A. (1954). *Modern Trends in Obstetrics and Gynaecology*, 20 (London: Butterworths)
41. Bedford, C. A., Challis, J. R., Harrison, F. A. and Heap, R. B. (1972). The role of oestrogens and progesterone in the onset of parturition in various species. *J. Reprod. Fert.*, Suppl. 16, 1
42. Thorburn, G. D., Nicol, D. H., Bassett, J. M., Shutt, D. A. and Cox, R. I. (1972). Parturition in the goat and sheep. *J. Reprod. Fert.*, Suppl. 16, 61
43. Csapo, A., Knobil, E., van der Molen, H. J. and Wiest, W. G. (1971). Peripheral plasma progesterone levels during human pregnancy and labour. *Amer. J. Obstet. Gynecol.*, **110,** 630

44. Johansson, E. D. and Jonasson, L. E. (1971). Progesterone levels in amniotic fluid and plasma from women. *Acta Obstet. Gynecol., Scand.*, **50,** 339
45. Runnebaum, B. and Zander, J. (1971). Progesterone and 20α-dihydroprogesterone in human myometrium during pregnancy. *Acta Endocrinol.*, **66,** Suppl. 150
46. Gorski, J., Toft, D., Shyamala, G., Smith, D. and Notides, A. (1968). Hormone receptors: Studies on the interaction of oestrogen with the uterus. *Rec. Prog. Horm. Res.*, **24,** 45
47. Shaaban, M. and Klopper, A. (1973). Changes in unconjugated oestrogens and progesterone concentration in plasma at the approach of labour. *J. Obstet. Gynaec. Brit. Cwlth.*, **80,** 210
48. Klopper, A. and Billewicz, W. (1963). Urinary excretion of oestriol and pregnanediol during normal pregnancy. *J. Obstet. Gynaec. Brit. Cwlth.*, **70,** 1024
49. Masson, G. M. and Klopper, A. (1972). Changes in plasma oestriol concentration associated with the onset of labour. *J. Obstet. Gynaec. Brit. Cwlth.*, **79,** 970
50. Shaaban, M., Jandial, V. and Klopper, A. (1973). Plasma oestrogen levels as an indication of probable response to artificial rupture of membranes. *J. Obstet. Gynaec. Brit. Cwlth.* (To be published)
51. Symonds, E. M., Fahmy, D., Morgan, C., Roberts, G., Gomersall, C. R. and Turnbull, A. C. (1972). Maternal plasma oestrogen and progesterone levels during therapeutic abortion induced by intra-amniotic injection of prostaglandin $F_2\alpha$. *J. Obstet. Gynaec. Brit. Cwlth.*, **79,** 976
52. Pinto, R. M., Leon, C., Mazzocco, N. and Scasserra, V. (1967). Action of oestradiol-17β at term and at onset of labor. *Amer. J. Obstet. Gynecol.*, **98,** 540
53. Pinto, R. M., Rabow, W. and Votta, R. A. (1965). Uterine cervix ripening in term pregnancy due to the action of oestradiol-17β. *Amer. J. Obstet. Gynecol.*, **92,** 319
54. Pinto, R. M., Fisch, L., Schwarz, R. L. and Montuori, E. (1964). Action of oestradiol-17β upon uterine contractility and the milk-ejecting effect in pregnant women. *Amer. J. Obstet. Gynecol.*, **90,** 99
55. Järvinen, P. A., Luukkainen, T. and Väistöl (1965). The effect of oestrogen treatment on myometrial activity in late pregnancy. *Acta Obstet. Gynecol. Scand.*, **44,** 258
56. Agüero, O. and Aure, M. (1971). Inutilidad de los estrogenos en la maduracion del cuello y en la induccion del parto. *Ginec. Obstet. Mexico*, **30,** 21
57. Klopper, A. and Dennis, K. J. (1962). Effect of oestrogens on myometrial contractions. *Brit. Med. J.*, **11,** 1157
58. Klopper, A., Dennis, K. and Farr, V. (1969). The effect of intra-amniotic oestriol sulphate on uterine contractions. *Brit. Med. J.*, **2,** 786
59. Dennis, K. J., Farr, V. and Klopper, A. (1973). The effect of intra-amniotic oestriol sulphate on abortion induced by hypertonic saline. *J. Obstet. Gynaec. Brit. Cwlth.* **80,** 41
60. Klopper, A., Farr, V. and Dennis, K. J. (1973). The effect of intra-amniotic oestriol sulphate on uterine contractility at term. *J. Obstet. Gynaec. Brit. Cwlth.*, **80,** 34

8
Steroid Hormone and Metabolite Binding in Reproductive Target Tissues

E. MILGROM and E. E. BAULIEU
Université de Paris-Sud

The figures of this text have been established on the basis of those illustrating the plenary lecture of E. E. Baulieu at the *4th International Congress of Endocrinology*, Washington 1972: 'A survey of the mode of action of steroid hormones' (in press in the *Proceedings of the Congress*, R. O. Scow, Editor, *Excerpta Medica*, Amsterdam). Some of them, as indicated in their captions, derive from published work.

8.1 INTRODUCTION

The hormonal signal emitted by an endocrine gland reaches most cells of the organism and yet only some of them—the target cells—respond. It is therefore necessary to postulate the existence of specific mechanisms whereby these cells recognise hormonal messages. It seems certain that such recognition can only be made through the means of macromolecules.

The existence of hormonal 'receptors' has thus been postulated for some time. These 'receptors' were expected to have a double polarity (Figure 8.1); on the one hand they should recognise and bind their specific hormone and

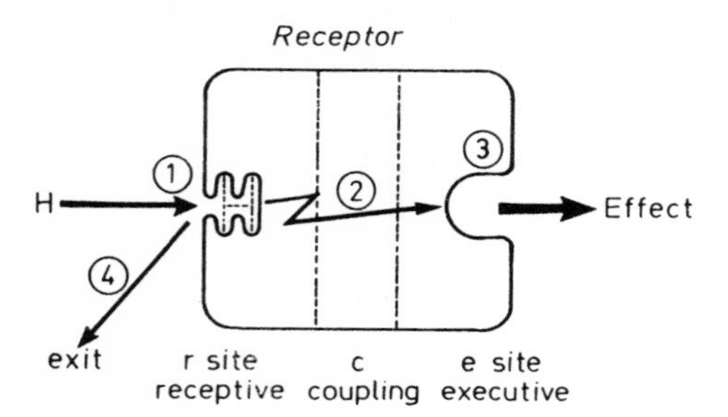

Figure 8.1 The *receptor*. Operationally, in the *receptor* there are three components: a receptive site *r* (1) 'specific' for a given hormone, of which the interaction provokes, through a coupling mechanism (2), a modification of an 'executive site' *e* (3), in charge of the initial response. The rest of the cellular events follows automatically whereas the hormone, having played its part, can leave (4). It is not known if the compounds are located on the same sub-unit (the *receptor* being composed of one or several such sub-units) or on different sub-units

on the other, because of their interaction with the hormone, become modified in such a way as to give rise to the chain of events which finally leads to the specific macroscopic effect of the hormone (growth, differentiation, secretion, etc.). In the case of the steroid hormones, the knowledge about these two functions of the 'receptor' has not developed simultaneously. The synthesis of radioactive compounds of high specific activity has made it possible to follow the fate of a steroid hormone under physiological conditions, i.e. at concentrations of the order of 1 nM at target level. Conversely, the present state of imperfection of knowledge concerning the control of gene expression in eucaryotic cells has strongly retarded the progress of studies on the transmission of hormonal information by the steroid–*receptor* complex to the cell machinery. In this respect there is a difference between the steroid and polypeptide classes of hormones. In the latter, the hormones are 'extracellular messengers' and interact with *receptors* situated on the surface of the cell membrane and presumably exert some, if not most, function by activation of adenyl cyclase; the next step is thereafter accomplished by cyclic-AMP inside the cell. Steroid hormones, however, enter the cells and their nucleus and thus one expects that the interaction with their specific *receptor* results in direct effects on gene expression for controlling the various cellular functions.

The literature devoted to steroid hormone receptors over the past 10 years or so is extensive and often contradictory. No attempt will be made in this chapter to provide an exhaustive summary of all the relevant papers, as these are the subject of several recent reviews[1-3]. However, in the presentation, will be formulated some general propositions which appear to these authors to represent the general consensus of workers in this field. It will also put forward criticisms of some generalisations which may have been rather too easily accepted in the absence of valid experimental evidence, and finally attention will be drawn to certain areas which appear to merit further research in depth. Naturally, such an approach will inevitably reflect personal bias.

8.2 CYTOSOL RECEPTOR

Glascock[4] and Jensen[5], using tritiated oestrogens of high specific activity demonstrated that target organs for these hormones were capable of concentrating and retaining them (Figure 8.2). These findings opened the way for the direct study of the binding macromolecules responsible for these effects.

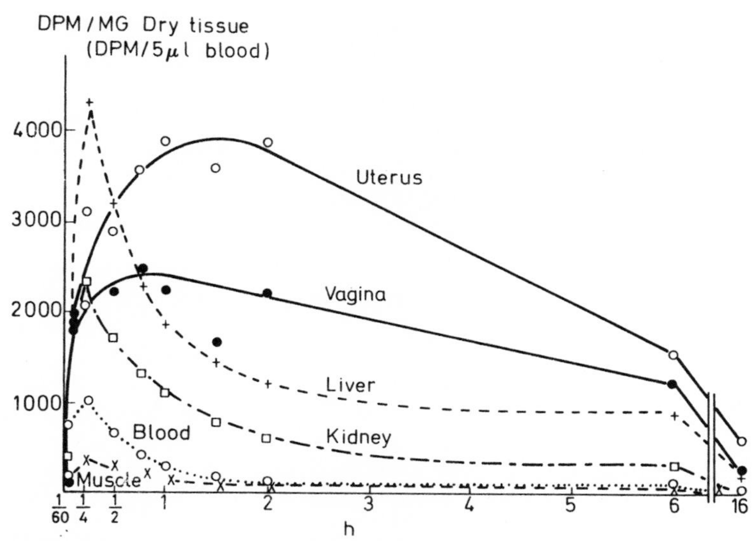

Figure 8.2 *Concentration and retention of hormone in target tissues.* The figure is taken from the work of Jensen and Jacobson[3], by courtesy of Pergamon Press: a 'physiological' dose of radioactive oestradiol has been injected at 0 time to a hormone deprived rat and radioactivity is measured in various tissues as indicated. Only target tissues of oestrogens concentrate and retain oestradiol significantly

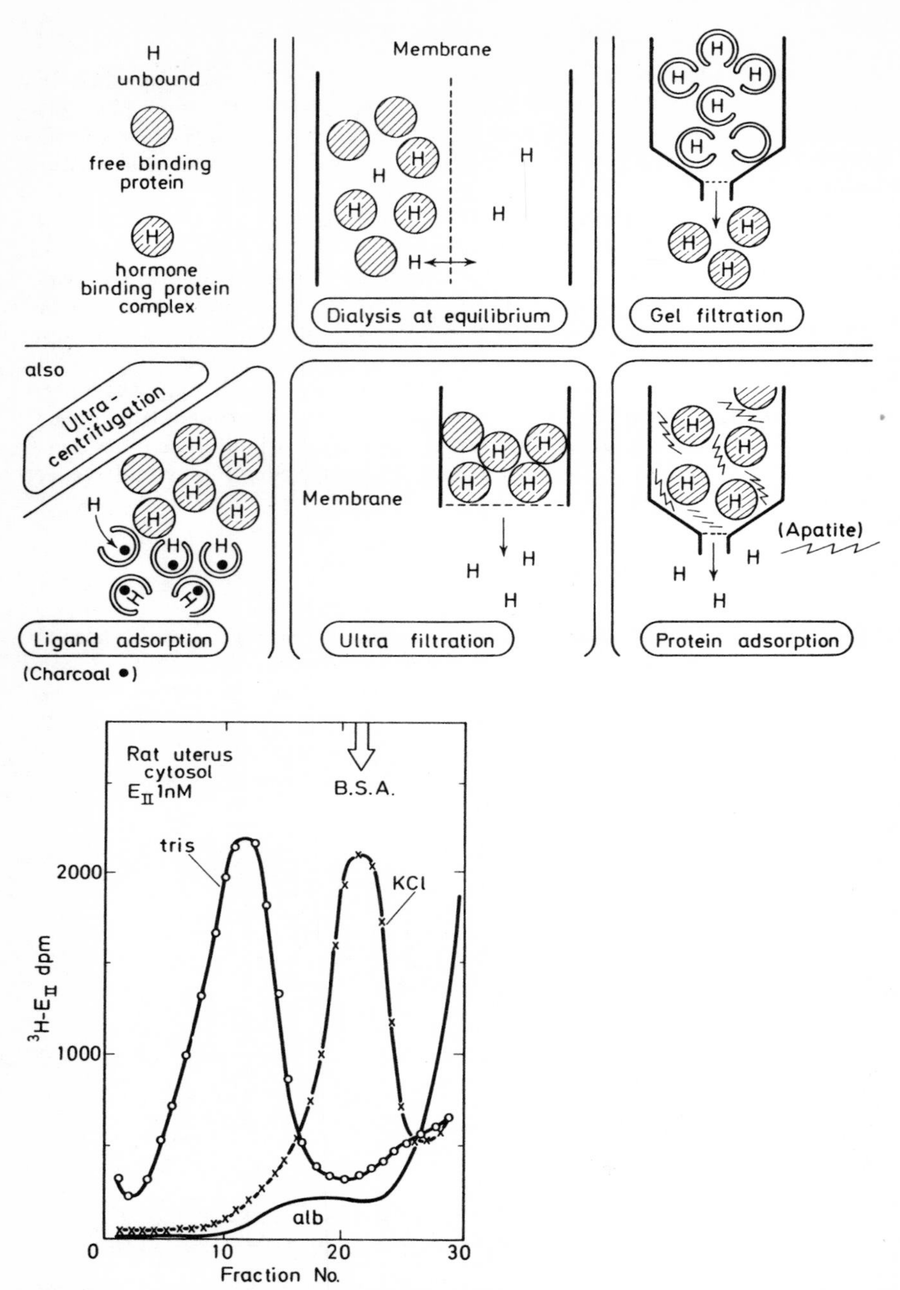

Figure 8.3 *Techniques for the measurement of binding at equilibrium.* Five techniques separating the hormone in unbound and bound forms, and the binding protein, are schematically represented on the upper figure.

On the lower graph, ultracentrifugation through gradient density medium compares the binding of oestradiol 1 nM to uterus cytosol *receptor* in salt and KCl 0.5 media, and to albumin (no salt).

Only dialysis and (with limitation) ultrafiltration do not modify the equilibrium

In animals deficient in a given steroid (e.g. in the case of sex steroids, a pre-pubertal or a castrate animal), specific binding proteins were demonstrated in the soluble fraction of tissue homogenate, operationally called 'cytosol'.

8.2.1 Methods of study[6]

The methodology is of prime importance in the study of *receptors* because their characterisation is based upon the definition of a non-covalent interaction between a hydrophobic small molecule, the steroid, and the hydrophobic region of the macromolecular *receptor*.

If one tries too hard, using high concentrations of protein and steroids, non-specific hydrophobic interactions are observed since any and every

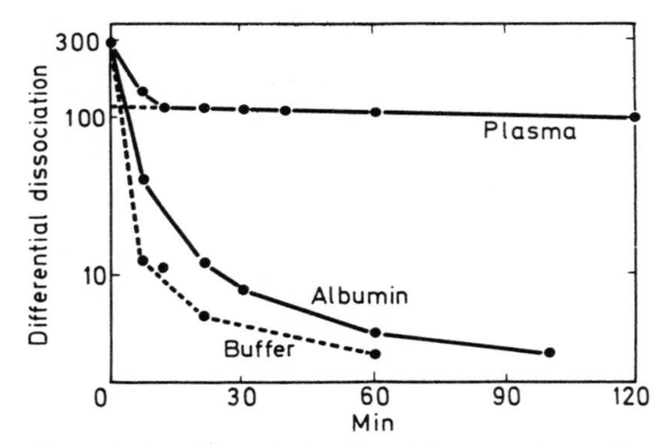

Figure 8.4 *Differential dissociation method.* The radioactive mixture is agitated for periods of time in the presence of an adsorbent (dextran coated charcoal) which takes up the unbound ligand. Hormone (radioactive cortisol) is removed rapidly in the case of a *buffer* solution and of fast dissociating non-specific (hormone albumin) complexes. When the adsorbent is added to a mixture of unbound, non-specifically and specifically bound hormone (cortisol in plasma is bound non-specifically to albumin, specifically to transcortin), the former is rapidly removed and after some time only the latter is left. The slow dissociation of the specific complexes can then be followed and extrapolation to the beginning of the adsorption permits the concentration of specific complexes initially present in the mixture, to be known[7].

protein binds every steroid, then one simply ends up by measuring concentrations of complexes which have no physiological significance. This sort of 'non-specific binding' gives therefore rise to 'positive' artefacts. Conversely, 'negative' artefacts can be produced because of the extreme fragility of the *receptors*; a single unintentional rise of temperature or an incubation unduly prolonged, is sufficient to conceal the existence of a binding which is of high physiological significance.

Those methods which do not alter the equilibrium of the binding reaction (Figure 8.3) yield results which are least open to question. However, ultrafiltration is little used because of the large volumes necessary, and dialysis at nanomolar concentration of the ligand involves at least 48 h equilibration at 0–4°C. Some *receptors*, e.g. those for oestrogens, maintain their binding properties over such a long period while other properties, for example sedimentation constant, may vary. Some other *receptors*, e.g. uterine for progesterone are most unstable and are rapidly inactivated.

Methods which disturb the equilibrium (Figure 8.3) as chromatography of the complex on small-pore Sephadex, adsorption of the ligand by dextran coated charcoal, or of the ligand–*receptor* complex by hydroxy-apatite also cannot be used without some reservations. The reason is that all these techniques give rise to a dynamic process, since separation of free ligand from steroid–*receptor* complexes will produce dissociation of the latter. Dissociation of the complexes occurs much more readily in the case of labile (non-specific) complexes and much slower in those of stable (specific) complexes. Hence it is possible to make more precise studies by following the time-course of the dissociation. This has led to the use of 'differential dissociation' techniques[7] which permit the evaluation of the concentration of the 'specific' (stable) steroid–*receptor* complexes (Figure 8.4). Similar considerations should be borne in mind when it is desired to make quantitative use of other methods which disturb equilibrium such as electrophoresis and sucrose-gradient sedimentation.

8.2.2 Binding parameters

Binding parameters of a hormone to a *receptor*, depending on the system studied, have some common properties, whereas other properties show wide variations. The physiological and experimental consequences of these variations may be very important. Thus the reported equilibrium binding association constant of various receptors varies from 10^8 to 10^{11} M^{-1}. Although published values are somewhat variable, it appears that the uterine *receptors* for oestradiol[1] and the prostatic r*eceptors* for androstanolone[1] have an affinity of about one order of magnitude greater than those of the uterus for progesterone[8] and of the liver for dexamethasone[9].

Since the rate constants of association are large in all these cases, the difference in equilibrium constant is related to variation in dissociation rates. At 0°C the half-life of the uterine oestradiol–*receptor* complex is several days[10] while that for the uterine progesterone receptor complex of the guinea-pig is of the order of an hour[11] and even less in case of the rat[12]. One very important experimental consequence of a relatively rapid dissociation rate is the possibility of exchanging ligand previously bound to the *receptor* with a second ligand, radioactive for instance.

Such exchange is desirable in order to measure the concentration of *receptor* sites in a situation in which a proportion of these is occupied by unknown amounts of endogenous hormone. Thus, for the nuclear *receptors* for oestradiol in rat uterus it is possible to exchange endogenous ligand with radioactive ligand at 37°C without inactivating the *receptor*[13]. In the case of the progesterone *receptors* in the cytosol of guinea-pig uterus, this can be carried out successfully by relatively short incubation (4 h) at 0°C[11]. The essential feature of these techniques is that they permit the study of changes in *receptor* concentration under diverse physiological and pathological situations.

The *in vivo* situation is different from these *in vitro* conditions since at low temperature, the rate of dissociation of hormone *receptor* complexes is much slower than at physiological temperature.

There are still a large number of problems remaining obscure in this field. For instance, the co-operative nature of the binding between hormone and *receptor* is still a matter of debate[3] and, in a number of systems, direct measurement of the equilibrium constant of association does not agree with that calculated indirectly from rate constants[10]. These considerations, as well as the biphasic character of dissociation curves[10,11] give rise to the suppositions that steroid–*receptor* complexes may well comprise a heterogeneous population, or that there may exist intermediate reaction products which are yet unknown.

8.2.3 Structural characteristics of receptor

All hormone *receptors* present in the cytosol appear to have certain physical properties in common. The one which has undoubtedly been the centre of most attention is the oligomeric character of the protein. In weak solutions of electrolyte, the *receptor* exists as an aggregate with a sedimentation coefficient of 6–10S according to the case. Under higher ionic concentrations (corresponding to 0.3–0.4 KCl or NaCl), it dissociates into sub-units of 3.5–5S. At physiological concentrations (corresponding approx to 0.15M) the sedimentation constant is intermediate (6S in the case of oestradiol *receptor*). The sedimentation of *receptor* also varies between certain limits, with dilution, suggesting a concentration–dependent aggregation–dissociation equilibrium[14]. Some workers interpret these results as indicating that in dilute ionic solution the *receptor* exists as a tetramer (of about 250 000 molecular weight), while at 0.3M it dissociates into a dimeric form[15]. These authors consider that all sub-units bind hormone and have similar molecular weights (as calculated from the sedimentation coefficient and from the Stokes radius determined by gel-filtration chromatography). On the contrary, Vonderhaar *et al.*[16] have provided arguments in favour of *receptor* heterogeneity: some sub-units could bind hormone while others not[16]. Another argument in favour of heterogeneity is the fact that, according to the physiological conditions, the same *receptor* may be present as an aggregate or in dissociated form even in dilute ionic solution[14]. The prostatic androgen *receptor* has a sedimentation coefficient of 8S if the animal has been castrated for 24 h whereas after 4 days the sedimentation coefficients is of 4S[17,18]. Also, the uterine progesterone *receptor* has a sedimentation coefficient of about 4S in the castrate or dioestrous animal while the coefficient is about 6.7S in the proestrous or the oestrogen-treated animal[14].

According to Schrader *et al.*[19], the progesterone *receptor* of the chick oviduct can be resolved into 2 components on DEAE cellulose; one of these components being able to bind DNA and the other to chromatin. Steroid hormone *receptors* have the properties of acidic proteins and they also appear to be inactivated by blockers of SH groups[3].

8.2.4 The problem of the purification of receptor proteins

Various checks have occurred in attempts to purify these macromolecules and these have resulted in little progress being made in their characterisation. Until now three major problems have been encountered: firstly, *receptor* is present in extremely low concentration (constituting 10^{-4} to 10^{-5} of the cytosol proteins of target organs), secondly, the macromolecules are extremely labile and thirdly they tend to form aggregates in the course of purification procedures.

In the face of these difficulties, some groups have tried without success up to now to make use of affinity chromatography for purifying oestrogen *receptor*. One explanation for their failure appears to be that oestradiol or its derivatives which have not reacted at time of binding to the support remain relatively firmly adsorbed to it in spite of multiple washings and hence are able, by competitive inhibition, to displace the radioactive steroid added for the measurement of *receptor* thus giving the false impression that the *receptor* has been retained on the column. Another difficulty lies in the intensity of binding of *receptor* to ligand, so that it is impossible to elute the protein after it has become bound to the column.

Partial purification has been carried out using conventional methods[20,21]. However, if after a series of purification steps, the *receptor* is still able to bind the hormone, it has occasionally lost its capacity for binding to the nucleus. This fact brings up another difficulty in purifying *receptor*: the entity ultimately isolated must have the potential not only to bind hormone but also to transmit the information it has thereby acquired.

8.3 NUCLEAR RECEPTOR

(1) *'Two-step' mechanism*[22]. If a target organ has not been exposed to the hormone, the *receptor* can be isolated after cellular fractionation from the cytosol alone. On the other hand, after contact with the hormone (whether injected *in vivo* or incubated *in vitro*) the vast majority (about 80%) of the steroid–*receptor* complexes is now located in the nuclear fraction (Figure 8.5). This phenomenon has been described in terms of a two-step mechanism: first, the hormone becomes bound to a *receptor* present in the cytosol and subsequently, during a temperature-dependent step, the complex is transferred to the nucleus.

Over the last few years, this proposition has been subjected to exhaustive experimentation without upsetting its principal tenets. The opinion originally prevailed that the role of the *receptor* was to provide transport of steroid into the nucleus. However the consensus now appears to be that steroids as such have no need of transportation to the nucleus and that, in fact, information is transmitted to this site by the steroid–*receptor* complex and not by the steroid action alone. In this scheme, after the hormone has become bound to *receptor*, the latter undergoes a conformational alteration which renders it capable of binding to a nuclear 'acceptor' and it is this new interaction which transmits the information initially carried to the cell by the hormone.

This conformational change of the *receptor* appears to take place in the cytoplasm. It is temperature dependent and becomes irreversible so that it retains its capacity for binding to the nucleus even after the temperature has been returned to 0°C[23]. In the case of oestrogen *receptor*, this transformation is accompanied by an alteration in coefficient of sedimentation (from 4S to 5S).

(2) A certain number of basic features of this theory are not based on sound experimental evidence.

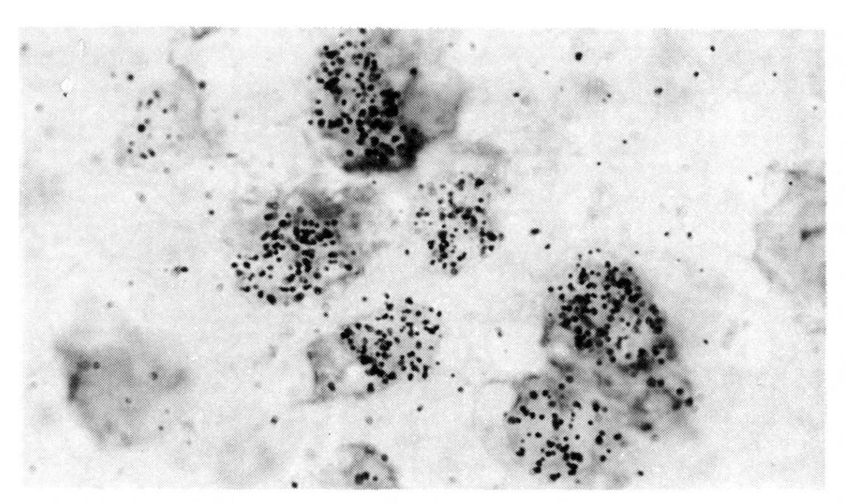

Figure 8.5 *Nuclear localisation of radioactive oestradiol in rat hypothalamus cells.* Preoptic median area Left upper corner: 3rd ventricule (Melle Warembourg, Laboratoire d'Histologie, Faculté de Médicine de Lille). The labelling is mainly found in the nuclear region

(a) The initial localisation of *receptor* to the cytoplasm is itself not absolutely certain. Artefacts may arise during homogenisation. If the *receptor* was weakly bound to a particular cellular structure (nucleus, microsome, membrane, etc.), it could become solubilised during this step of preparation. The only fact which can be stated with complete certitude is that, in the presence of the hormone, the *receptor* has a much greater affinity for the nucleus than in its absence.

(b) The nature of the eventual translocation of complex from cytoplasm to nucleus is studied *in vitro*, either by tissue incubation or by incubation of a cell free system. Such systems are very labile and these also can give rise to a variety of artefacts. In the case of incubation of tissues, the *receptor* becomes progressively inactivated as a function of time, while in the cell free incubations (cytosol + nucleus) it is not known if this system corresponds to a physiological situation since the specificity of the nuclear acceptor has not been unequivocally established (Clark has even shown that glass beads are capable of replacing nuclei of uterine origin in binding to the *receptor* complex).

(3) *The nature of the nuclear acceptor.* Among the numerous nuclear constituents there are two which appear to be possible candidates for the acceptor function, the acidic proteins[24] and the DNA[25-27]. Binding of the steroid *receptor* complex of these two types of macromolecules has been described. However, it is still uncertain as to whether such binding is similar to that observed *in vivo* and there is no theory which explains how information is transmitted as a result of such binding.

(4) *The problem of oestradiol binding to a non-histone protein of chromatin.* In the endometrium of the calf which has not been previously treated with hormone, there exists an acidic protein in the chromatin which specifically binds oestradiol[28], and this 'receptor' is distinguishable from all *receptors* previously described by its extreme affinity for ligand ($10^{14}M^{-1}$), its extremely low concentration (about 10 sites/cell) and the fact that it exists in the nucleus of an organ which has not been in contact with hormone (i.e. not in conformity with the two-step mechanism).

The role of this protein is unknown. However, the hypothesis has been put forward that the hormone may be transferred from 'classical' *receptor* to this protein in the chromatin.

8.4 ENTRY OF HORMONE INTO THE TARGET ISSUE

8.4.1 Role of transport in plasma

Steroids are present in plasma in the form of a reversible equilibrium between a free fraction and a fraction bound to protein. Changes in concentration of plasma binding proteins do not appear to be accompanied by any major modification of steroid action and it may be deduced from this that hormone activity is confined to the unbound moiety. Plasma proteins thus play the passive role of providing a rapidly available reservoir of hormone. However, this interpretation is not based upon indisputable experimental evidence.

A more active role has been suggested for plasma proteins by Keller *et al*[29]. A number of these plasma proteins have been recovered from certain organs in such concentrations as to exclude the possibility that they simply represent contamination of the tissue by plasma[30-32].

8.4.2 The passage across the cell membrane

Because steroids are small lipophilic molecules, it has been tacitly supposed until now that they pass easily across the cell membrane by simple diffusion.

However, studies in this field up to now have come up against two major difficulties. On the one hand, during *in vitro* incubation steroids are adsorbed onto the outside of cell membranes and thus can be falsely considered as located inside the cells[33]; on the other hand, any change in the binding of a hormone and hence every variation in *receptor* is echoed by like changes in the penetration of steroid into the target cell.

Taking into account these facts it was possible to show that the penetration of oestradiol into the uterus of the prepuberal rat exhibits the phenomenon of saturation (Figure 8.6) in conditions where the saturation of the *receptor*

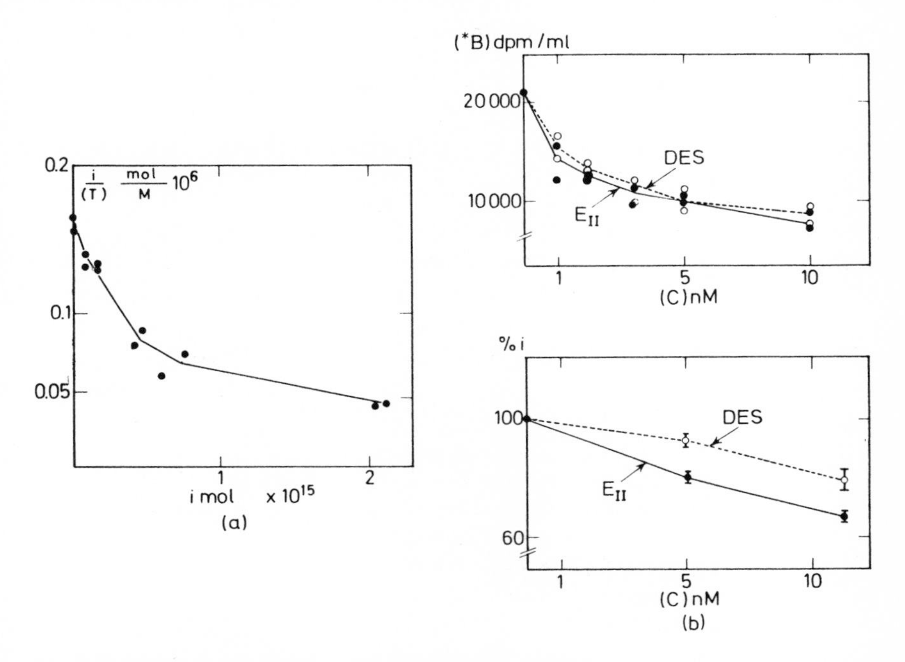

Figure 8.6 *Entry of oestradiol into immature rat uterus target cells saturability* (*a*) *and specificity* (*b*). a) The (Scatchard-like plot) graph correlates i, the incorporation of oestradiol per uterine horn per min and T, the total concentration of steroid obtained at 37 °C in 5 min experiments over a concentration of 0.1–50 nM of [^{3}H]-oestradiol. It shows the existence of a saturable process in the entry of oestrogen into the uterine cells. It has been established that this saturation is not due to the *receptor*

b) Competition between radioactive oestradiol and non-radioactive oestradiol (E_{II}) and diethylstilboestrol (DES) at 37 °C.

The incorporation i is measured in 5 min experiments with various concentrations of competitor C and [^{3}H]-oestradiol 1 nM i 100% is obtained in the absence of competitor. Binding B of [^{3}H]-oestradiol 0.5 nM to the cytosol *receptor* in the presence of various concentrations of non-labelled oestrogens. Unlabelled DES and E_{II} are equally effective as inhibitors of [^{3}H]-oestradiol binding to the *receptor* (6b upper part) whereas DES is a weaker inhibitor when the entry cells is studied (6b lower part)[35]

itself is not observed. This entry process has also a hormonal specificity different from that of the *receptor*. For example, diethylstilboestrol exerts a relatively slight competition against radioactive oestradiol for entry into the cells whereas it shows a very strong competition for binding to the *receptor*. This incorporation process has also different reactivity (when compared to that of the *receptor*), with compounds which block SH groups. It appears that the penetration of oestradiol into the cell does not appear to require the expenditure of energy. All these data suggest that at physiological concentration steroid enters the uterine cells by a protein mediated process ('facilitated diffusion')[34,35].

Another example of transport of steroids across cell membrane is the case of fibroblasts in which active transport permits the exit of glucocorticoids[36,37].

These examples show that variation of the passage of steroids through the cell membrane could well operate as a control mechanism in a certain number of systems.

8.5 LOCAL METABOLISM OF STEROID WITHIN THE TARGET ORGAN

In many target organs the active hormone undergoes metabolism.

It appears that, in general, two separate and distinct situations may arise (Figure 8.7); either the metabolite may be more active than the circulating

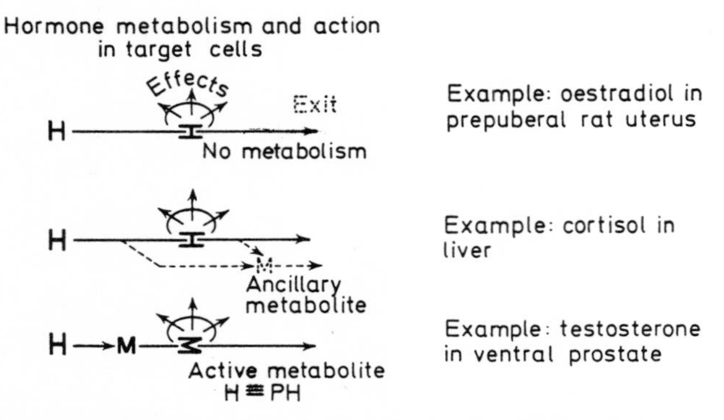

Figure 8.7 In the situations of the 2 upper lines, H is 'really' a hormone. In the third, the metabolic M is active and H becomes *de facto* a prehormone PH

hormone or on the contrary, it may be relatively or completely inactive. In the first case, the metabolic process activates a 'prehormone', while in the second inactivation is the dominant feature. The best known example of the former relates to testosterone and the prostate[38-41]. The steroid is metabolised into a number of compounds, the most important being dihydrotestosterone (androstanolone, 17β-hydroxy-5α-androstan-3-one) and 5α-androstane 3β and 3α, 17β-diols. Dihydrotestosterone is the metabolite that has received the most attention. It is bound to the soluble prostatic *receptor* with an affinity far greater than that of testosterone itself, and it alone seems to be able to promote the translocation of steroid *receptor* complex into the nucleus. There are some arguments indicating that androstane 3β, 17β-diol has an activity which differs from that of testosterone, which therefore is a prehormone for the prostate. On the other hand, in muscle and possibly in the kidney, testosterone itself is active and has an affinity for the *receptor* much stronger than that of dihydrotestosterone[42,43] (Figure 8.8).

In the case of progesterone, a theory has been put forward, analogous to that for testosterone, that 5α-dihydroprogesterone is the metabolite which acts on the uterine cells[44]. This does not appear to hold for many species, where progesterone is preferentially retained and is bound with higher affinity to its *receptor*[45]. As for oestradiol, it was originally considered not to

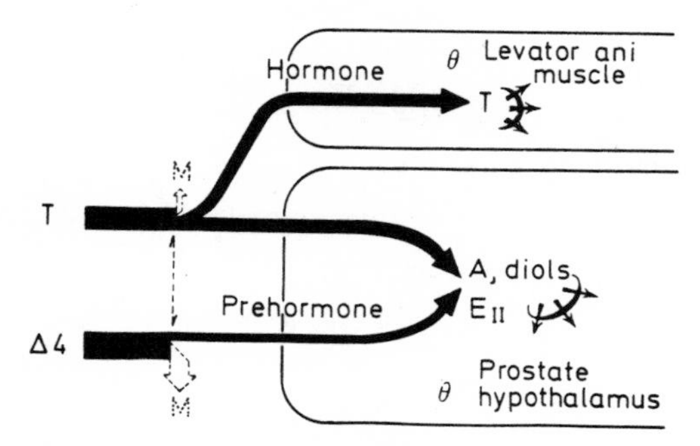

Figure 8.8 *Testosterone: an hormone and a prehormone*. In target organs (θ) such as *levator ani* and kidney, testosterone (T) is directly active as a hormone should be. In target organs such as ventral prostate and hypothalamus, the activity is partially mediated through the formation of metabolite(s) and testosterone is a prehormone. Androstenedione, Δ_4, is always a prehormone. Testosterone and androstenedone, have a peripheral detoxification metabolism into metabolites M, and are partly interconvertible.
A = androstanolone (dihydrotestosterone)
diols = androstanediols
E_{II} = oestradiol

be metabolised by the uterus[5]. However, recent studies on mature animals have shown that oxidative metabolism occurs (leading to oestrone)[46]. This is therefore a situation in which metabolism leads to a reduction in activity compared with the original hormone.

A more extreme case is that of the hepatic metabolism of cortisol where the process leads to inactivation of the steroid.

8.6 THE ROLE OF THE RECEPTORS

The definition of a 'receptor', as outlined in the beginning of this chapter, is purely operational and based upon capacity to bind hormone. Most authors have implied that *receptors* are indispensable for the action of hormones to the extent that they form a necessary link in the chain of transmission of information. The sole action of the hormone is to effect a transformation of *receptor* by the very process of becoming bound to it, and that it is this structural alteration which in turn transmits the information without further intervention by the hormone (Figure 8.9).

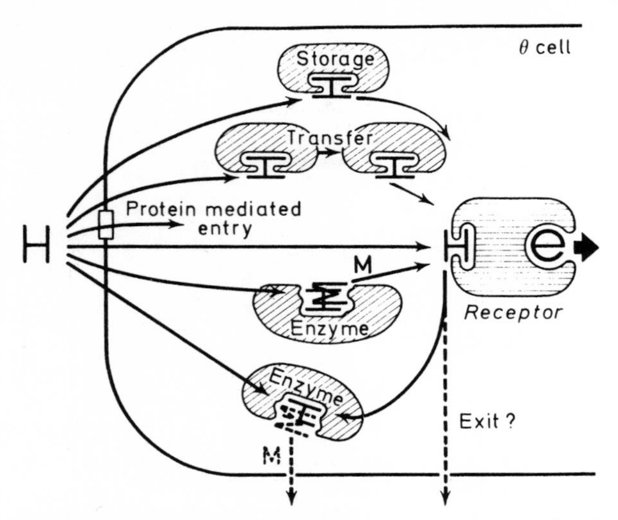

Figure 8.9 *Interactions of hormone H in a target θ cell.* Interaction of the hormone with specific proteins in target organs may correspond to entry, storage, transfer or enzymatic mechanisms. Finally they may also correspond to the interaction with the real *receptor*

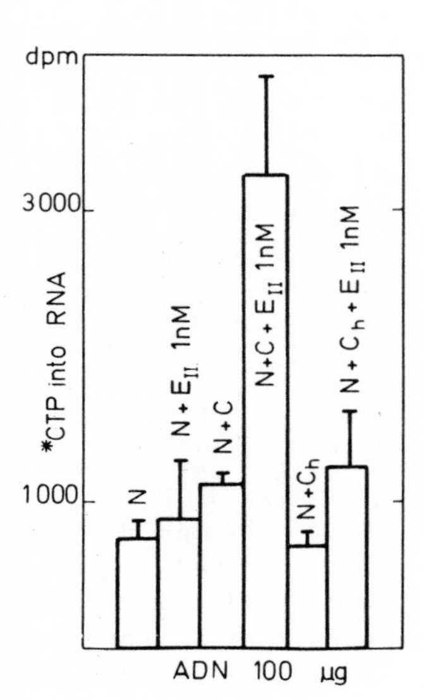

Figure 8.10 *Uterine nuclei RNA polymerase activation by oestradiol receptor complex.* Incorporation of cytosine triphosphate (CTP) into nuclear RNA. Calf endometrium nuclei eqivalent to 100 µg DNA) have been incubated alone or in the presence of oestradiol 1 nM, or cytosol, or of cytosol + oestradiol. Controls include incubation in the presence of heated cytosol or heated cytosol–oestradiol complex[47]

What are the arguments in favour of such a hypothesis? Can one eliminate other hypotheses? For example that the *receptor* may only exist to subserve transportation of the hormone, which is secondarily delivered to the 'true' *receptor*; or that the *receptor* hormone complex may be merely another form of reservoir of hormone etc. In favour of the more direct role of the *receptor* is the fact that there is reasonable parallelism between the physiological action and the affinity for *receptor* of a number of steroids. Also, the fact that *receptor* is only present in target cells constitutes another argument for the hypothesis.

Until now, only indirect arguments concerning the role of *receptors* have been adduced. More recently several authors have shown that, in acellular systems *in vitro*, the oestradiol–*receptor* complex can stimulate RNA polymerases in uterine nuclei[47-49] (Figure 8.10). However, the interpretation of these results is somewhat equivocal because of difficulties with reproducibility and also with their correspondance with the *in vivo* situation. Indeed, it has not been possible to elicit *in vivo* such a rapid stimulation of RNA polymerase and of such a magnitude.

This (as yet) unresolved question of how the steroid *receptor* complex is able to transmit information to the cellular apparatus is without doubt the most pressing problem in this area of research.

8.7 PHYSIOLOGICAL AND PATHOLOGICAL ALTERATION IN RECEPTORS

If a hormone exerts its effects by its interaction with a specific *receptor*, and if the intensity of such action is a function of the concentration of the complex so formed, it appears obvious that regulation of cell function could be brought about not only by alterating the level of circulating hormone, but also by changing the concentration of *receptor*.

The existence of the latter form of control system appears to have been well established in the case of sex hormones: after castration the concentration of prostatic *receptors* for testosterone falls, but the precise hormone responsible for this change has not yet been determined[18]. In the uterus of the prepubertal rat, injections of oestradiol lead first to the translocation of steroid *receptor* complex to the nucleus and later to *de novo* synthesis of more *receptor*[50].

The regulation of progesterone *receptors* in the guinea-pig uterus depends on the combined action of both oestradiol and progesterone[51]. In the castrate guinea-pig, a single injection of oestradiol leads to a perceptible increase in *receptor* by 6 h; this becomes maximal at about 20 h. This increase can be blocked by inhibitors of protein or of RNA synthesis injected at the same time as the oestradiol, which indicates that during this period, *receptor* is synthesised (or that there is synthesis of an activator). In the absence of progesterone the half-life of its *receptor* is about 5 days. After the injection of progesterone the concentration of *receptor* falls rapidly. It has been shown that this fall is neither the result of masking of sites by non-radioactive progesterone nor due to intranuclear translocation of *receptor*. Hence it appears not to be a question of a decrease in rate of synthesis but rather to a matter of 'inactivation' of binding protein already present. The nature and mechanism of this inactivation are as yet unknown, but may in some way be connected with an increased turnover when the *receptor* penetrates into the nucleus.

This endocrine control of progesterone *receptor* may explain its variations throughout the course of the oestrous cycle of the guinea-pig[14] (Figure 8.11).

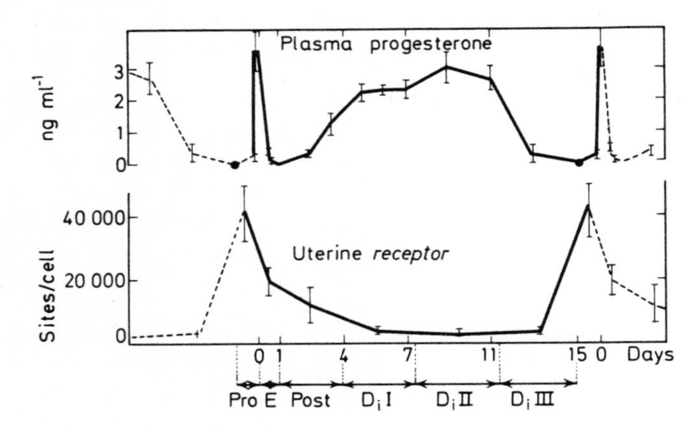

Figure 8.11 *Guinea-pig uterine cytosol receptor and plasma progesterone over the oestrus cycle.* E: oestrus, P: pro-oestrus, Post-: post-oestrus, Di: di-oestrus[14]. Plasma progesterone values are taken from Feder *et al.*[53]

The pro-oestrous peak in *receptor* concentration is probably due to the pro-oestrous peak in plasma oestradiol and the rapid fall in *receptor* concentration at oestrus and pro-oestrus appears to be due to the ovulatory secretion of progesterone as well as that secreted by the corpus luteum.

In some pathological states, 'non-receptivity' of genetic origin could provide an explanation for certain features of the disease. Thus a certain mutant strain of mouse (Tmf) exhibits a syndrome similar to human testicular feminisation. The kidneys of these animals are deficient in androgen *receptor*, which is present in the non-mutant strains[52].

Eventual progress in the study of the mechanism of action of hormones in general, and of *receptors* in particular, is bound up with overcoming two major difficulties. Firstly, methods must be devised for the preparation of *receptors* in purified form and secondly knowledge must be acquired as to the structure and mode of expression of the genome in eucaryotes.

Acknowledgements

The English version of this text has been prepared by Dr J. Goding to whom the authors are most grateful.

Many experiments reported in this text have been conducted by past and present members of our research group. Their names appear among the authors listed in the bibliography. The work has been made possible because of the support of the Ford Foundation, Population Council, Centre National de la Recherche Scientifique, Délégation Générale à la Recherche Scientifique et Technique, Fondation pour la Recherche Médicale Française and Roussel-UCLAF.

The manuscript has been typed by Michelle Vassal and Anne Atger.

References

1. Baulieu, E. E., Alberga, A., Jung, I., Lebeau, M. C., Mercier-Bodard, C., Milgrom, E., Raynaud, J. P., Raynaud-Jammet, C., Rochefort, H., Truong, H. and Robel, P. (1971). *Rec. Progr. Horm. Res.*, **27,** 351
2. Jensen, E. V. and DeSombre, E. R. (1972). *Ann. Rev. Biochem.*, **41,** 203
3. Raspé, A. (editor) (1971). *Adv. Biosciences No. 7, Schering Workshop on Steroid Hormone Receptors*, (Vieweg: Pergamon Press)
4. Glasscock, R. F. and Hoekstra, W. G. (1959). *Biochem. J.*, **72,** 673
5. Jensen, E. V. and Jacobson, H. I. (1962). *Rec. Progr. Horm. Res.*, **18,** 387
6. Baulieu, E. E., Raynaud, J. P. and Milgrom, E. (1970). *Acta Endocrinol.*, Suppl. 147, **64,** 104
7. Milgrom, E. and Baulieu, E. E. (1969). *Biochim. Biophys. Acta*, **194,** 602
8. Milgrom, E., Atger, M. and Baulieu, E. E. (1970). *Steroids*, **16,** 741
9. Baxter, J. D. and Tomkins, G. M. (1971). *Advan. Biosciences*, **7,** 331 (G. Raspé, editor) (Vieweg: Pergamon Press)
10. Truong, H. and Baulieu, E. E. (1971). *Biochim. Biophys. Acta*, **237,** 167
11. Milgrom, E., Perrot, M., Atger, M. and Baulieu, E. E. (1972), *Endocrinology*, **90,** 1064
12. Feil, P. D., Glasser, S. R. and Toft, D. O. (1972). IV Int. Congr. Endocrinol., Washington 18–24 June. *Excerpta Medica, Int. Congr.* Series No. 256, Abstr. 159
13. Anderson, J., Clark, J. H. and Peck, E. J. Jr. (1972). *Biochem. J.*, **126,** 561
14. Milgrom, E., Atger, M., Perrot, M. and Baulieu, E. E. (1972). *Endocrinology*, **90,** 1071
15. Puca, G. A., Nola, E., Sica, V. and Bresciani, F. (1971). *Advan Biosciences*, **7,** 97 (G. Raspé, editor) (Vieweg: Pergamon Press)
16. Vonderhaar, B. K., Kim, U. H. and Mueller, G. C. (1970). *Biochim. Biophys. Acta*, **215,** 125
17. Baulieu, E. E. and Jung, I. (1970). *Biochim. Biophys. Res. Commun.*, **38,** 599
18. Jung, I. and Baulieu, E. E. (1971). *Biochime*, **53,** 807
19. Schrader, W. T., Toft, D. O. and O'Malley, B. W. (1972). *J. Biol. Chem.*, **247,** 2401
20. Puca, G. A., Nola, E., Sica, V. and Bresciani, F. (1971). *Biochemistry*, **10,** 3769
21. Brecher, P. I., Chabaud, J. P., Colucci, V., DeSombre, E. R., Flesher, J. W., Gupta, G. N., Hurst, D. J., Ikeda, M., Jacobson, H. J., Jensen, E. V., Jungblut, P. W., Kawashina, T., Kyser, K. A., Neuman, H. G., Numata, M., Puca, G. A., Saha, N., Smith, S. and Suzuki, T. (1971). *Advan. Biosciences*, **7,** 75 (G. Raspé, editor) (Vieweg: Pergamon Press)
22. Jensen, E. V., Suzuki, T., Kawashima, T., Stumpf, W. E., Jungblut, P. W. and DeSombre, E. R. (1968). *Proc. Nat. Acad. Sci.*, **59,** 632
23. Jensen, E. V., Mohla, S., Gorell, T., Tanaka, S. and DeSombre, E. R. (1972). *J. Steroid Biochem.*, **3,** 445
24. O'Malley, B. W., Sherman, M. R., Toft, D. O., Spelsberg, T. C., Schrader, W. T. and Steggles, A. W. (1971). *Advan. Biosciences*, **7,** 213 (G. Raspé, editor) (Vieweg: Pergamon Press)
25. Baxter, J. D., Rouseau, G. G., Benson, M. C., Garcea, R. L., Ito, J. and Tomkins, G. M. (1972). *Proc. Nat. Acad. Sci.*, **69,** 1892
26. Musliner, T. A. and Chader, G. J. (1972). *Biochim. Biophys. Acta*, **262,** 256
27. Yamamoto, K. R. and Alberts, B. M. (1972). *Proc. Nat. Acad. Sci.*, **69,** 2105
28. Alberga, A., Massol, N., Raynaud, J. P. and Baulieu, E. E. (1971). *Biochemistry*, **10,** 3835
29. Keller, N., Richardson, U. I. and Yates, F. E. (1969). *Endocrinology*, **84,** 49
30. Milgrom, E. and Baulieu, E. E. (1970). *Endocrinology*, **87,** 276
31. Milgrom, E. and Baulieu, E. E. (1970). *Biochem. Biophys. Res. Commun.*, **40,** 723
32. Beato, M., Schmid, W. and Sekeris, C. E. (1972). *Biochim. Biophys. Acta*, **263,** 764

33. Williams, D. and Gorski, J. (1971). *Biochem. Biophys. Res. Commun.*, **45,** 258
34. Milgrom, E., Atger, M. and Baulieu, E. E. (1972)' *C.R. Acad. Sci.* (*Paris*), **274,** 2771
35. Milgrom, E., Atger, M. and Baulieu, E. E. (1973). *Biochim. Biophys. Acta*, in press
36. Gross, S. R., Arronow, L. and Pratt, W. B. (1970). *J. Cell. Biol.*, **44,** 103
37. Gross, S. R., Arronow, L. and Pratt, W. B. (1968). *Biochem. Biophys. Res. Commun.*, **32,** 66
38. Bruchovsky, N. and Wilson, J. D. (1968). *J. Biol. Chem.*, **243,** 2012
39. Wilson, J. D. and Gloyna, R. E. (1970). *Rec. Progr. Horm. Res.*, **26,** 309
40. Anderson, K. M. and Liao, S. (1968). *Nature* (*London*), **219,** 277
41. Baulieu, E. E., Lasnitzki, I. and Robel, P. (1968). *Nature* (*London*), **219,** 1155
42. Jung, I. and Baulieu, E. E. (1972). *Nature New Biology*, **237,** 24
43. Jung, I. and Baulieu, E. E. (unpublished observations)
44. Armstrong, D. T. and King, E. R. (1970). *Fed. Proc.*, **29,** 250, Abstr. No. 22
45. Milgrom, E., Atger, M. and Baulieu, E. E. (1971). *Adv. Biosciences*, **7,** 235 (G. Raspé, editor) (Vieweg: Pergamon Press)
46. Jutting, G., Thun, K. J. and Kuss, E. (1967). *Eur. J. Biochem.*, **2,** 146
47. Raynaud-Jammet, C. and Baulieu, E. E. (1969). *C.R. Acad. Sci.* (*Paris*), **268,** 3211
48. Arnaud, M., Beziat, Y., Guilleux, J. C., Hough, A., Hough, D. and Mousseron-Canet, M. (1971). *Biochem. Biophys. Res. Commun.*, **232,** 117
49. Mohla, S., DeSombre, E. R. and Jensen, E. V. (1972). *Biochem. Biophys. Res. Commun.*, **46,** 661
50. Sarff, M. and Gorski, J. (1971). *Biochemistry*, **10,** 2557
51. Milgrom, E., Luu, M. and Baulieu, E. E. (1973). *J. Biol. Chem.*, in press
52. Gehring, U., Tomkins, G. M. and Ohno, S. (1971). *Nature New Biology*, **232,** 106
53. Feder, H. H., Resko, J. A. and Coy, R. W. (1968). *J. Endocrinol.*, **40,** 505

9
Recent Progress in the Study of Male Reproductive Physiology: Testis Stimulation; Sperm Formation, Transport and Maturation (Epididymal Physiology); Semen Analysis, Storage and Artificial Insemination

T. D. GLOVER
University of Liverpool

9.1 INTRODUCTION

To attempt a comprehensive review of the literature on male reproduction in a single chapter of a book would be unrealistic, so in the account that follows many deserving and even distinguished papers may not be cited. I, therefore, ask any authors who might feel that their work has not been fairly treated to be indulgent and understanding of the difficulties. Here, a generalised picture is presented, in the hope that some perspectives might be reached and some of the outstanding and unresolved problems might be recognised.

Essentially, the male reproductive system in mammals is designed to produce mature spermatozoa, to store them until they are needed and then to discharge them in a convenient fluid into the female so that they may ascend the female tract. For this, each male mammal has a site of sperm production, an organ for sperm maturation and storage, and a group of other accessory organs to contribute to the seminal plasma or fluid portion of the ejaculate. This is the basic theme (Figure 9.1) although differences in the occurrence and arrangement of the organs varies widely between species.

With the evolutionary development of internal fertilisation, copulation has been a necessary innovation and the urge and ability to copulate are, to a great extent, due to the activity of androgens. In adult males, testosterone is the main androgen, although in some prepubertal mammals, androstenedione appears to predominate[1] and doubtless plays a role in sexual behaviour patterns that may be seen before puberty. There are probably many species differences, however, because in some animals the testes of the new-born are active in synthesising testosterone[2].

Obviously, androgens are vital to male reproductive activity and they also maintain structure and function in the accessory organs. The production

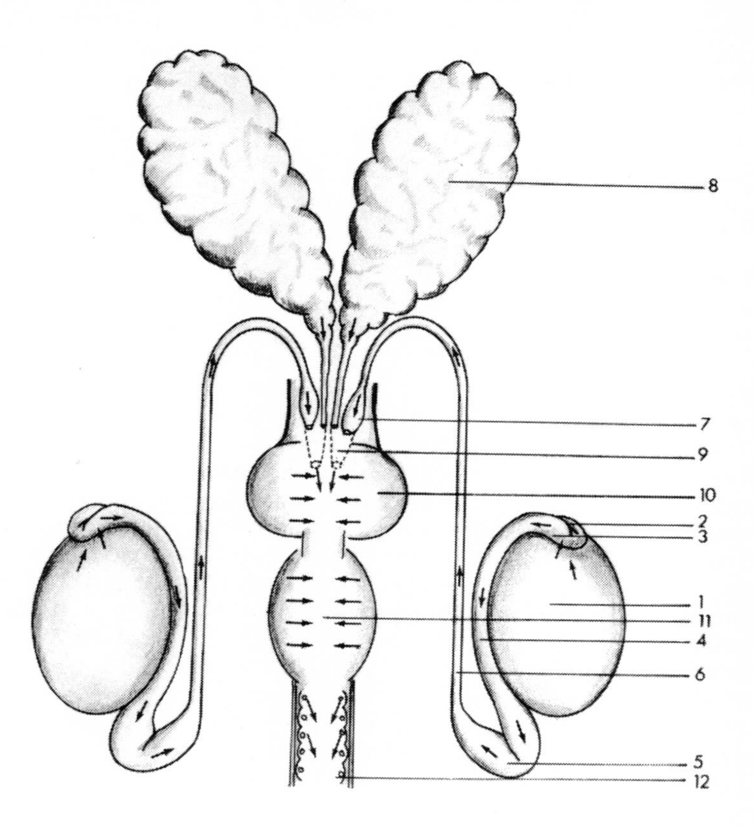

1. Testis
2. Initial segment
3. Caput epididymidis
4. Corpus epididymidis
5. Cauda epididymidis
6. Ductus deferens (vas)
7. Ampulla ductus deferens
8. Seminal vesicle
9. Ejaculatory duct of man and some other species of mammals, but not typical
10. Prostate emptying into pelvic urethra
11. Bulbo-urethral glands
12. Other urethral glands

Figure 9.1 Generalised diagram of a mammalian male reproductive tract illustrating the direction of flow of spermatozoa and fluid

and release of these steroids by the testes are themselves governed by protein hormones of the adenohypophysis (gonadotrophins) and the entire process appears to be self-regulating by virtue of feed-back mechanisms. In addition, spermatogenesis, and perhaps the release of spermatozoa also, are hormonally controlled.

9.2 THE TESTIS

9.2.1 Biological aspects of gross structure and development

The mammalian testes serve both an exocrine and endocrine function, releasing spermatozoa and fluid into the seminiferous tubules and elaborating male sex hormones which pass into the blood stream.

The testes develop from the urogenital ridge and much of their tissue is mesodermal in origin. However, it is now generally accepted, after considerable controversy, that the germinal epithelium develops from 'gonocytes' that migrate from the yolk sac endoderm to the site of gonadal development. The basis of the differentiation of an asexual gonad into a testis or an ovary and the manner in which the accessory organs subsequently develop are complicated subjects and it is not considered appropriate to discuss these in detail here, it is simply noted that the action of hormones or other chemical substances appears to be superimposed upon, and indeed to modify, a fundamental genetic imprint. Exactly how this happens is as yet unclear, but one point is certain, that maleness in mammals is essentially a matter of suppressing a natural tendency by the agonadal foetus to become female[3]. At some point in the development of the mammalian embryo, differentiation of the gonad occurs and in the male, testis cords develop from peripheral epithelium, a tunica albuginea ultimately develops between the cords and the epithelium, the cords then elongate and from this time on, development appears to proceed from the deeper parts of the gonad. Development of a testis is thus recognised by medullary development in the early gonad, in contrast to cortical development in a gonad that is destined to become an ovary.

That the seminiferous tubules eventually come to possess germ cells that are transformed into spermatogonia and that these occur alongside entirely different and highly specialised Sertoli cells, is something of an embryological mystery and it warrants more careful examination. But outside the seminiferous tubules, the interstitium, which is composed largely of the Leydig cells has quite a separate mesenchymal origin.

Development of the genital ducts follows these changes in the gonad, so that the testis comes to join the mesonephric (Wolffian) duct through efferent ductules, the seminal vesicles grow out from the mesonephric duct, and the prostate, bulbo-urethral and other urethral glands form in associaton with the urethra itself.

It is well known that in the majority of mammals, the testes migrate caudad and eventually come to lie outside the abdomen in a scrotal sac. Thus, testicular tissue is surrounded first by a relatively inelastic coat (the tunica albuginea), is enclosed in a double-layered peritoneal sac (the visceral and parietal layers of tunica vaginalis) and is contained within thin scrotal skin. This skin is often pigmented, is rather sparsely endowed with characteristically coarse hairs and is always lined by a layer of connective tissue called the tunica dartos. The dartos is interesting because it contains plain muscle that allows the scrotal skin to wrinkle and when the muscle contracts it helps to bring the testes closer to the abdomen. It is a well known observation that this occurs in cold conditions when presumably temperature receptors in the scrotum are stimulated, and that the reverse is true when the environmental temperature is high. Retraction of the testes themselves is primarily accomplished, however, by the action of the striated external cremaster muscle which lies on the surface of the spermatic cord and is usually regarded in most species as being a flange of the internal abdominal oblique muscle. The smooth internal cremaster muscle that is found in the mesorchium of several species is, by contrast, probably of little functional significance.

The overall design of the scrotum and its contents indicates beyond doubt that the testes require an especially precise thermoregulation, so that when the ambient temperature rises or the individual becomes febrile, maximum heat may be lost by radiation and other means, but where circumstances demand, heat loss may also be prevented. It thus seems that the scrotum is exceptionally sensitive to changes in environmental temperature and in addition to its sensory fibres carrying normal sensations[4], it has been shown that they are also associated, in the ram at least, with a definite thermoregulatory reflex[5,6]. This whole subject has recently been fully reviewed by Waites[7].

The need for delicate thermoregulation of the testes is confirmed by its unique vascular architecture. In most scrotal mammals, the testicular artery is excessively coiled as it approaches the gonad[8,9], and later it takes a superficial subalbugineal course[9]. Coiling of the artery at a point where it is surrounded by a complex of veins allows blood flowing to the testis to be precooled by a countercurrent heat exchange mechanism[7,10,11], and results in a slow, virtually non-pulsatile flow of blood in the testis itself[12]. The superficial course of the artery on the testis will allow the blood to be cooled further. As Setchell[11] points out, this peculiar vascular pattern depends to some extent upon the location of the testis, because in some eutherian mammals in which the testes are permanently maintained in the abdomen, the testicular artery is straight and penetrates the testis parenchyma directly[13]. The testicular artery of the rock hyrax and the elephant demonstrate this well and lend still more credence to the pampiniform plexus being associated with the thermoregulation of scrotal testes.

The testicular artery of man is relatively straight compared with that of some mammals such as the ram, and it is interesting that this corresponds to a striking variation between the two species in abdominal and scrotal temperature differences. In ruminants, the difference in temperature between the abdomen and scrotum may be as much as a 8 °C or even 13 °C[8], whilst in man, it is only about 2.2 °C[14]. Has the human testis become partly adapted to temperature change, therefore, or has it only just avoided being abdominal? Distinct stages in the evolution of testicular descent seem to be displayed by different species, so that in some mammals the testes are entirely abdominal and occupy the anterior part of the cavity, in others they are situated near

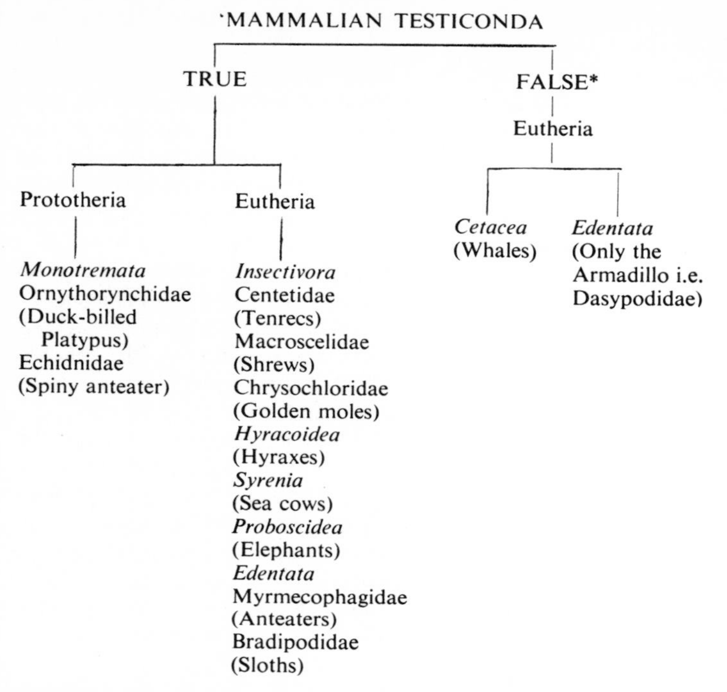

Figure 9.2 Diagram to illustrate mammalian species with abdominal testes (so-called 'testicond mammals')

the brim of the pelvis and yet in others they lie in the inguinal canal. It must be taken that the abdominal position of testes is the primitive condition, since it is the most common position in vertebrates and from the evolutionary standpoint scrotal testes may be regarded as being more advanced.

Owen[15] believed that inguinal testes, such as occur in the whale, represent a secondary withdrawal of the organs into the abdomen from a former scrotal position, but it is not necessary to postulate this, since an equally valid explanation is that inguinal testes represent an intermediate stage of descent. The testes of the armadillo lie just within the internal inguinal ring (personal observations) and in other edentates such as the sloths (*Bradipodidae*) the testes lie somewhat more anteriorally in the abdomen and might, therefore, be considered as being rather more primitive[16]. More physiological studies of the male reproductive system of mammals with abdominal testes (so-called testiconda) should be rewarding, because it is still not clear how the testes of some mammals function normally in the abdomen whilst in other species they are unable to do so.

It is also interesting that there has been considerable divergence in the evolution of testicular descent, because there are a number of eutherian mammals that, like the Prototheria, have abdominal testes and yet metatherian mammals, such as the marsupials, have scrotal testes.

On the other hand, some lagomorphs become spontaneously cryptorchid during periods of sexual quiescence, but it is likely that this is an effect of hormonal change rather than a direct cause of sterilisation. It may well be, in fact, that heat has been overrated generally as a cause of sterility in animals and man.

9.2.2 The interstitial tissue

9.2.2.1 Structural features

The interstitial tissue of the testis consists of a rather orderly network of blood vessels and lymphatics arranged around the seminiferous tubules and a more intense and randomly orientated network of connective tissue fibres intermingled with a variety of cells including macrophages and fibroblasts and perhaps some mast cells also. But in the majority of species, the most conspicuous interstitial cells are the Leydig cells.

The advent of electron microscopy has permitted a more detailed study of the structure of the interstitial tissue of the testis and particular attention has been focused on the Leydig cells[17-20]. These cells occur in clumps around the blood vessels and tight junctions bind them together. The nuclei of Leydig cells are characteristically round and under the light microscope the cytoplasm has a dusty appearance. The usual organelles are present in the cytoplasm but an extensive smooth endoplasmic reticulum is striking and occurs together with an abundance of mitochondria that are possessed of tubular cristae. Various lysosomal bodies, lipid, and lipofuchsin pigment are also to be seen. In man, the Leydig cells are often seen to have a peculiar crystalloid structure in their cytoplasm[21]. This crystal, known as Reinke's crystal occurs in the form of a hexagonal prism, and although its function is unknown it is of interest in its specificity to man. The role of organelles in relation to Leydig cell function is not absolutely clear either, and this applies to all species, but there is considerable evidence that the development of the smooth endoplasmic reticulum (ER) is associated with steroid biosynthesis. Hypophysectomy and LH replacement comprise the main experimental techniques that have been used in the study of this problem, which has been reviewed by Christiansen and Gillim[22].

Obesrvations on the cytology of interstitial cells in seasonally breeding male mammals have added support to the view that smooth ER and an abundance of mitochondria are important in androgen biosynthesis, since the development of each appears to be optimal at the height of the mating season. This has been shown in the rock hyrax[23] and in the grey squirrel[24]. Towards the end of the breeding season in the hyrax, both glycogen and lipid accumulate in the Leydig cells and the smooth ER becomes vesiculated. It is possible, therefore, that steroid precursors accumulate at this time due to an arrest of androgen synthesis. Correspondingly, the testosterone content of sexually quiescent animals may be as low as 20 ng as compared with 15 000 ng or more during the mating season and presumably these conditions are related to the activity of gonadotrophins.

9.2.2.2 *Pituitary control of Leydig cells*

There is no firm evidence from recent work to refute the now classical contention that LH or ICSH is the pituitary hormone that controls the Leydig cells[25,26], or that these cells are the main source of androgen production. Moreover, the Leydig cells produce more testosterone than any other steroid, at least in adults, so that although several steroids are produced by the testis[27], other androgens, and in those species where they occur, oestrogens also, are produced in much smaller quantities. Dihydrotestosterone, for instance, represents only a fraction of the total androgen in blood and much of this probably results from a breakdown of circulating testosterone[28]. This might also hold true for levels of oestradiol recorded in the blood stream[29].

Hypophysectomy results in diminished cholesterol in the Leydig cells[30] and reduced 3β-steroid dehydrogenase activity[31] and these events may represent a reduction in steroid synthesis. Reduced glucose-6-phosphate dehydrogenases also suggest relative inactivity of the cells under these circumstances, but information is still required for a full biochemical explanation of the control of Leydig cell function. It is known that there are sensitive receptors for the binding of LH or ICSH in the testis[32], that adenyl cyclase is activated[33] and it is also possible that cyclic AMP is involved in the formation of pregnenolone from cholesterol. But much hidden information remains, and it seems likely that biochemical studies at the cytological level will form a most important area of research in testicular endocrinology over the next few years. More needs to be learned, for example, about enzymatic activity in the Leydig cells.

It is also important to know precisely how different cells in the testis respond to particular hormones, because some confusion has arisen from an earlier inability to purify hormones, and particular responses may in some cases have been due to contamination either with another exogenous hormone or with endogenous hormone. An observation might thus be the result either of a synergistic or antagonistic action of two or more hormones. It has been claimed in rabbits that FSH is a synergist to LH in the stimulation of Leydig cells[34] and in some rodents prolactin appears to be included in the control of Leydig cell function[35,36].

If, as one hopes, purified pituitary hormones become progressively more accessible, work in this field should be made easier in the future, and a large number of puzzling problems should be resolved.

9.2.2.3 *Androgens and pituitary regulation*

It is now accepted that androgens inhibit the release and probably the synthesis of LH, although the manner in which they do so is incompletely understood. Dihydrotestosterone (DHT) and other substances with androgenic activity can all exert this effect just as well as testosterone and so the view that androgens act on hypothalamic receptors only after their conversion to oestrogen[37] is perhaps questionable, since DHT and several other androgens cannot be aromatised. But androgens do suppress LH either directly or indirectly by inhibiting an LH releasing factor or hormone, i.e. LH-RF, or LH-RH. This is a decapeptide, the action and control of which has been reviewed in some detail by Davidson[38] and by Martini[39].

It seems that, in contrast to FSH-RH, LH-RH is synthesised in the supra-chiasmatic and arcuate-ventromedial areas of the brain. This was shown by hypothalamic deafferentation and implants of protein synthesis inhibitors such as cyclohexamide[39]. The factor controls not only the release but also the synthesis of LH or ICSH and the amygdala, cortex, hippocampus and diencephalon, all appear to influence the neurones concerned in the secretion of the releasing factor. Acetylcholine might be the mediator, for Fiorindo and Martini (cited by Martini[39]) have shown that prostigmine increases LH release. But dopamine may also release LH[40], so it looks as if both cholinergic and adrenergic mechanisms play a part. Much yet needs to be done in this area, though the mechanism of LH release would seem to be simpler than that of FSH release.

One of the most interesting of recent findings in connection with this subject is that in some species LH release, in contrast to FSH release, is not constant but episodic[41,42] (Figure 9.3). In rabbits, LH levels may be increased by coitus[43] and in bulls, coitus and even the sight of a cow, stimulates the release of LH. The maximum level of LH is rapidly followed by elevated concentrations of circulating testosterone[44]. This effect is illustrated in Figure 9.4. In rabbits, testosterone is released rhythmically and in man there appears to be a distinct diurnal pattern of testosterone release[42].

It is evident though, from experiments on animals, that LH begins to decline before testosterone reaches its peak in the circulation. This is also apparent in seasonally breeding male mammals[45] and raises the thought as

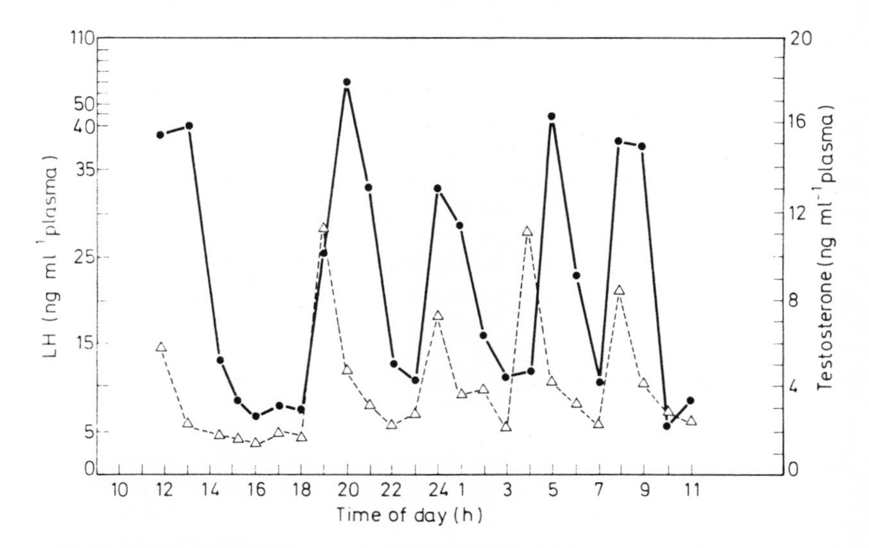

Figure 9.3 Fluctuations in the concentration of luteinising hormone (LH) (△- - -△), and testosterone (●——●) in the peripheral blood of a bull (see text). (From Katongole *et al.*[41], by courtesy of *the Journal of Endocrinology* and Cambridge University Press.)

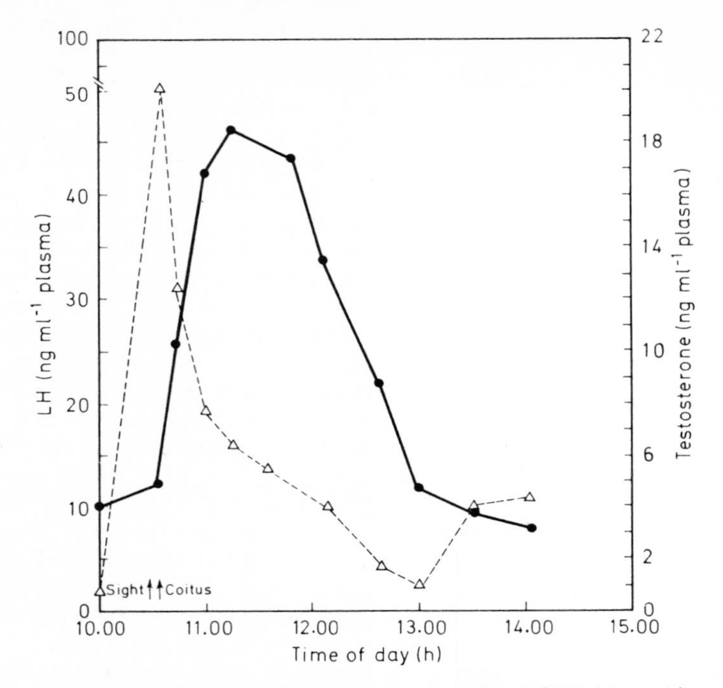

Figure 9.4 The effect of coitus on concentrations of LH (△- - -△) and testosterone (●——●) in the peripheral blood of a bull (see text). (From Katongole *et al.*[41], by courtesy of *The Journal of Endocrinology* and Cambridge University Press.)

to whether testosterone feed-back only operates at a certain threshold, and, therefore, whether LH release may not be refractory to very high levels of testosterone. Here again, studies on seasonally breeding male mammals should be valuable because similar results have been obtained in a variety of species including the red deer and the roe deer[46,47].

Finally, there is the question of whether androgens influence the synthesis and release of FSH as well as LH. This is not firmly established as a general principle and it demands further work in a variety of species.

9.2.3 The hormonal control of spermatogenesis

Spermatogenesis consists of mitotic and meiotic activity in spermiogenesis and a phase of maturation in which spermatids are transformed into spermatozoa. This is the phase known as spermateliosis and no cellular division is involved. In the past, the study of spermatogenesis has suffered particularly from the limitations of light microscopy and from the failure of research workers to describe the process in quantitative terms.

More recently, however, a number of papers have been published on the different aspects of the fine structure of germ cells[48-51] and interesting descriptions of the structure and activity of the nuclei, Golgi bodies, endoplasmic reticulum and mitochondria have helped to clarify details of the process of spermatogenesis.

In addition, since the now standard works of Roosen-Runge[52], Leblond and Clermont[53] and later of Oakberg[54], it has been recognised that at certain given levels in the seminiferous tubule a number of generations of germ cells develop concurrently and that new generations are intimately related to and synchronised with the earlier ones. Consequently, distinct cell associations occur and when such an association is complete, so that the process is ready to start again in that region of the tubule, a so-called seminiferous epithelial cycle is considered to have been completed. The stages of this cycle have been classified in a variety of ways so that Roosen-Runge and Giesel[55], described eight stages, whilst Leblond and Clermont[53], using acrosomal development as the basis of their method, described fourteen stages. Slight differences between species may exist, but recognition of the different cell associations in the cycle has permitted quantitative assessments of spermatogenic activity to be made and has thus obviated a major obstacle to accurate interpretation.

In addition to the spermatogenic cycle itself having been described however, it has also been observed that spermatogenesis proceeds in the seminiferous tubules in a wave moving from the centre of the testis to the periphery or in the reverse direction, and that as a result, successive periods or stages of the wave may be revealed by serial sections through the testis. Unfortunately, the duration of each stage is not constant, so that a length of tubule showing one particular stage of spermatogenesis might be much shorter than the length of tubule showing another stage. If, of course, a particular stage is especially long, several tubules as seen on an histological section may appear to be at the same stage. Cleland[56] and Hochereau[57] have also shown in rats and bulls respectively, that there is some modulation in the spermatogenic stages in any given tubule, so that any one stage cannot automatically be expected to follow immediately a preceding stage. The position is worse in human testes[58] where cellular associations in the seminiferous tubule appear to be quite chaotic. These aberrations of the general theme make interpretation more difficult and it must also be remembered that spermatogenesis is a dynamic process and any given stage in a wave will soon move on to a subsequent stage.

Although the rate of spermatogenesis might vary, it is a continuous process except in seasonally breeding mammals, and a constant supply of cells is ensured by the production of both type A_0 and type A_1 spermatogonia which are respectively reserve and renewing stem cells. In the rat, type A spermatogonia undergo four divisions and produce stem cell and intermediate type spermatogonia[53], whilst in the bull three generations of type A spermatogomia arise with new stem cells originating after the second one[59]. Doubtless other species differences exist also. In any case, the total number of spermatozoa arising from these divisions is less than expected, and this is due to some degeneration of cells taking place during spermatogenesis. Very little degeneration occurs during meiosis but mostly during the division of type A spermatogonia.

Mitoses may be seen as type A spermatogonia turn into type B spermatogonia but they are rarely to be observed in type B spermatogonia, because in these cells metaphase is of relatively short duration. The number of spermatogonia or divisions during a cycle is important in assessing spermatogenic activity quantitively and this has been determined by estimating the percentage of cells that incorporate tritiated thymidine and are, therefore, likely to be synthesising DNA[60]. The percentage of cells undergoing mitosis that also become labelled has, in addition, been used to assess gonadal divisions[60].

When type B spermatogonia divide, DNA synthesis or replication occurs in telophase and not, as might be expected, in interphase, but it takes some hours to complete and thus resting preleptotene spermatocytes have considerably less DNA than leptotene spermatocytes. This early replication is doubtless associated with the syncitium that may be seen to occur between type B spermatogonia as they give rise to spermatocytes.

Spermateliosis has been studied with the electron microscope[61-63] and among the more interesting structural changes that occur there is the extrusion of the chromatoid body from the spermatid nucleus, its contribution to the so-called ring centriole, changes in the spermatid nuclear membrane and the production of the residual body during spermiation or sperm release from the Sertoli cell[64].

The kinetics of spermatogenesis in a number of different mammals have been reviewed in detail by Courot, Hochereau-de Reviers and Ortavant[65] and by Clermont[66], and it is apparent that different generations of germ cells develop in a co-ordinated fashion to produce the characteristic cell associations seen in a spermatogenic cycle. Although it seems likely that Sertoli cells are in some way involved in achieving this coordination, details of the underlying mechanism are still unknown.

Much more work is needed before spermatogenesis can be completely understood and the chemical changes taking place during the various cellular divisions stand out as needing more investigation. Already a combination of cytochemical analyses and autoradiographic studies has yielded valuable information[67]. Work of this kind has shown, for instance, that, unlike oogenesis, male meiosis is characterised by the absence of ribosomal RNA synthesis. RNA is, therefore, associated with the chromosomes at this stage of spermatogenesis and is released into the cytoplasm at diakinesis or early metaphase, thus contributing to the cytoplasmic RNA of the spermatid. Nuclear protein synthesis also ceases at the same time as meiotic RNA is eliminated from the nucleus, although an arginine-rich histone is synthesised at the end of spermiogenesis (Figure 9.5). These events might well be extremely important factors in the regulation of nuclear activity in spermiogenesis and further work on the chemistry of chromosomal activity, particularly during the meiotic phase of spermatogenesis, is bound to be instructive.

An important development in the study of spermatogenesis is the interest taken in testicular enzymes and it may be predicted that work in this field will develop and be shown to be increasingly valuable. The subject has been concisely reviewed by Bishop[68] who has drawn attention to the presence of a number of enzymes in the testis including phosphatases and a variety of dehydrogenases. Some emphasis has been placed by this worker on the occurrence of sorbitol dehydrogenase in the testis and certainly this is interesting, but since its activity appears to be limited to the later stages of spermatogenesis, lactic dehydrogenase (LDH), which is active throughout

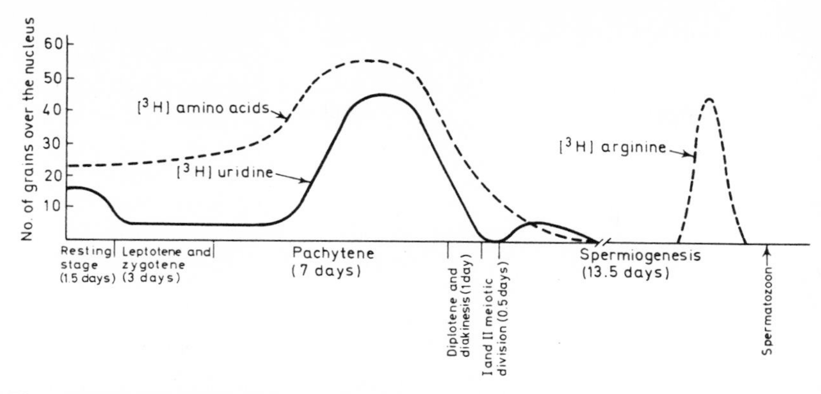

Figure 9.5 Diagrammatic description of the pattern of (^{3}H) uridine and (^{3}H) amino acids incorporation into RNA and nuclear protein, respectively, during spermatogenesis in the mouse, 1 h after administration of the radioactive precursor. (From Monesi,[67], by courtesy of the Editor, *Journal of Reproduction and Fertility*.)

the process, is probably of greater interest in relation to the control of spermatogenesis. LDH and its constituent isozymes, certainly seem worthy of closer scrutiny in relation to the spermatogenic process.

9.2.3.1 Pituitary hormones and spermatogenesis

A knowledge of intracellular events is ultimately necessary to an understanding of how hormones influence and control spermatogenesis. We are some way yet from such an understanding and a number of factors conspire to make the interpretation of hormonal effects difficult. First, a hormone may act differentially on different germ cells and the response of the germinal epithelium to any given hormone may differ in different species. Also, since hypophysectomy is frequently involved in studies of this subject the completeness or otherwise of the operation must be taken into account. This applies equally to the interpretation of clinical data. Nevertheless, and in spite of differing opinions, some basic tenets seem to be emerging.

The concept of FSH being the controlling factor in spermatogenesis[25,26]. has to some extent been overtaken by more recent observations. Indeed, it has been suggested that ICSH may largely be involved in controlling the process[69] although Woods and Simpson[70], whilst confirming this, believed also that FSH may serve as a synergist to ICSH in stimulating spermatogenesis. Lostroh[71] felt that both FSH and ICSH were needed for spermatid formation in the initiation of spermatogenesis following its arrest by hypophysectomy and Berswordt-Wallrabe and Neumann[72] also felt that FSH alone was an insufficient stimulus for spermatogenesis to be restored after hypophysectomy. Also, in hypophysectomised lambs, Courot[73] has shown a synergistic effect of the two hormones, though Steinberger[74] has suggested that they may act consecutively rather than synergistically.

On the other hand, Heller and Nelson[75] found that in man, HCG was incapable of stimulating spermatogenesis and it is known that it may even be detrimental to the human testis. But, paradoxically, in the presence of some endogenous hormones, HCG may give rise to complete spermatogenic cycles[76].

FSH-rich preparations such as human menopausal gonadotrophin (HMG) have proved more successful[77] in man, as also has a combination of HMG and HCG[78], but the possibility of contamination of one hormone with another confuses the issue, and Clermont and Harvey[79] believe that the stimulatory effects of FSH are due to Leydig cell stimulation resulting from the contamination of preparations with ICSH. Using purified gonadotrophin, Courot[80] reached a similar conclusion in his studies of hypophysectomised lambs.

Undoubtedly, therefore, gonadotrophins are vital in the control of spermatogenesis, but the specific roles of FSH and LH are not clarified in every species and it is particularly important that in man, the situation be established soon since both gonadotrophins are widely used in the treatment of male infertility. But since ICSH contamination is recognised in the most widely used preparations of gonadotrophin, it is not surprising that considerable attention has also been focused on the influence of androgens on spermatogenesis.

9.2.3.2 Steroids and spermatogenesis

That testosterone can maintain spermatogenesis is now well established, but dosage is important. Thus, Moore and Price[81] illustrated that in rats, low doses of testosterone suppress spermatogenesis, and Ludwig[82] showed that this is due to pituitary suppression. When high doses of testosterone are used, however, it seems that the effects of reduced ICSH resulting from pituitary suppression are overcome by a direct effect of the androgen on spermatogenesis. Steinberger and Nelson[83] have shown that testosterone can maintain spermatogenesis in animals whose pituitary gonadotrophins have been suppressed by treatment with oestrogen and there is some evidence that spermatogenesis can actually be re-initiated by it following hypophysectomy[84]. Similarly, if spermatogenesis is partly arrested by pituitary suppression, testosterone administration may allow the process to be completed[85,86].

In man, Heller and co-workers[87] found that testosterone may cause atrophy of the testis and these workers described the now well-known 'rebound phenomenon' following the end of treatment. But the effects of testosterone in human beings are not comparable to those obtained in laboratory animals, because it is doubtful if it is possible to administer high enough doses to man in order to make a valid comparison. If this were not so, then surely testosterone would already be available as a contraceptive agent for the human male.

It might be that the hormonal control of spermatogenesis is mediated through the Sertoli cells in the seminiferous tubules and there is evidence that these fascinating cells are involved in some kind of hormonal activity. For some time, Sertoli cells have been considered as sources of oestrogen and Lacy[88] is of the opinion that they may also produce androgen. Some of the features of the fine structure of these cells are suggestive of steroid synthesis, and include an extensive smooth endoplasmic reticulum, numerous mitochondria and several lipid droplets surrounded by annulate lamellae. Also, lipid appears to accumulate in the Sertoli cells when spermatogenesis is inhibited and this may well represent the accumulation of steroid precursor such as cholesterol. In addition, resorption by the Sertoli cells of residual bodies (cytoplasmic masses pinched off from the spermatozoa during spermiation) is interesting, especially in view of these bodies also having a high lipid content. The function of Sertoli cells, therefore, remains a mystery and their study will be made the more difficult if they are to be found in different stages of activity in one and the same tubule. Johnsen[89], for instance, has found that in human biopsy specimens there are two types of Sertoli cells and it is evident that much still needs to be learned about them.

An important contribution to the study of the hormonal control of spermatogenesis has been made by the use of an organ culture system developed by Steinberger and Steinberger[90]. This and subsequent work has led these authors to the view that testosterone is definitely required in diakinesis following the prolonged meiotic prophase of spermatogenesis and that it is also required for reduction division in the production of secondary spermatocytes. The steroid might also be needed at other stages of spermatogenesis but at present its role here, versus that of gonadotrophins is controversial. Other hormones, some as yet unidentified, may also be involved in the control of spermatogenesis. Steinberger[74] has mentioned growth hormone, and steroids other than androgens may also play a part. Oestrogens, for example are known to suppress spermatogenesis in the rat and hamster, exerting their action via the pituitary, whereas the inhibitory action of progestagens on spermatogenesis appears to depend upon the preparation that is used. The subject has been succinctly reviewed fairly recently by Steinberger[74].

9.2.3.3 Testicular feed-back

Since testosterone exerts a regulatory effect on the synthesis and release of LH, it seems reasonable to expect that a similar feed-back mechanism associated with spermatogenesis might exist for controlling the production and release of FSH. Such a mechanism is established, for it is well known that when spermatogenesis is impaired, gonadotrophin production increases even when the Leydig cells are functioning normally.

Johnsen[91] believes that in man, spermatogenic feed-back only operates when spermatozoa are present in the seminiferous tubule and that only the last stages of spermatogenesis, including spermatids and spermatozoa, are involved in the mechanism. Moreover, this worker found a correlation between the level of gonadotrophins and the number of spermatozoa in the ejaculate, which may be regarded as supporting evidence. Paulsen[76] claimed that an increase in gonadotrophin occurring in oligospermic men was mainly due to FSH, and this hormone has been shown to accumulate in the seminiferous tubules[92]. However, de Kretzer *et al.*,[93] found no correlation between various degrees of oligospermia and plasma FSH titres. De Kretzer (personal communication) believes that spermatozoa and spermatids are not necessary for testicular feed-back, because in many cases of impaired spermatogenesis there is damage to spermatogonia as well as to spermatids. In some cases, such spermatogonial damage may be recognised electron microscopically, but missed under the light microscope.

Exactly how the germinal epithelium exerts a regulating influence on gonadotrophins is still unknown, therefore, and requires some reappraisal. Johnsen[91] has implicated the lipid rich residual bodies in the mechanism and put forward the hypothesis that these bodies might contain a gonadotrophin inhibitory substance, namely 'inhibin'. If this explanation is true then the Sertoli cells must also be involved in the process, since Lacy[94] has shown that residual bodies are in essence phagocytosed by Sertoli cells (although this worker believes that the residual bodies induce steroidogenesis in the Sertoli cells). Even if the residual bodies are not involved in pituitary feed-back, however, the Sertoli cells could still be important in the mechanism, for their intimate association with the germ cells strongly suggest that they play some part in the control of spermatogenesis. But Setchell[95] has recorded an inhibitory factor in testicular fluid and has suggested the possibility that this may gain access to the circulation by being absorbed in the head of the epididymis.

Further work on testicular feed-back mechanisms is to be encouraged and hopefully an interest in 'inhibin' will be rekindled.

9.2.4 Fluid secretion and the blood–testis barrier

Knowledge of the exocrine function of the testis has been greatly enhanced by the discovery that the rete testis and seminiferous tubules may be cannulated and that fluid may be obtained therefrom. This is possible in rams[96] and rats[97] as well as in other species[98] and the procedure is illustrated in Figure 9.6. Study of the chemical composition of fluids from the rete testis and the seminiferous tubules is made somewhat difficult, because, in spite of the tubuli recti (straight termination of the seminiferous tubules) having a valve-like arrangement as they open into the rete testis[99] there is probably interchange of fluid between the two regions. Nevertheless, analysis of rete testis fluid has been most informative in terms of the secretory activity of the testis, since the chemical composition of the fluid can be compared with that of blood plasma.

Fluid of the rete testis has less sodium, bicarbonate, calcium and magnesium than blood plasma and it also has considerably less phosphate, but it has more potassium and more chloride[100].

There is virtually no fructose or glucose in the rete testis fluid although there is very much more inositol and ascorbic acid than is to be found in blood plasma. There is very little citric acid or glycerylphosphorylcholine, however, and less lactate than in the blood plasma.

A number of amino acids including glycine, alanine, aspartic acid and glutamic acid are to be found in rete testis fluid in higher concentrations than in testicular lymph or blood plasma and glutamine and asparagine can be expected to follow a similar pattern. It is suggested by Setchell[101] that these amino acids might be synthesised in the seminiferous tubules from glucose.

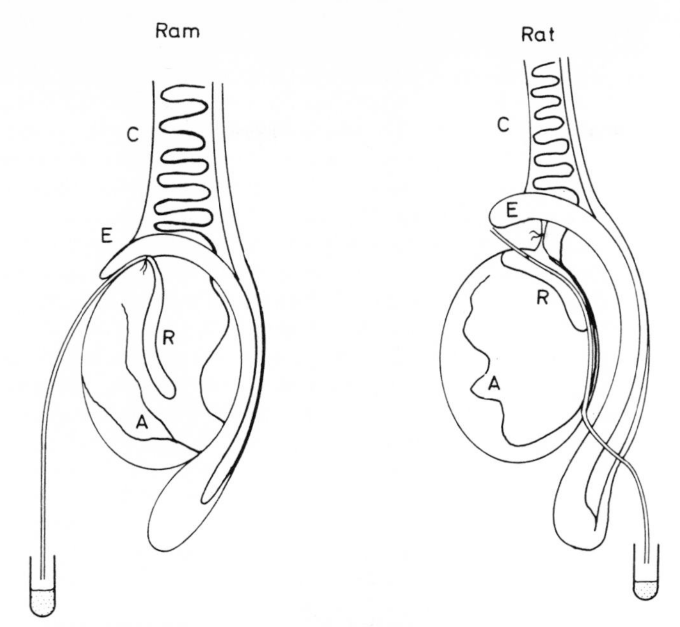

Figure 9.6 Diagrams of the anatomy of the excurrent ducts of the ram and rat and the procedures adopted for collection of rete testis fluid. R, rete testis, central in the ram and lateral in the rat; A, testicular artery; E, caput epididymidis; C, vascular cone (region of the pampiniform plexus)—in these species this comprises the coiled testicular (internal spermatic) artery surrounded by many small veins. In the ram, the caput epididymidis must be reflected and a cannula is placed in the rete testis through the efferent ducts and tied in place. In the rat, a tube with a side-hole is passed through the rete and the side-hole manoeuvred until fluid flows from the catheter (Modified from Setchell and Waites[98], by courtesy of the Editor, *Journal of Reproduction and Fertility*.)

There are other nitrogenous compounds in the rete testis fluid also and urea is present in similar concentrations to those found in blood plasma and a number of proteins are also present. Serum proteins occur in lower concen-minal segment[155]. In the golden hamster, these three regions roughly correspond to the initial segment, the head and body of the epididymis combined, and the tail of the epididymis[157], but this by no means applies to every species of albumin, and, in contrast, the larger molecule of α_2 macroglobulin does not appear to pass readily into the fluid. Globulin also seems to have some difficulty. There seems, therefore, to be some selection of proteins that is not entirely related to molecular size.

Testosterone is also found in the fluid of the rete testis in fairly high concentrations but its origin and fate are as yet unknown and this and other steroids in the fluid offer themselves as interesting subjects for study in the future.

Available evidence suggests that the fluid of the rete testis is actively secreted. Among other features, there is more chloride in the fluid than can be accounted for entirely by filtration and large molecules such as inulin do not get into the fluid. Hence, the passage of substance into the seminiferous tubules and rete testis has attracted considerable attention in recent years.

Dyes do not easily penetrate the rete testis and seminiferous tubules from the blood and this lack of permeability has been studied in some detail by Kormano[103], and a blood–testis barrier comparable to the blood–brain barrier has been described by Setchell[102]. It appears that water, urea, ethanol and bicarbonate pass freely into the rete testis fluid, but that inorganic ions such as sodium, potassium and chloride, together with creatinine and galactose pass in only slowly. Inulin and (^{51}Cr)-chromium EDTA completely fail to penetrate the tubule and permeability to albumin is low. It would seem that this permeability barrier is very much involved in the secretion of fluid by the seminiferous tubules and there is evidence that it is influenced by gonadotrophins and by temperature[100].

Morphological studies represent a major step towards an explanation of how substances penetrate the seminiferous tubules and Clermont[104] in a study of the investments of the tubules described a layer of contractile cells on the outside of the basal lamina, being separated from it by amorphous material. A similar structure was described in the rat and mouse by Lacy and Rotblat[105] and Ross[106] respectively. The cells have been termed 'interlamellar cells' by Clermont and 'myoid cells' by Fawcett *et al.*[107]. The cells are characterised by an abundance of myofilaments and by several pinocytotic vesicles which occur on both the inner and outer surface of the cells. Boundaries between the cells appear to vary, so that sometimes they are short and direct whilst in others they run obliquely. Dym and Fawcett[108] have shown, however, that the extracellular space between myoid cells is often occluded by tight junctions, particularly when the intercellular space runs obliquely. These workers have elegantly demonstrated, moreover, that when the testis is perfused with fixative containing lanthanum nitrate, much of the lanthanum fails to penetrate the seminiferous tubules and the barrier appears to be due to the tight junctions between the myoid cells.

A second barrier exists between Sertoli cells[109]. Understandably, there are no specialised cell membrane attachments between Sertoli cells and germ cells, because presumably, desmosomes or other such specialisations would prevent the constant movement of germ cells towards the lumen of the tubule[110]. In fact, it is probable that cytoplasmic projections of the Sertoli cells that extend around the spermatogonia successively retract or disappear during spermatogenesis and allow the germ cells to move luminally as they become spermatocytes.

However, special junctional complexes occur between adjacent Sertoli cells[109,111]. Periodic narrowing of the intercellular space forms a series of gap junctions which are recognised also by bundles of fine filaments in the subjacent cytoplasm and cisternae of endoplasmic reticulum lying parallel to the cell surface. These junctions serve as a barrier to markers larger than lanthanum and occlusion is made still more effective by tight junctions that occur in addition to each gap junction or nexus. These tight junctions result in there being two distinct compartments of the seminiferous epithelium, first, the

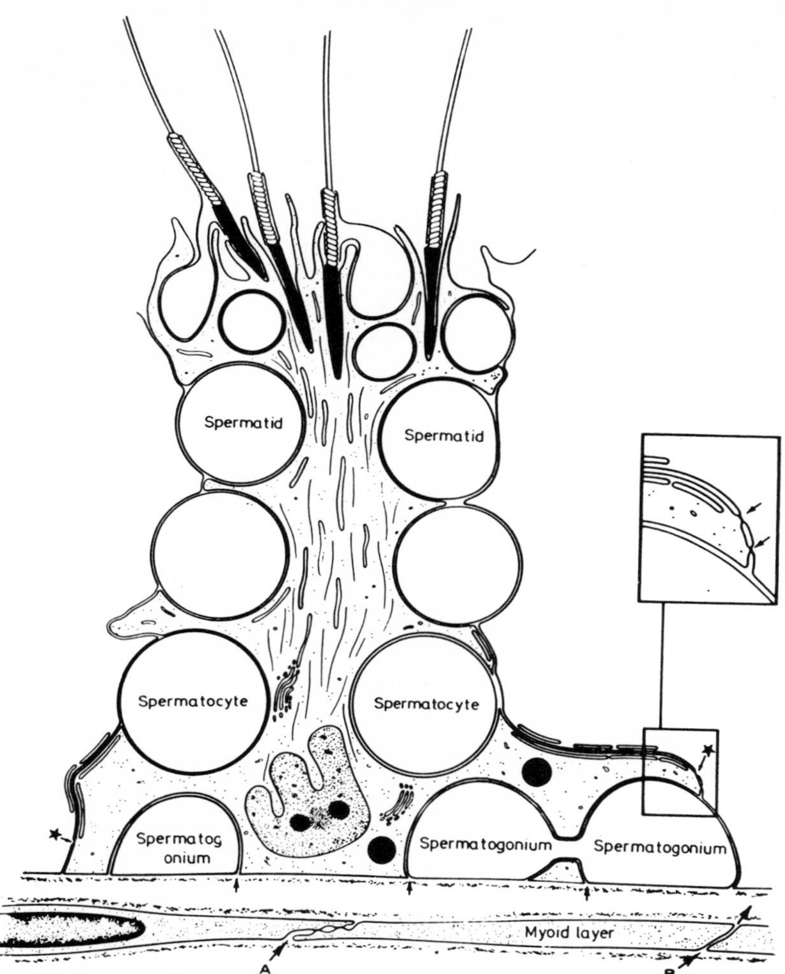

Figure 9.7 A diagram depicting the localisation of the blood-testis barrier and the compartmentalisation of the germinal epithelium by tight junctions between adjacent Sertoli cells. Note the germ cells and their relationship to a columnar Sertoli cell. The primary barrier to substances penetrating from the interstitium is the myoid layer. The majority of cell junctions in this layer are closed by a tight apposition of membranes as indicated at A. Over a small fraction of the tubule surface, the myoid junctions exhibit a 200 Å wide interspace and are therefore open as depicted at B. Material gaining access to the base of the epithelium by passing through open junctions in the myoid layer is free to enter the intercellular gap between the spermatogonia and the Sertoli cells. Deeper penetration is prevented by occluding junctions (stars) on the Sertoli–Sertoli boundaries. These tight junctions constitute a second and more effective component of the blood–testis barrier. In effect, the Sertoli cells and their tight junctions delimit a *basal* compartment in the germinal epithelium, containing the spermatogonia and early preleptotene spermatocytes, and an *adluminal* compartment, containing the spermatocytes and spermatids. Substances traversing open junctions in the myoid cell layer have direct access to cells in the basal compartment, but to reach the cells in the adluminal compartment, substances must pass through the Sertoli cells. (From Dym and Fawcett[108], by courtesy of the Editor, *Biology of Reproduction* and Academic Press)

basal compartment containing spermatogonia and early leptotene spermatocytes, and secondly, the adluminal compartment with spermatocytes and spermatids (Figure 9.7). It is apparent from this arrangement that substances which succeed in penetrating the myoid cell layer will immediately gain access to the basal compartment, but they may still be barred from reaching the adluminal compartment with its associated spermatocytes.

This fascinating subject of the blood–testis barrier requires further investigation, particularly in relation to the kinetics of spermatogenesis, the site of action of hormones and the passage of drugs into the seminiferous tubules and rete testis.

9.3 THE EPIDIDYMIS

9.3.1 Functional activity

9.3.1.1 Sperm maturation

The mammalian epididymis is an extremely long, coiled duct that is attached to the testis through excurrent ducts, or ductuli efferentes, and extends over the surface of the gonad to join the ductus deferens distally. Spermatozoa must, therefore, be transported through the epididymis before they can be discharged from its distal end at ejaculation, and during periods of sexual rest, spermatozoa accumulate at this distal end which is known as the tail of the epididymis or cauda epididymidis. Not surprisingly, therefore, this region of the duct appears to have become adapted to preserve spermatozoa for fairly long periods. During their passage through the epididymis, spermatozoa undergo several changes in both structure and function which collectively are said to reflect a process of maturation. Maturation really begins in the testis because the spermatozoa spend some time in the seminiferous tubules before and during their transport into the epididymis. But testicular spermatozoa appear to be functionally inadequate and further maturation changes in both morphology and function take place in each of these independent cells as they migrate through the epididymis. These changes result in the spermatozoa acquiring the ability to fertilise an ovum if placed in the female tract and a capacity for vigorous motility if removed from the epididymis and exposed to oxygen or to a glycolisable sugar such as fructose.

Perhaps the most striking morphological change in epididymal spermatozoa is the migration of the cytoplasmic droplet (remnant of spermatid cytoplasm) from the neck of the spermatozoon to the end of the midpiece, an event which is accompanied by changes in the fine structure of the droplet[112,113]. Changes in the size and shape of the acrosome have also been described in the rabbit[114] and the guinea-pig[115], although evidently these are not apparent in the human spermatozoon as it passes through the epididymis[116]. The acrosome of rabbit spermatozoa appears to contract down over the nucleus during maturation[114,117] and increased specific gravity in the spermatozoa has also been reported[118,119]. It would seem, therefore, that during transit through the epididymis, the spermatozoa undergo a process of dehydration. Lindahl *et al.*[120] also claimed that the light-reflecting capacity of the spermatozoa changes during their maturation. In addition, more disintegrated spermatozoa are found in samples taken from upper regions of the epididymis than in those from the tail of the epididymis[121,122] and increased capacity for motility *in vitro* is a well known feature of maturing spermatozoa. Other changes, such as occur in Feulgen-positive material[123] and in the relative resistance of the spermatozoa to cold shock[124] also characterise the maturation process. Evidence that maturation of epididymal spermatozoa is accompanied by a change in their membrane characteristics was produced by Ortavant[125] and later by Glover[126] and Amann and Almquist[122] with the observation that mature and immature spermatozoa showed a difference in response to differential staining. This change is confirmed by an alteration in the electrophoretic properties and autoagglutination of the spermatozoa during their passage through the epididymis[127,128]. Recently, recorded differences in the binding of positively charged particles of colloidal ferric oxide to the plasma membrane of spermatozoa at different levels of the epididymis provides more evidence of a change in the cell surface during sperm maturation[129] and new techniques such as freeze fracturing for the examination of cell membranes should prove useful in further investigations of the subject.

There is little doubt that spermatozoa mature in the epididymis, but it cannot be claimed categorically that each cell does so at exactly the same rate. It has been shown in rabbits that the bulk of the spermatozoa attain their ability to fertilise by the time they reach the lower part of the body of the epididymis[130-132], but there might nevertheless be some that mature earlier or later. Even if the time of maturation is standard, it is unlikely that each spermatozoon travels through the epididymis at the same rate[133]. In order to be certain of what happens, it is important to know whether the process of maturation in the spermatozoa is entirely temporal in which case, given a consistent and congenial environment, the spermatozoa would mature by virtue of an intrinsic ability to do so, or whether a specific or constantly changing environment is an equally or more important factor. In the past, opinions have differed about this, but more recently, Gaddum and Glover[134] put forward the view that both factors were involved in epididymal maturation, because they found that whilst spermatozoa trapped in the head of the epididymis by duct ligation were actively motile when exposed to the air, and showed some morphological signs of maturation, their individual movement was not normal. Subsequently, Gaddum[135] has shown that the nature of movement in immature spermatozoa differs from that of mature forms, so it looks as if immature spermatozoa, blocked by epididymal occlusion, do not mature completely. This is confirmed by tests of their fertilising capacity[136,137]. But the site of ligation, that is whether it is low or high, does appear to be important, because very high ligation prevents any serious signs of maturation from appearing in the spermatozoa at all.

Immature spermatozoa taken from the testis, or upper parts of the epididymis of a mammal are usually either immotile or only weakly vibratile, but they can be invigorated by adjusting their environment. Thus, the addition of physiological saline may stimulate an immature spermatozoon so that it becomes actively and progressively motile. This has been seen, if not reported in print by a number of workers. It rather suggests that the environment might be the all important factor in the acquisition of progressive motility by spermatozoa but the precise relationship between motility and fertilising capacity remains unclear. It might, in fact, be argued from existing evidence, that the development of a capacity for motility *in vitro* relates mainly to the environment and that its connection with sperm maturation is largely fortuitous. Clearly, the significance of motility in relation to fertilising capacity needs to be better established. This means that more work is needed on metabolic systems in mature and immature spermatozoa.

9.3.1.2 Sperm transport and storage

In a wide variety of mammalian species the lumen of the tail of the epididymis is larger than that of any other area of the duct and this is mostly due to the accumulation of spermatozoa and probably fluid also, at this site. It is controversial whether the tail of the epididymis is to be regarded as a reservoir for spermatozoa, but there is no doubt that many spermatozoa are stored here even if there is some constant overspill into the ductus deferens. From the point of view of the ejaculate, spermatozoa in this storage area are very important, because unlike the rest of the duct, the tail of the epididymis contracts together with the ductus deferens in seminal emission. This is due to its sharing the same nerve supply as the ductus deferens[138]. In rabbits, stimulation of the hypogastric nerves elicits contraction of both the tail of the epididymis and the ductus deferens, but does not appear to affect the head of the epididymis or the body of the epididymis[138]. It would seem, therefore, that upper regions of the duct have their own separate autonomic innervation, although species differences are known to exist[139-141]. On anatomical and physiological grounds, therefore, the frequency of ejaculation would not necessarily be expected to influence contractile activity in the head of the epididymis or the body of the epididymis, and there is evidence in rabbits that it does not affect the rate of transport of spermatozoa through these regions of the duct[142]. This is not surprising in view of the fact that only a small part of the tail of the epididymis is emptied at ejaculation. However, in man, the tail of the epididymis is relatively small compared with that of many other mammals and it might just be that, in contrast to these other species, frequent ejaculation in man effects the rate of sperm transport. More detailed studies on the innervation of the human epididymis would be helpful in relation to this problem, which is a rather important one because from a clinical point of view, it is helpful to know whether or not the tail of the epididymis can be completely emptied by very frequent ejaculation. In man it looks as if it is possible, but with frequent ejaculation, progressive fatigue of the seminal emission reflex must also contribute to low sperm counts. This applies also in the reverse direction so that after long periods of sexual abstinence high sperm counts may be expected, although under these circumstances, other factors, such as greater sexual excitement, give rise to general systemic effects also[138] and these may affect the quality of the ejaculate.

But what happens to spermatozoa stored in the male tract when coitus is not taking place regularly? Many of them are probably eliminated by means of masturbation or spontaneous emissions, and doubtless some are flushed from the urethra by urine. It has, in fact, been claimed that in rams the number of spermatozoa in the urine corresponds to sperm output by the testis[143]. However, it would be unwise to generalise about all species or individuals, and an accurate quantitative estimate of sperm output by the testis is, in any case, rather difficult to make, unless it is measured directly. Nevertheless, estimates of daily sperm output in rams made by Voglmayr *et al.*[96] after cannulation of the rete testis correspond quite closely to values obtained by Lino *et al.*[143] from exhaustive ejaculation tests. Amann and Almquist[144] however, claimed that unwanted spermatozoa may be destroyed and absorbed in the tail of the epididymis of bulls and phagocytosis of sperms in this area has subsequently been claimed[145]. It has also been claimed that phagocytosis occurs in the human epididymis[146, 147], but there is no substantial evidence that it is associated with periods of sexual rest. The possibility cannot be entirely discarded, however, for there is evidence presented later in this chapter that massive evanescence of spermatozoa can occur when the hormonal climate of the epididymis is changed and individual hormone levels might well fluctuate considerably even in one and the same individual.

The situation in man is again obscure, but in animals it would seem that if under normal circumstances any spermatozoa are removed from the epididymis, the process is inefficient. In vasectomy there is a striking swelling of the epididymis over a period of time and in the rabbit, prolonged vasectomy results in the epididymis, especially the tail of the epididymis, occupying infinitely more space in the scrotum than the testis. In rams and rats spermatocoeles are almost inevitable following vasectomy and in the golden hamster the tail of the epididymis often ruptures altogether.

It is interesting that in man, apart from occasional spermatocoele formation, these affects do not seem to occur following vasectomy, although some clinicians have observed transitory enlargement of the epididymis. This suggests that spermatozoa and epididymal plasma are removed more efficiently from the human epididymis following vasal occlusion, or alternatively that the flow rate of fluid and spermatozoa from the human testis is much lower than that in animals. However, whether any given spermatozoon is ultimately passed out whole from the urethra during long periods of sexual rest, or is destroyed within the epididymis, it will first lose its fertilising capacity, because the viability of spermatozoa in the tail of the epididymis cannot be preserved indefinitely[148,149]. If the tail of the epididymis is isolated from the rest of the duct by ligation, contained spermatozoa eventually disintegrate, but even after long periods, a large number of disintegrated forms still remains in the lumen. This is further evidence that any mechanism for the removal of excess spermatozoa, if it exists at all, is not very efficient.

The disintegration of spermatozoa in the tail of the epididymis may be hastened and augmented by increasing the local temperature[150], but even then, mature spermatozoa appear to be more resistant than immature ones. In artificially cryptorchid rabbits, for instance, mature spermatozoa in the tail of the epididymis do not disintegrate until well over a week has elapsed[151] even though they lose their fertilising capacity much more rapidly[152]. By contrast, many immature forms are decapitated within a day of cryptorchidism.

It is interesting that in most animals the main sign of sperm degeneration is decapitation and yet in man a decapitate spermatozoon is relatively uncommon and does not appear as a characteristic outcome of increased body temperature, or local application of heat to the scrotum. In man, these conditions appear rather to evoke oligospermia[153]. Could it be, therefore, that in man, when the temperature of the scrotum is raised, mature spermatozoa are destroyed more rapidly and more completely in the epididymis than they are in most other mammals? Much more comparative information is needed on the physiological characteristics of the epididymis before such a question may be answered. Nevertheless, it seems that increased temperature curtails the survival time of epididymal spermatozoa in scrotal mammals, but it is not known yet whether the effect is directly on the spermatozoa themselves or whether it results from a heat response of the epididymal cells. There is plenty of evidence that the cells lining the epididymis play a role in the preservation of spermatozoa but their relative significance in different pathological conditions needs to be elucidated.

9.3.2 The evolution and structure of the epididymis

The evolution of the epididymis clearly demonstrates that the duct is much more than a simple conducting tube, because it has evolved, together with the ductus deferens, from the primitive mesonephric duct. The take-over by the vertebrate testis of a duct from the urinary system must surely be significant, bearing in mind especially, that in some vertebrates such as some of the teleosts, it has not occurred. Thus, in some teleost fishes the sperm duct is short, structurally simple and entirely separate from the urinary duct[154].

The beginnings of the take-over are to be seen in lung fishes (Diptera) where the testis joins to the urinary duct through the caudal end of the kidney, but the final stages of take-over only appear to have been achieved, as in amniotes, with the advent of a metanephric kidney possessing its own excurrent duct. An intermediate position is occupied in this context by some sharks and urodeles where the metanephros, in spite of having its own duct, utilises the mesonephric duct also. It seems, therefore, that in many primitive vertebrates, sperm storage, and sperm maturation if it occurs, take place within the testis, but that in the course of evolution, these functions have been acquired by the mesonephric duct. Thus, in the salamander, the toad (*Bufo bufo*) and in some snakes, spermatozoa are stored for considerable periods in ampullae or in a large central canal of the testis. In mammals, by contrast, there is no room for such storage, because the seminiferous tubules enter into a relatively small rete testis.

Recently, Glover and Nicander[155] have outlined this evolutionary trend and attempted to draw up anatomical homologies in the excurrent duct system of a variety of vertebrates. The work illustrates how the extratesticular sperm store (cauda epididymidis) has come to lie close to the testis as a result of testicular descent, since in animals with abdominal testes this sperm store is remotely situated from the gonad. This is confirmed by observations on true testicond mammals in which the testes lie close to the kidneys[13,156]. Moreover, although sperm ageing may simply be a continuation of sperm maturation, the two processes, as far as the bulk of the spermatozoa are concerned, take place in distinct and separate regions of the epididymis and, whilst sperm maturation is a fairly rapid process, sperm ageing is not.

Thus, whilst dynamic changes take place in spermatozoa as they traverse upper regions of the epididymis, further changes appear to be halted or restrained as the spermatozoa enter the sperm store. Such a situation demands a sophisticated system and marked regional differences in the histological structure of the epididymis indicate that a functionally complex system exists.

As many as eight distinct regions have been described in histological accounts of the epididymis, but recently it has been suggested that from a functional standpoint, the mammalian epididymis may be regarded as having three main regions, namely an initial segment, a middle segment and a terminal segment[155]. In the golden hamster, three regions roughly correspond to the initial segment, the head and body of the epididymis combined and to the tail of the epididymis[157], but this by no means applies to every species and failure to recognise sharply defined species differences has led to confusion. In the guinea-pig, for example, the so-called body of the epididymis is, in reality, the initial segment as structurally represented in other species. Therefore, a terminology based on microscopical features is more meaningful than the conventional terms of head, body and tail of the epididymis which appear to be used purely for convenience and may not refer to the same functional areas in different species.

The initial segment is the most proximal part of the epididymis and is recognised by having high epithelium with long straight stereocilia (Figure 9.8(1) and Figure 9.9(1). The middle segment, according to this terminology, is complicated and consists of three main subregions. The first of these, consists of very vacuolated cells, whilst in the next (intermediate) part vacuoles are less profuse (Figure 9.8(2) and Figure 9.9(2)). In this region also, however, spermatozoa in the lumen are very concentrated (Figure 9.8(2) and Figure 9.9(2)). This suggests that fluid absorption occurs in the first part of the middle segment and there is plenty of supporting evidence for this. The distal part of the middle segment shows only moderate vacuolation of the cells and a lower concentration of spermatozoa in the lumen (Figure 9.8(3) and Figure 9.9(3)). In the terminal segment, which in most species includes the two limbs of the tail of the epididymis, the epithelium and the stereocilia are much lower than in other regions (Figures 9.8(4), 9.8(5), 9.9(4) and 9.9(5)) whilst the lumen is larger and packed with spermatozoa. Most of this area serves as the sperm store.

There are four clearly defined cell types in the epididymal lining. These comprise mainly the principal cells and basal cells, but two other types may be present, namely, the so-called clear cells and halo cells. These are most apparent in the epididymis of rats and hamsters, and whilst halo cells are present in the rabbit epididymis also, clear cells are not to be seen in this species.

The principal cells are the tall columnar cells that are found throughout the extent of the epididymis (Figure 9.10(1)). They possess microvilli of variable length and type and different degrees of pinocytosis are to be seen taking place in these cells, indicating that absorption is characteristic activity of the epididymis. Thus, pinocytotic vesicles and multivesicular bodies are present at most levels of the duct. Other features of the principal cells

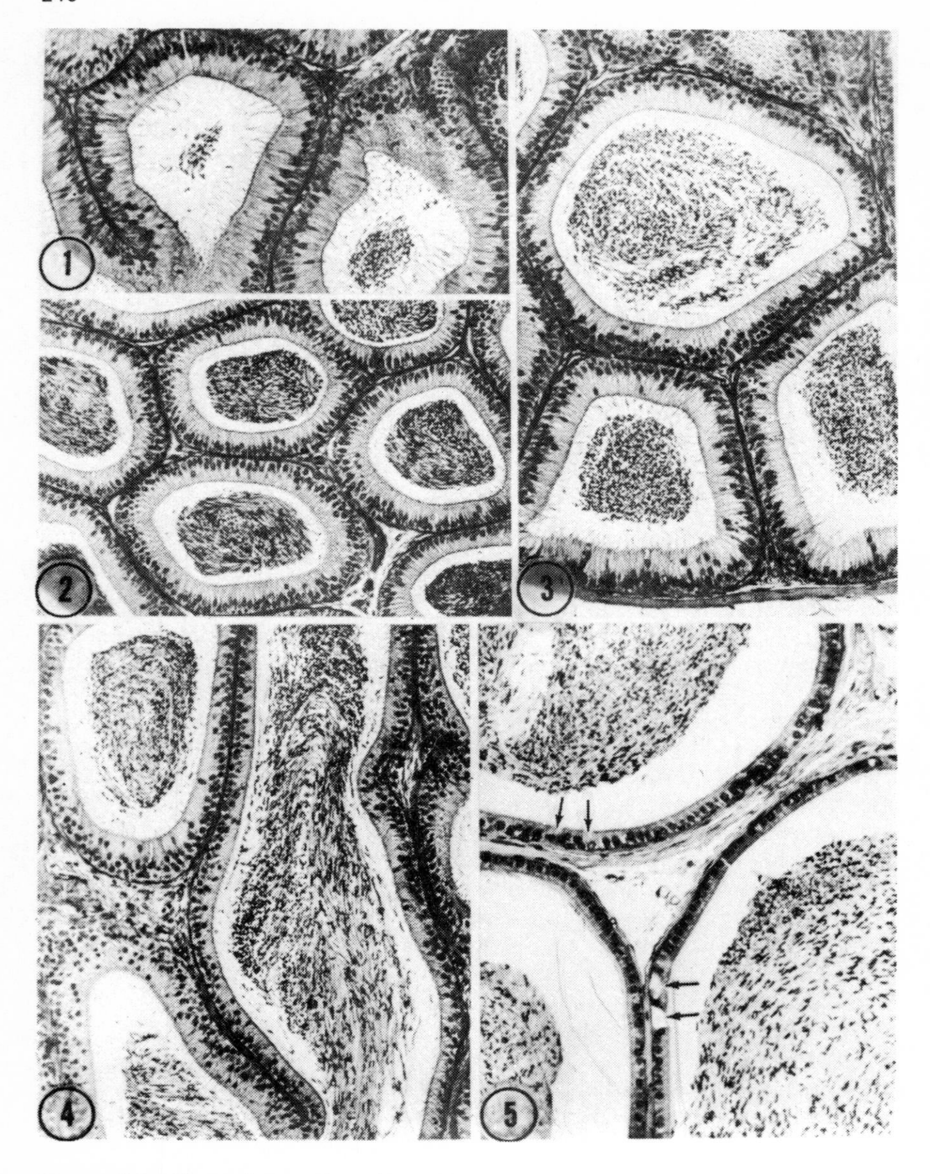

Figure 9.8 Photomicrographs of paraffin sections of a hamster epididymis. Bouin fixation, H and E. × 160 (Reduced $\frac{2}{3}$rds on reproduction)

(1) Initial segment
(2) Middle segment, intermediate part
(3) Middle segment, distal part (proximal corpus)
(4) Terminal segment, proximal part
(5) Terminal segment, distal part (middle cauda)

Arrows show a number of clear cells (see text). (From Nicander and Glover[157] by courtesy of Cambridge University Press.)

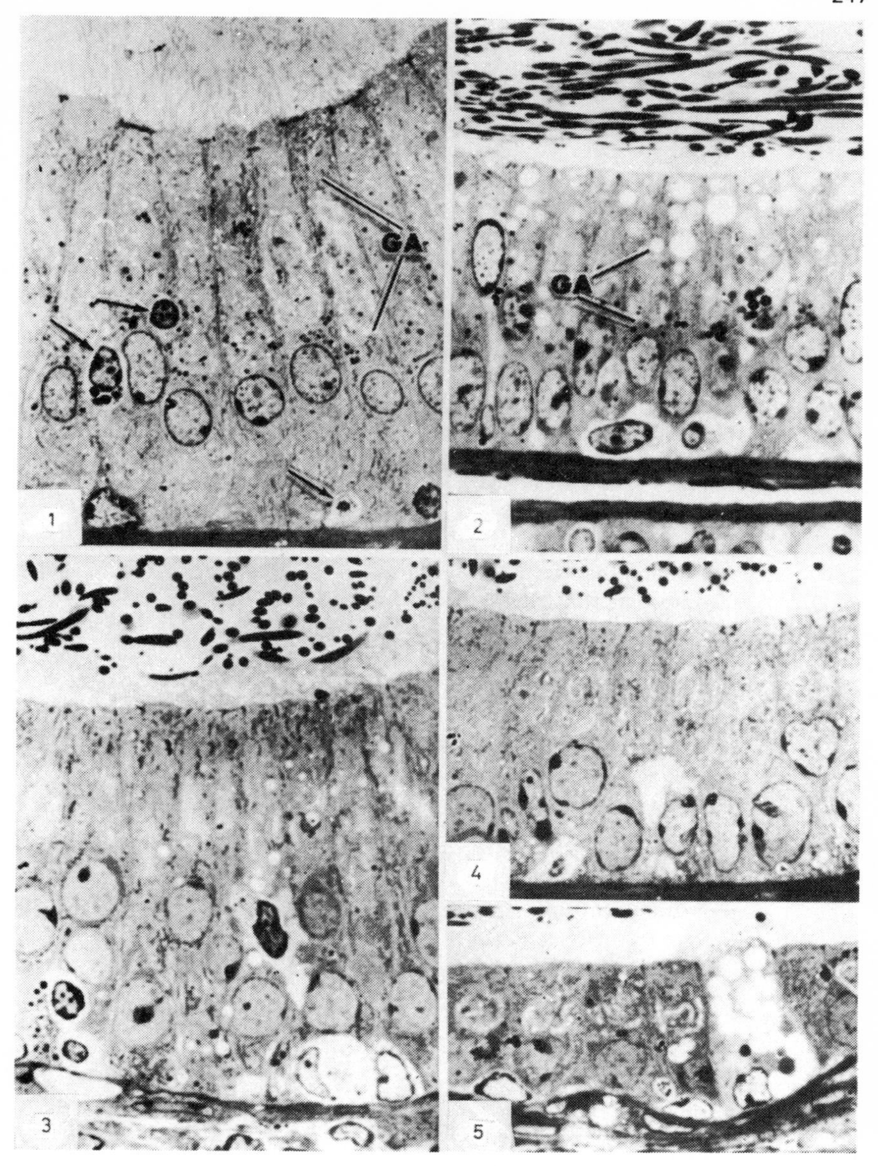

Figure 9.9 Photomicrographs of Epon sections from a hamster epididymis. Fixed by vascular perfusion with glutaraldehyde, followed by osmium tetroxide. Toluidine blue (see text). × 1 200 (Reduced ⅔rds on reproduction)

(1) Initial segment, with long, coarse stereocilia, large Golgi area (GA), columnar cell nuclei at a high level, and numerous supranuclear granules. Three small light cells (halo cells) indicated by arrows.

(2) Middle segment, intermediate part. Note densely crowded sperm in lumen, low stereocilia and mainly apical vacuoles. The Golgi apparatus (GA) is of moderate size.

(3) Middle segment, distal part, with few vacuoles but numerous mitochondria in apical cytoplasm. Moderate amounts of spermatozoa in the lumen.

(4) Terminal segment, most anterior part. Lower columnar cells with apparently homogeneous cytoplasm.

(5) Terminal segment, distal part (middle cauda). Low epithelium with dense cytoplasm (From Nicander and Glover[157], by courtesy of the Editor, *J. Anatomy*, and Cambridge University Press.)

indicate that they are metabolically very active and there is a preponderance of euchromatin in lobulated nuclei and a highly developed Golgi apparatus. Lysosomes and other cell inclusions, whilst occurring variably in different species, are usually present. Mitochondria are fairly randomly distributed throughout the cytoplasm, and although rough ER is extensive, many free ribosomes are invariably to be seen in the cytoplasm also. Large dense bodies are often to be seen in a supranuclear position and are lysosomal[158]. Definite morphological evidence of secretion by the principal cells is so far lacking, which makes distinct areas of selective secretion difficult to define and thus strengthens the view that the epididymis is primarily an organ of absorption. Certainly, the cells of the head of the epididymis absorb both fluid and particulate matter[159,160], but as mentioned later, chemical analyses show that some secretion also occurs.

Apically situated tight junctions between principal cells (Figures 9.10(1) and 9.10(2)) are probably associated with absorption of fluid, since they occur in other fluid absorbing organs such as the gall bladder. And it is also interesting that the cisternae of smooth ER are sometimes seen to form an open channel with the lumen.

Clear cells may occasionally be seen interposed between the principal cells, particularly in the terminal segment of rodent epididymides. These cells were originally named by Reid and Cleland[161] on the grounds of excessive vacuolation and they are the so-called 'holocrine cells' of Martan[162]. However, as Hamilton[163] points out, better fixation methods have now shown these cells to contain densely packed lipid droplets and a rather amorphous nucleus. Pinocytosis may be seen taking place at the basal border of these cells. They often contain supranuclear vacuoles and an abundance of supranuclear lysosome-like bodies may also be demonstrated (Figure 9.10(2)).

The epithelium of the epididymis is pseudostratified, but true apical cells that do not extend to the basement membrane have been described in the epididymis of the rat by Reid and Cleland[161]. These cells appear to be different from principal cells but may be their precursors. The observations of Suzuki (personal communication) on the epididymis of the rabbit and hamster indicate that apical cells actually reach the basement membrane, albeit perhaps by tenuous cytoplasmic extensions (Figure 9.11(1)).

Deeper in the epithelial lining of the epididymis, halo cells are periodically to be seen from time to time (Figure 9.11(2)). These cells have some resemblance to lymphocytes, but like basal cells they do not have many organelles. They often contain dense lysosome-like bodies and even if they are not lymphocytes they may perform a similar function. A possible objection to this interpretation of their function is that they are unable, it would seem, to pass out into the lumen because of the tight junctions between the principal cells.

The basal cells of the epididymis are exceptionally interesting. They rest on the basement membrane of the duct, and by contrast with the principal cells they have few organelles or cell inclusions (Figure 9.11(3)). Mostly, basal cells are in contact with principal cells and appear to have a special association with them in that the two interdigitate in a complex manner and exhibit definite adhering junctions at the cell surface[163]. When basal cells are in contact with halo cells these junctions are not to be seen and the plasma membrane surfaces are smooth.

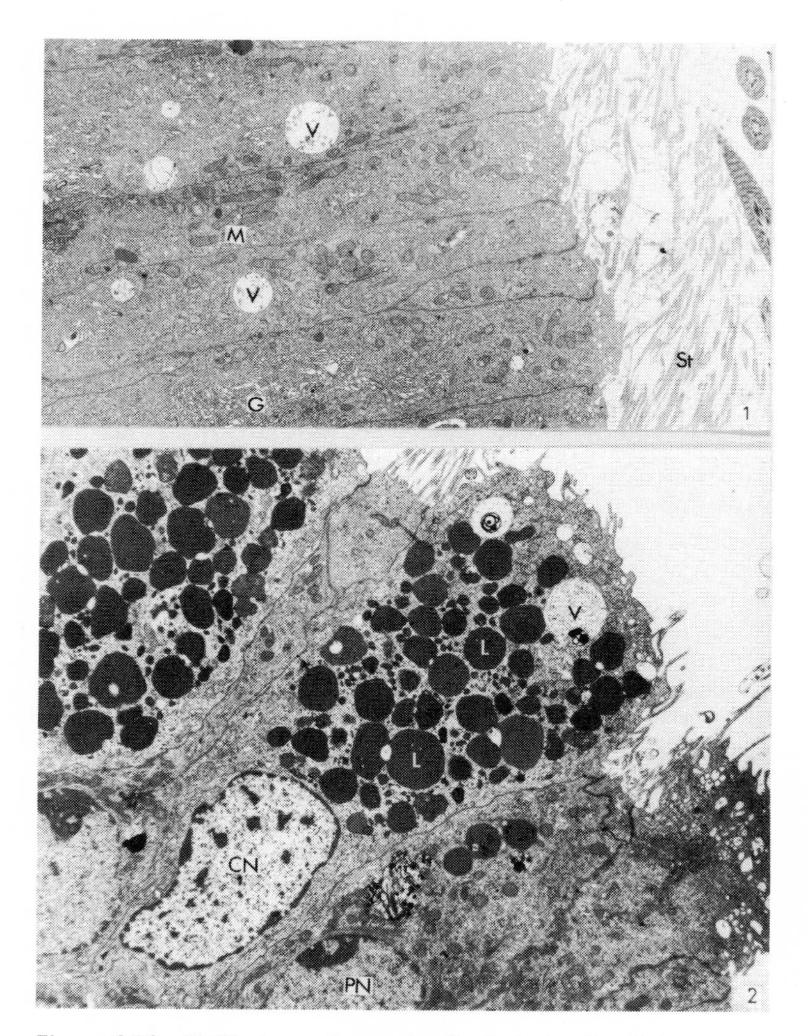

Figure 9.10 (1) Electron micrograph of principal cells of the hamster epididymis. V, Vacuoles; M, Mitochondria; G, Golgi cisternae. × 7 500 (2) Clear cell in the epididymis of the hamster. V, Vacuoles; L, Lysososme-like bodies; CN, clear cell nucleus; PN, Principal cell nucleus. × 7 500 (Reduced ½ on reproduction.)

In material obtained from a vasectomised man whose testis and epididymis had been exposed to x-irradiation, we have observed that the basal cells in the head of the epididymis (intermediate part of the middle segment) contained granules with a lot of lipofuschin pigment (Figure 9.11(4)). In the same material, similar granules were seen in the principal cells and Suzuki[164] believes that these are associated with the absorption of luminal debris and other material that may be passed from the principal cells to the basal cells. If this is true, the basal cells may effectively act as scavengers.

Loose areolar tissue surrounds the tubule of the epididymis, binding its convolutions together and containing a number of wandering cells. Plain muscle surrounds the tubule also, as might be expected, and continues as a massive muscular wall in the ductus deferens. Spontaneous contractions in upper regions of the epididymis are more marked and more rapid than those in the tail of the epididymis[135, 165, 166]. This might be expected on the grounds of differences in functional demand between the two areas bearing in mind that only the tail of the epididymis contributes to the emission reflex, but that higher levels of the duct are concerned with sperm transport.

The blood supply to the epididymis has been studied in great detail by Kormano[167,168] and Kormano and Penttilä[169,170]. These workers demonstrated the distribution of capillaries arising from arteries running in the connective tissue septa between the epididymal convolutions, and they showed that there is more permeability of blood vessels in the head of the epididymis than in other regions. Hamilton[160] believed that this may be related to fenestrations in the capillary wall, but it should be rewarding to examine vascularity and blood flow in a wider variety of species, since there are quite striking species differences in the structure of the epididymis. Detailed information about vascularity, blood flow and capillary permeability are extremely important to an understanding of epididymal physiology, because the facility with which substances pass from the blood stream to the lumen of the duct is clearly of great significance. Moreover, the manner in which epididymal cells handle materials passed to them is important and has led to an interest in the chemical composition of epididymal plasma, since this must be intimately connected with the problem.

It is important to bear in mind when trying to interpret the chemistry of epididymal fluid that whilst the basic contribution is from the testis, materials are added or withdrawn as the fluid traverses the epididymis. Thus, the chemical composition of the plasma at different levels of the epididymis should be instructive in terms of the activity of epididymal cells, particularly when compared to the composition of testicular plasma. Spermatozoa themselves might affect the composition of the plasma, however, particularly if they degenerate and some intracelluar constituents leak out.

A further problem in attempting to study epididymal plasma is that only very small quantities of fluid are available for analysis, that the fluid is difficult to collect without contamination by blood, lymph and epididymal cells, and that resort has often been made to *postmortem* material. Chemical estimations on postmortem material are bound to be quantitatively suspect, but even so, definite patterns in the chemistry of epididymal plasma have been established and appear to be remarkably constant between species.

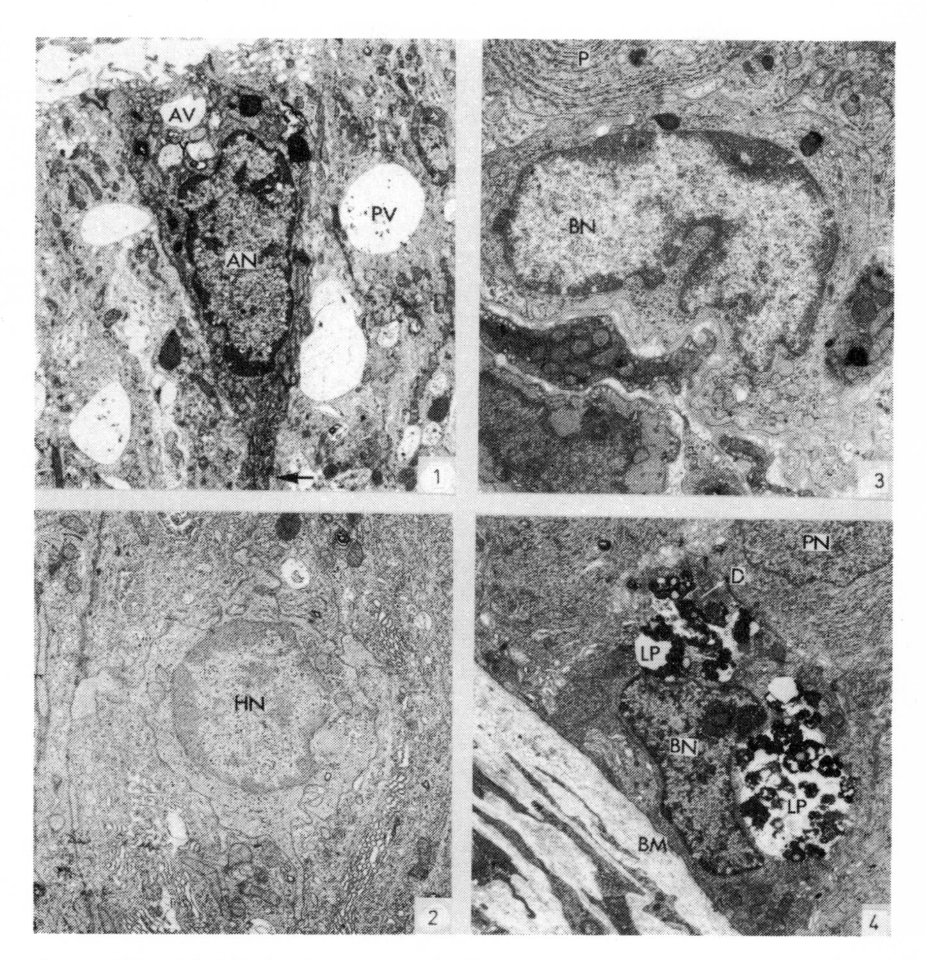

Figure 9.11 (1) Apical cell of hamster epididymis. AV, Apical cell vacuole; AN, Apical cell nucleus; PV, Principal cell nucleus. Arrow pointing to cytoplasmic extension (see text). × 7 500
(2) Halo cell in hamster epididymis in between principal cells. HN, Nucleus. × 15 000
(3) Hamster basal cell. P, Rough ER of principal cell; BN, Basal cell nucleus. × 15,000
(4) Human basal cell (x-irradiated). PN, Principal cell nucleus; D, Desmosome, BN, Basal cell nucleus; BM, Basement membrane; LP, Lipofuchsin body. × 7 500 (Reduced ½ on reproduction)

(These electron micrographs were prepared by Miss F. Suzuki. Staining with lead citrate and uranyl acetate.)

Changes in the concentration of inorganic ions are perhaps outstanding in this context.

Scott *et al.*[171] showed a progressive decrease in the concentration of sodium and chloride ions from the rete testis to the tail of the epididymis in rams and a comparable effect has been demonstrated in bulls and boars[159,172-174] and in rats[175]. The sharpest decline in sodium ions evidently occurs in the head of the epididymis, but, by contrast, potassium increases in concentration in the head of the epididymis and only declines afterwards, except in the rat where it continues to become more concentrated right through to the tail of the epididymis[175]. The concentration of divalent ions such as calcium and magnesium changes little throughout the epididymis and of particular interest is the fact that in spite of substantial changes in sodium, potassium and chloride, the osmotic pressure of epididymal plasma remains relatively constant throughout the length of the duct. Similar values have thus been obtained for the osmotic pressure of rete testis fluid and fluid from the tail of the epididymis respectively[175]. However, there is a discrepancy between the osmotic pressure calculated from total ionic strength and that obtained by direct measurement. The calculation assumes, however, that sodium, potassium, chloride and bicarbonate are the main ions concerned in osmotic regulation and on this assumption the discrepancy is apparently greater with epididymal plasma taken from the tail of the epididymis than with rete testis fluid. It seems, therefore, that other substances must be involved in the maintenance of osmotic pressure in the epididymal fluid, particularly towards the distal extremity of the duct, but although several substances, including glycerylphosphorylcholine, present themselves as possible candidates for such a role, the true situation is as yet unknown. It is, nevertheless, worth while examining organic constituents of the plasma and their possible relationship to sperm maturation and sperm preservation.

Glycerylphosphorylcholine or GPC occurs in progressively higher concentrations in tubular plasma between the testis and the tail of the epididymis[159] but since seminal plasma contains much less GPC, the concentration in ejaculated semen is relatively low[176]. The precise mechanism of GPC synthesis is unknown, but it is accepted that it is produced in the epididymal epithelium and is not the result of breakdown of phospholipids in the spermatozoa[177]. The precursors and enzymes concerned in this synthesis are still unknown, but Martan and Risley[178] attempted to explain GPC synthesis and secretion by proposing a so-called 'holocrine secretory cell cycle' involving the clear cells. Unfortunately, this interesting suggestion has not received support in that in some species no clear cells are to be found[179] and in any case, cell turnover would involve replacement, and frequent mitoses would be anticipated. Such mitoses are not evident and indeed the cell population of the epididymis appears to be particularly static. Clearly, more work is needed on the synthesis and function of G.P.C. It may contribute to the maintenance of the osmolarity of the epididymal plasma, but until its degree of dissociation in the epididymal lumen is known, it is not possible to judge this.

The protein content of tubular fluid also increases between the rete testis and the cauda epididymis in rams and bulls[174]. Setchell *et al.*[180] found that the concentration of glutamic acid followed a similar pattern and Setchell *et al.*[181] implied that some of this amino acid was synthesised in the epididymal cells.

Hamilton[163] has speculated that this amino-acid might be the result of glutamine metabolism and might be derived from the formation of glucosamine-6-phosphate from fructose-6-phosphate. This possibility is suggested by a high level of epididymal sialic acid which represents amino sugars that might be bound to polysaccharide, lipid or protein. Sialic acid is undoubtedly associated in some way with the metabolism of epididymal cells and accordingly appears to be related to the level of androgen, upon which the metabolism of these cells is so dependent[182].

Details of carbohydrate metabolism in the epididymis are not yet available, however, and more experiments in this field should provide valuable information about the metabolism of the cells. A number of relevant enzymes have been identified, but most attention has so far been focused on determining the distribution of these enzymes by histochemical methods and the location of PAS-positive material, particularly that which is resistant to diastase.

As might be expected, species differences exist, but the distribution of glycogen, for instance, seems to be fairly extensive, having been reported in the basal cells of the human epididymis[183] and in the principal cells of certain regions of the epididymis in other species[184-187]. Manelly[188] suggested that PAS-positive particles were secretory granules and might represent a source of mucoprotein for assisting the transport of spermatozoa through the epididymis, but Nicander[189] is of the opinion that the particles are associated with absorption.

Fouquet and Guha[190] have demonstrated glycogen synthetase and phosphorylase in the epididymis of the hamster and Allen[191] has shown that in the epididymis of the mouse, glucose-6-phosphatase is variably distributed throughout the length of the duct. The presence of glucose-6-phosphatase is especially interesting in that in other tissues it is concerned with the release of sugar into the blood stream. The question thus arises as to whether this applies in the epididymis or whether the enzyme is associated with the release of carbohydrates into the lumen of the duct. The occurrence of glycosidases, including β-galactosidase and β-*N*-acetylglucosaminidase, is diffuse within the cells of the epididymis and it seems that the presence of β-glucuronidase is variable[192].

Allen and Slater[193] have studied the lactate dehydrogenase–diphosphopyridine–diaphorase system in the epididymis of mice and whilst the enzymes seem to be homogeneously distributed in the cytoplasm, there is a marked apical concentration of lactate dehydrogenase in the head of the epididymis. Lactate, succinate, and malate dehydrogenases have also been reported in the head of the epididymis in bulls[194] and Niemi[195] has described the presence of succinate and β-hydroxybutyrate dehydrogenases in the apical cytoplasm of the ductus deferens in rats.

Since lactate dehydrogenase and β-hydroxybutrate dehydrogenase are involved in glycolysis and in the oxidation of fatty acids respectively, glucose-6-phosphate dehydrogenase is concerned in pentose cycle activity, and succinate and malate dehydrogenases are vital enzymes in the Kreb's cycle, it is clear that a whole new area of research into the metabolism of epididymal cells is available for exploration. Allen and Slater[196] have also recorded the presence of esterases in the epididymis of the mouse and the activity of nearly all these enyzmes is affected by testosterone. The presence of hydroxysteroid dehydrogenases in the epididymal cells[197], therefore, is particularly interesting in view of their association with steroid metabolism. 3-β-hydroxysteroid dehydrogenase, for example, is needed in the conversion of pregnenolone and dehydroepiandrosterone to progesterone and androstenedione respectively, and Moniem[198] has described the distribution of this enzyme in the epididymis of the rat. Taken together with evidence of steroid synthesis by the cells of the epididymis *in vitro*[199-201], it seems possible that the epididymis may synthesise steroids *in vivo*, but even if it does, the physiological significance of this activity is at present obscure.

Acid phosphatase and β-glucuronidase are localised as discrete granules and may be regarded as being lysosomal[202]. Their respective association with the metabolism of phosphate and mucopolysaccharides suggests that they might be involved in the metabolism of substances absorbed from the lumen of the duct.

Hydrolytic enzymes such as alkaline phosphatase, adenosine triphosphatases and adenosine monophosphatase occur in the subepithelial connective tissue of the epididymis and in endovascular structures. Judging from their physiological activity in other parts of the body, it is reasonable to assume that these enzymes are involved in active transport between the vascular structures and the epithelial cells lining the duct. The occurrence of alkaline phosphatase activity in the apical parts of epididymal cells similarly suggests active transport between the cells and the contents of the lumen also.

Lipid metabolism in the epididymis is extremely obscure although the presence of carnitine[203] is interesting and should be investigated further, particularly if it is associated with phospholipids and hence, with cell membranes.

It is difficult to be sure of the origin of all the enzymes and other organic substances occurring in the epididymal plasma and this emphasises the difficulties of interpreting the chemical composition of this fluid. The enzyme glutamic-oxaloacetic transaminase (GOT) appears to arise mostly from the spermatozoa, especially if they degenerate[204]. It has been suggested that an increase in LDH following castration is related to degenerating spermatozoa[205] and yet, GPC appears to have its origin in the epididymal cells. Inter-relationships between the epididymal cells, the epididymal plasma and the spermatozoa are, however, important, and, therefore, the control of metabolic activity in the epididymal cells, and its ultimate influence on the spermatozoa, need to be studied.

9.3.2.1 *Hormones and the epididymis*

That the epididymis is dependent upon androgen for its metabolic activity is well known and so it is not surprising that the maturation and viability of spermatozoa are also androgen-dependent[206,207]. Normally, the androgen is testicular in origin, but there is some evidence that pituitary-controlled extratesticular androgen might also be involved[207].

Niemi and Tuchimaa[208a] have shown radioactively labelled testosterone to be mostly concentrated in the wall of the epididymis but Back, D. J.[208b], has shown that the concentration of radioactivity in the lumen of the cauda epididymidis is greater than that in the cells lining that area. Thus, the exact mechanism of action of the hormone is as yet unknown, but it seems that, as in several other steroid dependent tissues, it acts by becoming attached to 'primary receptors' in the cells[209-211]. However, the active form of hormone in the epididymal cells is not yet established and although it is known that in a number of androgen dependent tissues, testosterone is reduced to dihydrotestosterone (DHT), there is still no unequivocal evidence that in the epididymis DHT is the active form. But the finding of Lubicz-Nawrocki[212] that DHT is equipotent or even more potent than testosterone in maintaining the viability of epididymal spermatozoa suggests that DHT might well be an intermediate metabolite of testosterone in the cauda epididymidis. This applies even more to 5α-androstanediol which has been shown to be much more potent than testosterone in maintaining sperm viability in the epididymis[212].

Thus, quite a lot is known already about the hormonal control of epididymal function and recently, considerable advances in this subject have been made. Important gaps still remain, however, and include the passage of hormones from the bloodstream to the cells at different levels of the epididymis and the manner in which epididymal cells utilise androgen. No doubt the biosynthesis of enzymes is governed by androgen[213] so that various secretion products will in their turn be under its control. The synthesis of messenger RNA might also be influenced by androgen[214] and it is interesting that in the epididymis of the golden hamster, Moniem[215] has shown that removal of the testes results in a loss of histochemically detectable RNA from the epididymal cells.

Another interesting role of androgen is to stimulate or at least to regulate contractile activity in the epididymis[216]. Rhythmical contractions of the testes have been observed[217] and these would appear to be in part due to activity of the testicular capsule[218] but also to contraction of the seminiferous tubules themselves[219,220]. Hormonal control of these latter contractions has not been fully explored although Niemi and Kormano [221]have demonstrated the sensitivity of the contractile elements to oxytocin. These testicular contractions presumably serve to flush out the spermatozoa and fluid into the epididymis via the efferent ductules, but Macmillan[222] showed that the passage of spermatozoa through the epididymis is not dependent upon a *vis a tergo* from the testis and, therefore, the androgen dependent contractions of the epididymis would appear to be responsible for the transport of spermatozoa through the duct.

Finally, it has been shown that when androgen is withdrawn by removal of the testes, epididymal spermatozoa not only lose their viability and rapidly degenerate but may completely disappear from the lumen[205,207]. This also occurs in some seasonally breeding mammals when androgenic levels decline at the end of the mating season[23] and in these mammals it appears to result from autolytic rather than phagocytic activity. This is not to say that phagocytosis such as has been reported in the human epididymis by Phadke[147] may not occur, but it does not seem to be the main means for eliminating epididymal spermatozoa *in situ*. Available evidence suggests, therefore, that very little dissolution of spermatozoa occurs in the epididymis provided the levels of circulating androgen are adequate. Both species and individual differences may be expected, but the situation poses an interesting question

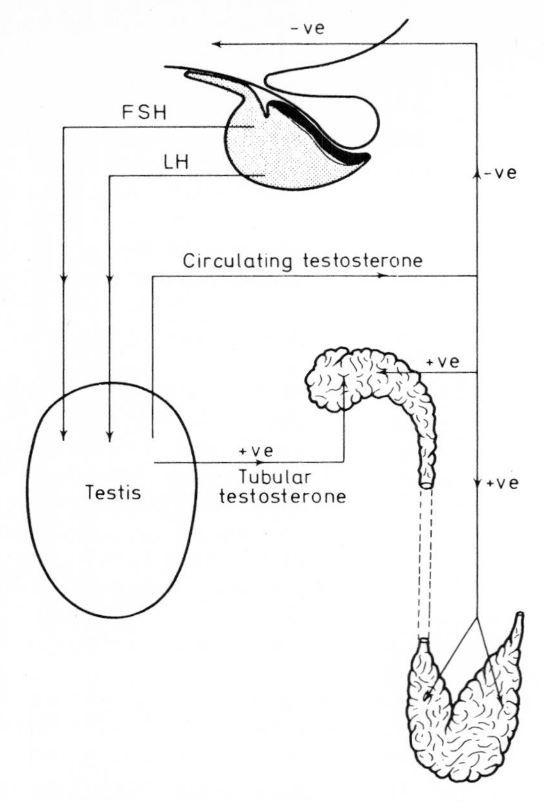

Figure 9.12 Diagram illustrating how the epididymis receives testosterone from two different sources, and the indirect role of the pituitary on epididymal function

in relation to vasectomy, where presumably hormone levels remain normal.

Although circulating androgen is obviously indispensable to epididymal function, testicular fluid, which is reasonably rich in testosterone, might also participate. It has been shown by Reid[223] that development of the caput epididymidis, unlike that of the cauda epididymidis, is dependent, upon a direct connection with the testis. Moreover, it is known that proximal regions of the epididymis and the efferent ductules are primarily absorptive and it is possible that cells of these upper regions take up testosterone together with other substances. In the cauda epididymidis it is doubtful if the epididymal plasma has much testosterone in it, although the position might be different in early pubescent animals[224] and no firm data exist in this connection. It seems that whilst the cauda epididymidis is primarily dependent on circulating androgen the cells of the caput epididymidis might require tubular testosterone in addition (Figure 9.12). Tubular testosterone may, of course, be important to the spermatozoa themselves, particularly in the early stages of maturation, but there is as yet no conclusive evidence that androgens act directly on spermatozoa in the epididymis.

9.4 OTHER ACCESSORY ORGANS OF REPRODUCTION

Secretions of the epididymis undoubtedly contribute to the fluid portion of an ejaculate, but the main volume is made up of secretion from the prostate and bulbo-urethral (Cowper's) gland, and particularly, in those species where it exists; from the seminal vesicle. Fluid, especially precoital fluid, is also added by urethral glands, and in stallions, the glands of the ampulla of the ductus deferens contribute most of the 'tail-end' portion of an ejaculate[225]. The anatomy and physiology of these accessory organs differs widely between species and the well known and comprehensive review of Mann[177] is recommended for information on the subject.

Nevertheless, it is worth while emphasising in the present context that these glands are all sensitive to circulating steroid and sometimes they appear to respond to them in rather a special way. It is not surprising, therefore, that in addition to hormonal studies on the epididymis, investigations have been undertaken recently to determine how the cells of the prostate and seminal vesicles deal with steroid that is delivered to them in the blood stream.

Injected testosterone has been shown to be converted into dihydrotestosterone (DHT) in the prostate, seminal vesicles and preputial glands of the rat[226]. In the prostate this occurs in the cell nuclei and attempts have been made to characterise the binding protein[227, 228]. *In vitro* studies have shown that prostatic nuclear chromatin selectively binds 5α-DHT[229] and whilst testosterone metabolism yields DHT, androsterostanediol and androsterone, only DHT is retained by the nuclei[229].

In the rat, testosterone is converted more readily to DHT in prostatic cells than in the cells of the seminal vesicles[230] and this sort of information may be most valuable when trying to interpret the response of accessory glands to various hormonal conditions. More detailed information on this fascinating topic will surely be anticipated with interest.

The penis and scrotum are to be included in external accessory organs of reproduction and in different species there are a number of specialised accessory reproductive glands including preputial glands and such secretory structures as the dorsal gland on the back of the rock hyrax. Many of these glands are also sensitive to steroid, but since they constitute such a large subject in themselves they will not be considered here.

9.5 THE ANALYSIS OF SEMEN

9.5.1 General appearance of ejaculates

The analysis of semen is usually undertaken for the purpose of predicting an individual's fertility, but it may also be of value in diagnosing abnormalities of the male genital tract. It is noticeable, however, that over the last 15 years, few new methods of semen appraisal that are of immediate value to the clinician have come to light, although well known tests have been modified from time to time. The appraisal of semen in domestic animals has been reviewed in great detail by a number of authors[231], but since it has been recognised in man that the husband is frequently the infertile partner in a childless marriage, and since both AIH and AID have been more widely used in recent years, assessment of semen quality in the human being has become increasingly important. The following remarks will, therefore, be confined mainly to the problem of semen analysis in man.

It is important to be aware at the outset, that the fertility of a man cannot be predicted as accurately as that of many domestic animals. Several semen characteristics in bulls, for instance, have been correlated with fertility[232] but this cannot be achieved in man with the same precision since adequate control data are almost impossible to obtain and since there is much more variability in semen quality even among apparently fertile males. Thus, many assumptions have to be made when using semen quality as an indicator of the fertility of a man, whilst a statement about abnormal function of his male reproductive organs may be made with more certainty.

Functional status of the reproductive organs may be revealed in the general physical characteristics of an ejaculate and this applies to both human and animal semen. Unlike the semen of most animals which is odourless, however, human semen has a characteristic tarry smell which is due to the organic base 'spermine'[177]. The ejaculate is greyish in colour and contains gelatinous material which liquefies on standing at room temperature. Usually this liquefaction takes about 20 min to complete and the significance of the process in semen appraisal will be mentioned later.

Typically, a man ejaculates between 2 and 4 ml of semen, although occasionally much larger volumes are produced. Voluminous ejaculates are to be viewed with some caution though, because they are often associated with a low sperm concentration and even low total sperm numbers. It is not uncommon (personal observations) for complete azoospermia to be associated with a large volume of ejaculate.

Occasionally, of course, there may be contamination of an ejaculate with urine and discoloration of semen samples may arise also from other causes such as infections of the reproductive tract, blood pigments (icteric patients may discharge yellow-coloured semen) and detritus. Sometimes, freshly ejaculated semen appears streaky due to the presence of cellular debris and microscopically, human semen is well known to contain prostatic casts. However, a good quality ejaculate will appear as fairly clear fluid once liquefaction is complete. After this time the motility of the spermatozoa is optimal and after measurement of ejaculate volume, microscopic examination may next be undertaken.

9.5.2 The spermatozoa

An early test in the analysis of semen is the test for motility of the spermatozoa. In the semen of bulls or rams, where the sperm concentration is between a million and four million per μl, motility may be assessed quantitively by measuring impedance change frequency (ICF). This is not possible with relatively dilute samples and so the microscopic observation of sperm motility is still undertaken in the analysis of human semen. Various methods have been described for assessing motility and that of Harvey[233] is quantitatively the most useful. However, motility is frequently scored arbitrarily and this is not entirely valueless with an experienced observer, particularly since the assessment usually involves observations on the quality of sperm movement as well as an estimate of the number of motile spermatozoa. MacLeod[234] for instance, uses the term 'motility index' in combining the percentage of motile spermatozoa with the rate of forward progression of the cells.

Nevertheless, there is plenty of evidence from animal experiments that on its own, sperm motility is a poor criterion of potential fertilising capacity. For example, sperm motility may be retained following fairly low dose x-irradiation, yet the ability to fertilise may be lost. Clinically, therefore, poor motility is more meaningful than good motility, because asthenospermia (defined by Eliasson[235] as less than 50% motile spermatozoa) at least indicates that something is wrong.

That some motile spermatozoa may be infertile is perhaps not surprising, because situations might be envisaged where midpiece activity is normal, but where the nuclear DNA of the spermatozoa or the lipoglycoprotein complex of the acrosome is adversely affected. It has been shown, for instance, that linkage between DNA and nuclear proteins may be altered and aberrations in the amino acid composition of the nuclear proteins may occur[236]. Such changes may not necessarily influence motility.

In human semen, optimal motility is only seen when liquefaction of the sample has taken place and, therefore, when there is a failure of liquefaction, the addition of a trypsin-like agent to the semen may be required before proper examination of the spermatozoa can be made.

The whole question of sperm motility has been admirably reviewed by Nelson[237], but although several papers on the subject have appeared since, it is evident that much more research is required before the mechanisms of motility, and therefore, the significance of motility, can be fully understood.

The concentration of spermatozoa in semen only acquires significance when several samples are examined, because any one sperm count may reflect the efficiency of emission and ejaculation rather than sperm production. By examining a number of samples from each patient, however, and by ensuring that a constant period of sexual abstinence is maintained before the samples are produced, a general picture of a patient's sperm production may be obtained.

Counting the spermatozoa in diluted samples using a haemocytometer slide is the usual method of assessing sperm concentration in human semen. Absorptiometric methods would be much quicker and probably more accurate, but their development in this context is hampered by the variable presence of crystals and cellular debris as well as a tendency for various diluents to cause clumping of the spermatozoa. Nevertheless, more efficient methods are needed and more work in this field is demanded. Similar difficulties are experienced with the use of electronic cell counters[238,239] although some of the problems can doubtless be overcome by suitable adjustment of the diluents.

A sperm concentration of less than 30 000 000 per ml in a human ejaculate is usually taken as constituting oligospermia, but the significance of sperm concentration over total sperm number is not clear. Pregnancies have occurred when the husband's sperm count has been less than 10 000 000 per ml and it might be emphasised that if semen volume is large, the total number of spermatozoa in an ejaculate might be considerable. Thus, low sperm counts may not necessarily signify a quantitative deficiency in sperm output but may rather indicate excessive production of seminal plasma. Invariably, however, so-called 'low density' samples are not very potent and when artificial insemination is prescribed for the condition it is necessary to use only the 'high sperm' fraction. This subject will be discussed again later.

Sperm concentration is markedly affected by the frequency of ejaculation, so that when a semen sample is examined, it is important that the period of sexual abstinence before the sample is produced is known and preferably standardised. Without this information a semen analysis is essentially valueless. Frequent ejaculation, it has been claimed, may almost empty the sperm reserve in the tail of the epididymis[240] since the sperm count first declines sharply but later stabilises at a low level. If this is true, then the phenomenon provides a means of assessing sperm production as opposed to sperm output, but since the efficiency of emission also decreases with repeated stimulation, interpretation of these effects must be made with care. The subject is an important one, however, and should be examined more closely.

In domestic animals there is an inverse relationship between the fertilising capacity of an ejaculate and the incidence of certain morphologically abnormal spermatozoa. But in man, the significance of sperm abnormalities is more difficult to ascertain, because a wide variety of morphological forms of spermatozoa appears to be characteristic. These different forms have been described by a number of workers and have fairly recently been reviewed by Freund[241] who believes that some standard classification is needed. This would certainly facilitate easier communication between different workers and might avoid some confusion, but it is difficult to understand how it would assist interpretation. For example, sperm tails coil in very variable fashion, so to classify them all as 'coiled tails' might cause confusion and might obscure differences in the origin of the abnormality.

It is interesting, however, that in man many aberrations in sperm structure occur in the head of the cell and are, therefore, presumably mainly of spermatogenic origin, but more attention should be paid both to the origin and the cause of these abnormalities. Increased interest in the cytogenetic aspects of human male fertility is, therefore, encouraging, and experiments where chromosomal activity is related to sperm morphology are particularly useful (see, e.g., Skakkebaek and Beatty[242]). Human spermatozoa differ markedly in size even in one and the same ejaculate, so that microsperms and macrosperms are easily identifiable. Why should this occur especially in man, what is its significance in terms of fertility and what is the origin of the discrepancy? None of these questions has been answered. Moreover, new methods of scanning, counting and measuring the size of cells are now available and yet they do not appear to have been adequately adapted to the examination of spermatozoa. It is to be hoped that methods will be improved in the future, because MacLeod[243] has shown that from the point of view of practical techniques available for routine diagnostic purposes, there has been little advance since the pioneer studies of Moench[244]. However, from the research point of view, the electron microscope has proved a valuable innovation and much more is now known about the basic structure of the human sperm cell. Very recent techniques such as the binding of colloid to the sperm cell surface[129] are likely to prove useful in understanding further the development of spermatozoa and their reaction to the environment. The histochemical demonstration of glycoprotein binding[245] also has promise and the microscopic observation of proteolytic activity in spermatozoa is also most interesting[246].

For many years differential staining has been widely used in the appraisal of semen for determining the proportion of so-called 'live' spermatozoa. Nigrosin eosin appears to have been the most popular differential stain, largely, one suspects, because it is relatively easy to use. In bulls, however, the stain has proved useful in assessing fertility[232] and in several other species it has been shown that differential counts on nigrosin eosin smears are reasonably reproduceable[247]. This stain does not appear to have been widely used in the appraisal of human semen, because opinions differ as to its validity with human spermatozoa. There is no doubt that human spermatozoa respond differentially to nigrosin eosin, but it is difficult to give a meaningful interpretation of the results. However, Eliasson and Treichl[248] have described a method for estimating the percentage of living cells by using supravital staining and this may prove to be a more useful technique to human seminologists.

The manometric measurement of oxygen uptake by spermatozoa has long been used experimentally as a measure of their activity, but more recently the oxygen consumption of human spermatozoa has been measured polarographically using Clark electrodes[249]. Normally, oxygen uptake by human spermatozoa is very low and Eliasson[250] has shown that this is due to a factor in the seminal plasma that appears to depress oxidative metabolism in the mitochondria. Thus, in human spermatozoa oxygen uptake is increased after washing in Ringer or upon transfer to a salt solution[251,252].

Anaerobic and aerobic fructolysis may also be used as measures of metabolism in ejaculated spermatozoa, but, like oxygen uptake, these rates express activity of the spermatozoa rather than their fertilising ability. Although good quality semen samples contain vigorously motile spermatozoa, therefore, motility in itself may be regarded as a separate entity from the mechanism of fertilisation. The midpiece of a spermatozoon is the centre of the cytochrome system and in addition to the coenzyme ATP it has a variety of glycolytic and respiratory enzymes. The isozyme of lactate dehydrogenase (LDH_{IV}) is, for example, sperm-specific. The midpiece is also rich in lipids, particularly plasmalogen and lecithin. Thus, this area of the spermatozoon represents the source of energy for movement and the enzyme adenosine-triphosphatase (ATPase) which occurs in the tail fibrils is likely to be responsible for the transfer of energy from ATP to contractile elements in the flagellum.

By contrast, the instruments of fertilisation lie in the acrosome, which contains enzymes such as hyaluronidase, proteinases, esterases and glycosidases which are released for penetration of the envelopes of the egg.

Since much yet needs to be learned about the chemistry of the interaction of spermatozoa and their fluid environment, chemical tests on the semen must at this stage have a particular place in the diagnosis of malfunction in accessory organs rather than in the prediction of fertility, and on occasions, it is important to distinguish between the two.

9.5.3 The seminal plasma

It is now well-known that in a variety of mammalian species the accessory sex organs contribute to the composition of ejaculated semen in a characteristic manner. Thus, in man, integrity of function in the seminal vesicle may be assessed by determining the content of fructose in the ejaculate as well as inositol, prostaglandins and other substances including enzymes. On the other hand, prostatic function may be revealed by the content of citric acid, particular proteins and acid phosphatase. Thus, in prostatitis, for example, the citric acid content of ejaculated semen is invariably reduced. The possible value of various other seminal constituents as tools in the diagnosis of reproductive dysfunction should also be investigated. In this context, GPC, sialic acid, carnitine, acetyl carnitine, glutamic acid and hypotaurine are of interest. Electrophoretic analysis of proteins is also recognised as being of increasing importance in the diagnosis of abnormal reproductive function in the male, particularly in relation to prostatitis[253].

Enzymes in semen may also yield very valuable information about secretory activity in the accessory organs. Infections of the prostate, for example, not only result in decreased concentration of zinc, cholesterol and citric acid in the semen, but also cause a decrease in acid phosphatase activity[254]. By contrast, the LDH fraction of semen is elevated in prostatistis, seminal fluid has an additional band to the usual five LDH isozymes and this has been called LDH_x. It is an extremely interesting isozyme in that it appears to be associated with the spermatozoa themselves, and might, therefore, be used for estimating enzyme leakage.

The liquefaction of human semen *in vitro* is a complex enzymic process, which has been discussed in some detail by Mann[177], but it is important that more details of the exact mechanism involved are determined, because failure of liquefaction is a surprisingly common defect in human ejaculates. The coagulum may be digested with trypsin, however, and it is interesting that Hirschhäuser and Kionke[255] have demonstrated a specific inhibitor for trypsin in human seminal plasma. These workers have suggested that this substance inhibits proteinase on the acrosome of spermatozoa and thus prevents their penetration into cells other than the ovum. This is an unusual idea and bears further consideration.

9.6 THE STORAGE OF SEMEN AND ARTIFICIAL INSEMINATION

9.6.1 Deep freezing spermatozoa

On the face of it, it might seem difficult to separate the subject of sperm preservation from artificial insemination, because stored semen is usually used when insemination is indicated. However, it is important that the success of semen storage can be tested *in vitro* and it is this aspect of the subject that will be mainly dealt with here.

Although it is well known that cells in general are very susceptible to freezing and thawing, the introduction of protective substances has enabled mammalian spermatozoa to be stored at low temperatures for long periods without any adverse effects on their fertilising capacity or ability to produce viable embryos.

Since the finding of Polge *et al.*[256] that animal spermatozoa could be protected against the effects of very low temperatures by the addition of glycerol, an enormous amount of work has been carried out on the subject of deep freezing spermatozoa. Many of the earlier papers have been reviewed by Parkes[257] whose account should be consulted for an assessment of the original literature. The research has meant that for some time now the semen of animals has been successfully stored at low temperatures for long periods and predictably, attention eventually turned to the possibility of storing human spermatozoa by similar means. The literature in this field is now also quite extensive.

The usual technique used for freezing human spermatozoa is essentially the same as that used with other species, although doubtless each laboratory will have its own special modifications. In Liverpool, we use a method based on that given to us by Jackson (personal communication) and the method is described here as a general guide.

The semen diluent is prepared by adding 7.0 ml of egg yolk to 18 ml of citrate/fructose/glycerol buffer. This buffer consists of 3.2 g trisodium citrate (Analar), and 2.2 g fructose, made up to 100 ml with distilled water. 28 ml glycerol (Analar) are added to this mixture and the pH is adjusted to 7.4 if necessary. The buffer is stored in 18 ml aliquots at -20 °C. Next, the volume of the semen sample is measured after it has been allowed to liquefy and the sample is transferred to a 10 ml conical flask. An equal volume of the semen diluent is then measured into a graduated centrifuge tube and is added to the semen at a rate of ten drops every 3 min, and the two fluids are thoroughly mixed between the addition of each drop. 0.8 ml aliquots of the diluted sample are next transferred to 1 ml ampoules which are afterwards sealed and incubated in a refrigerator at 5 °C for 30 min. The ampuoles are then clipped into the top of a cane and allowed to stand in liquid nitrogen vapour in a sealed vacuum flask for 30 min. Finally, the ampoules are quickly transferred to the bottom of the cane and the cane is placed into the liquid nitrogen refrigerator. Each of these steps is illustrated in Figure 9.13.

The semen is thawed by rapid removal of the ampoule from liquid nirogen and immersion in water at 35 °C for 2 min. It is then usually left for a further 5 min at room temperature before insemination.

Semen was at first frozen by means of solid carbon dioxide after slow cooling, and samples were taken down to -79 °C, although Bunge[258] has suggested that human semen stores better at -85 °C. Sherman[259] however, described the use of liquid nitrogen for low temperature preservation of spermatozoa and the use of this technique has now become common practice for long term storage. This is not to say that other substances such as isopropyl alcohol[260] which might cool to -70 °C, might not be feasible for short periods of storage.

The history and development of research on frozen human semen has been surveyed by Sherman[261], who explained some of the technical improvements in storage methods that were introduced about 10 years ago, and this worker recommended that further experiments be undertaken to improve diluents and protective substances and to refine the methods of cooling that were then in use. Since then, work in this field has developed considerably, but no substance appears to have superceded glycerol as a practical protective agent. Richardson and Sadlier[262] showed that the toxicity to spermatozoa of glycerol, dimethysulphoxide, ethylene glycol, methyl formamide and methyl acetamide, was about the same, but that dimethyl formamide and dimethyl acetamide were more toxic when used in the same concentrations, namely 2.5%–10%. Dimethyl sulphoxide has been widely applied for preserving cells against the effects of freezing and it has been used in the freezing of animal semen[263] but Zimmerman *et al.*[264] found that with human semen, glycerol produced better results as far as post-thaw motility and the percentage of live spermatozoa were concerned.

Although actively motile spermatozoa are not necessarily fertile, a good semen sample contains a large proportion of them. It is thus important in freezing semen to achieve good post-thaw motility and the criterion is widely used as a test for the success of freezing.

Differential staining has also been used as a test (Zimmerman *et al.*[264]), but it often proves unreliable with thawed spermatozoa (personal observations). There is absolutely no evidence from *in vitro* studies[265] that sperm DNA is adversely affected by current methods of freezing and a very large number of normal babies have been born from artificial insemination with semen previously stored at low temperatures. An abnormal baby resulting from the use of stored semen may, therefore, be regarded as a statistical fortuity and to reflect either an abnormality of the original semen sample or some defect in the mother. Nevertheless, with AID, it is obviously prudent to choose donors whose ability to produce normal offspring is already proven.

Freezing does cause damage to some spermatozoa, however, so that the percentage of motile spermatozoa is usually reduced and the fertilising ability of samples may also be affected[266]. Several inseminations are almost always necessary to achieve a pregnancy with stored semen, but it is difficult at present to assess how far this is due to the adverse effects of freezing and thawing, since the same situation often seems to apply when fresh semen is inseminated[267,268]. The ability of thawed human spermatozoa to penetrate cervical mucus is also progressively impaired by periods of freezing[269].

Attention might well be turned, therefore, to the possible injury caused by freezing to metabolic activity in the spermatozoa, to the function of the acrosome and to the sperm surface. Ackermann[270] produced evidence that exposure to low temperatures may influence sperm hyaluronidase activity, which is normally fairly constant, and he held the view that glycerol preserves this activity in a cold environment. Heeley[271] showed also that the ultrastructure of spermatozoa in some domestic animals was damaged by freezing. The spermatozoa of chinchillas and rams were cited as examples, and it was claimed, that freezing caused gross deformity in the acrosome of boar spermatozoa.

A further problem is that there are species differences in the ability of

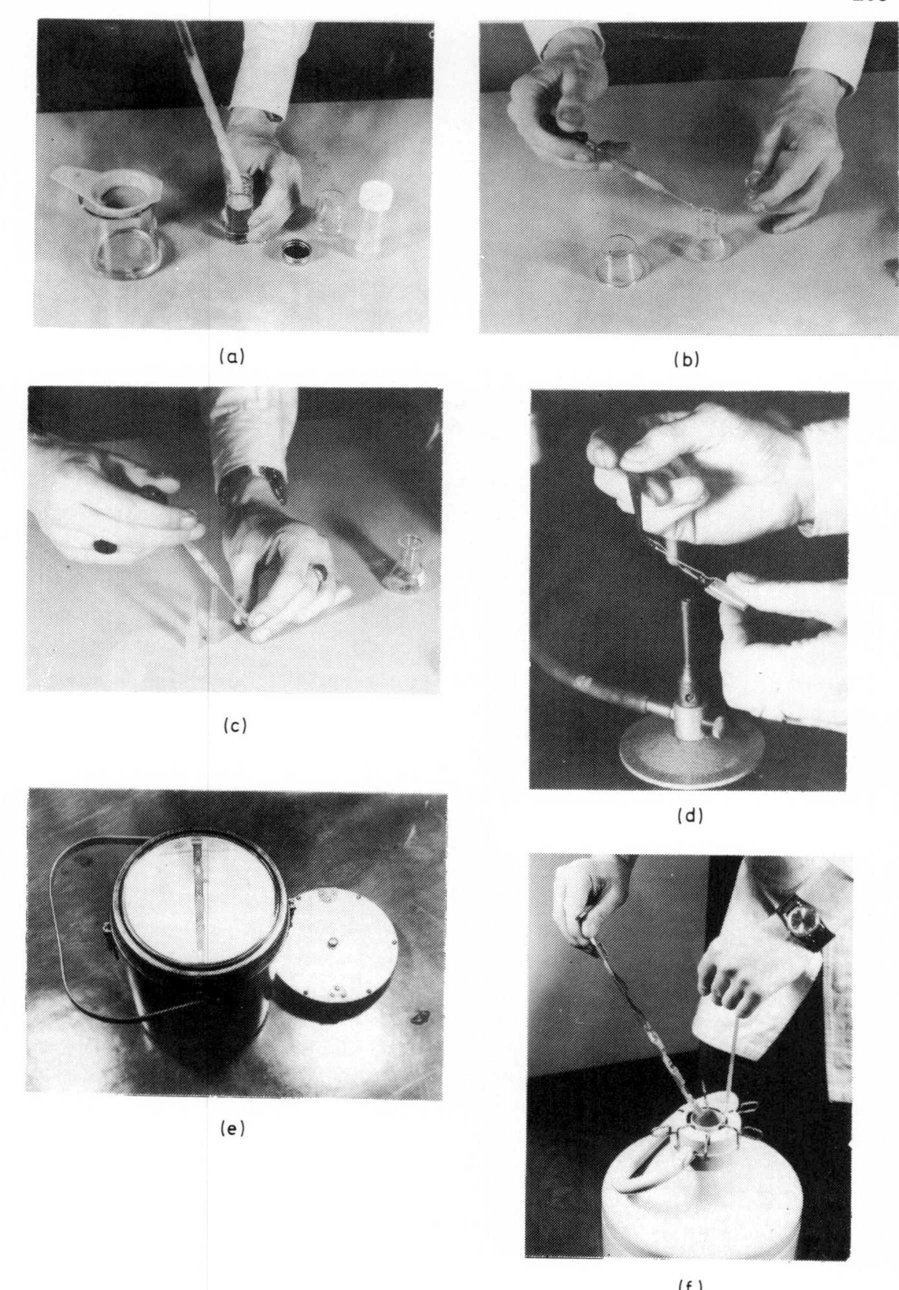

Figure 9.13 (a) The preparation of diluent by addition of egg yolk to the buffer
(b) Dropwise addition of diluent to semen sample
(c) Aliquots of diluted semen being placed in ampoules
(d) Sealing of ampoule containing sample, care being taken to keep ampoule as near upright as possible so as to avoid heating the mixture
(e) Vacuum flask containing cane with samples in it representing when closed, the second stage of cooling, after an initial stage in a refrigerator at + 5.0 °C. About 2 inches of liquid N_2 is placed in the bottom of the vacuum flask
(f) The last stage of cooling and storage. The cane is quickly transferred to liquid N_2 in an immersion storage tank (N_2 refrigerator (see text).)

spermatozoa to withstand the effects of freezing. It has been shown, for example, that in different species of primates survival of spermatozoa after freezing varies considerably[272]. So with the complicated genetic history of different human beings, variation in the resistance of human spermatozoa to freezing is predictable and has been borne out in the light of experience. Thus the semen of different men responds variably to low temperature storage and unfortunately different ejaculates from one and the same person sometimes vary in their resistance to freezing.

Undeniably, therefore, this whole subject needs looking at much more closely with particular attention being paid to the control of media and to the entire question of interactions between spermatozoa, diluents and temperature. Freund and Wiederman[260] made an approach to this subject and Norman *et al.*[273] studied the prolonged survival of human spermatozoa in chemically defined media. This sort of work needs to be expanded and a thorough study of cryobiology in spermatozoa, particularly human spermatozoa, is sorely needed, since work in this field has, for a variety of reasons, tended to be piecemeal. Obviously, if the demand for artificial insemination in man is to become more widespread, safe and efficient methods of semen storage must be ensured.

9.6.2 Artificial insemination

The indications for artificial insemination in man, the ethics of AID and the actual techniques employed are not the province of the biologist, since they form the background to clinical decisions. But looking into the picture from the outside as it were, the potential value of artificial insemination in certain conditions is clearly apparent. The greatest call for AID is likely to be in cases of unresolved azoospermia in the husband, but rhesus incompatibility and predicted disease arising from recessive genes are also situations where AI may be indicated.

Requests for semen storage for purposes of AIH (Artificial Insemination Husband) are on the increase, since vasectomy has become more popular as a means of birth control. Opinions differ as to the desirability of storing the semen of patients before vasectomy and it seems that it is not generally encouraged, although again this is not for the biologist to decide. The storage of semen before vasectomy is certainly quite feasible on technical grounds.

There may be indications for storing human semen from patients who have been subjected to unilateral orchiectomy for seminoma, when the contralateral testis appears to function normally, but where it might have to be sacrificed later in the course of treatment.

Finally, there is the question of concentrating successive ejaculates in cases of oligospermia. Available data offer little guide as to the value of attempting artificial insemination with these cases, except that it must surely be a last resort and the portents are that it will mostly yield disappointing results. The plight of an infertile couple may demand that AIH with bulked semen be undertaken, but it would seem commonsense to ensure beforehand that the couple have been trying to reproduce naturally for a period of several years.

The indications are that the high sperm fraction should always be used with artificial insemination, particularly when ejaculates of large volume and low sperm density are used[274]. A method of obtaining such partitioned ejaculates has been described by Harvey and Jackson[275] and by Harvey[276]. Generally, the first portion of an ejaculate to be discharged is the high sperm fraction, although in some 6% of cases this sequence is reversed[274].

The whole subject of artificial insemination in man has been fully reviewed by Behrman[277,278] and these reviews should be consulted for an overall interpretation of the subject.

Acknowledgements

My sincere thanks are due to all members of the staff of the Unit of Reproductive Biology in Liverpool who have been both helpful and patient. Without their assistance, it is doubtful if this chapter would ever have been completed. I wish to thank especially, Mrs June Kelly, for her help, with the manuscript and the Ford Foundation for making possible our own contribution to the subject.

References

1. Lindner, H. R. and Mann, T. (1960). Relationship between the content of androgenic steroids in the testis and the secretory activity of the seminal vesicles in the bull. *J. Endocrinol.*, **21**, 341
2. Steinberger, E. and Ficher, M. (1968). Conversion of progesterone to testosterone by testicular tissue at different stages of maturation. *Steroids*, **11**, 351
3. Jost, A. (1960). Hormone influences in the sex development of bird and mammalian embryos. *Mem. Soc. Endocrinol.*, **7**, 49
4. Hodson, N. (1970). The nerves of the testis, epididymis and scrotum. *The Testis*, Vol. 1, 47 (A. D. Johnson, W. R. Gomes and N. L. Vandemark, editors) (New York: Academic Press)
5. Waites, G. M. H. (1962). The effect of heating the scrotum of the ram on respiration and body temperature. *Quart. J. Exp. Physiol.*, **47**, 314
6. Waites, G. M. H. and Voglmayr, J. K. (1963). The functional activity and control of the apocrine sweat glands of the scrotum of the ram. *Aust. J. Agric. Res.*, **14**, 839
7. Waites, G. M. H. (1970). Temperature regulation and the testis. *The Testis*, Vol. 1, 241 (A. D. Johnson, W. R. Gomes and N. L. Vandemark, editors) (New York: Academic Press)
8. Harrison, R. G. (1948). Vascular patterns in the testis, with particular reference to *Macropus*. *Nature, London*, **161**, 399
9. Harrison, R. G. (1949). The comparative anatomy of the blood supply of the mammalian testis. *Proc. Zool. Soc. London.*, **119**, 325
10. Setchell, B. P. and Waites, G. M. H. (1969). Changes in the permeability of the testicular capillaries and of the 'blood-testis barrier' after injection of cadmium chloride in the rat. *J. Endocrinol.*, **47**, 81
11. Setchell, B. P. (1970). Testicular blood supply, lymphatic drainage and secretion of fluid. *The Testis*, Vol. 1, 101 (A. D. Johnson, W. R. Gomes and N. L. Vandemark, editors) (New York: Academic Press)
12. Waites, G. M. H. and Moule, G. R. (1960). Blood pressure in the internal spermatic artery of the ram. *J. Reprod. Fert.*, **1**, 223
13. Glover, T. D. (1973). Aspects of sperm production in East African mammals. *J. Reprod. Fert.*, **35**, 45

14. Badenoch, A. W. (1945). Descent of the testis in relation to temperature. *Brit. Med. J.*, **2,** 601
15. Owen, R. (1856). *Comparative Anatomy and Physiology of Vertebrates.* Vol. 2 (London: Longmans Green)
16. Glover, T. D. (1968). *VIth International Congress of Animal Reproduction and Artificial Insemination. The production of spermatozoa in some species of 'Mammalian Testiconda'.* 273 (Paris, 1969)
17. Fawcett, D. W. and Burgos, M. H. (1960). Studies on the fine structure of the mammalian testis. II. The human interstitial tissue. *Amer. J. Anat.*, **107,** 270
18. Christensen, A. K. (1965). The fine structure of testicular cells in guinea pigs. *J. Cell. Biol.*, **26,** 911
19. Christensen, A. K. and Fawcett, D. W. (1966). The fine structure of testicular interstitial cells in mice. *Amer. J. Anat.*, **118,** 551
20. Kretser, D. M. de (1967). The fine structure of testicular interstitial cells in men of normal androgenic status. *Z. Zellforsch.*, **80,** 594
21. Nagano, T. and Ohtsuki, I. (1971). Reinvestigation of the fine structure of Reinke's crystal in the human testicular interstitial cell. *J. Cell. Biol.*, **51,** 148
22. Christensen, A. K. and Gillim, S. W. (1969). The correlation of fine structure and function in steroid-secreting cells with emphasis on those of gonads. *The Gonads*, 415 (K. W. McKerns, editor) (New York: Appleton Century Crofts)
23. Millar, R. P. (1972). Reproduction in the rock hyrax (*Procavia capensis*) with special reference to seasonal sexual activity in the male. *Ph.D. Thesis, University of Liverpool*
24. Pudney, J. A. S. (1972). Cytological and biochemical studies on the testis of the American grey squirrel *Scuirus carolinensis* (Gmelin) during the annual reproductive cycle. *Ph.D. Thesis, University of London*
25. Greep, R. O., Fevold, H. L. and Hisaw, F. L. (1936). Effects of two hypophyseal gonadotrophic hormones on the reproductive system of the male rat. *Anat. Rec.*, **65,** 261
26. Greep. R. O. and Fevold, H. L. (1937). The spermatogenic and secretory function of the gonads of hypophysectomized adult rats treated with pituitary FSH and LH. *Endocrinology*, **21,** 611
27. Lipsett, M. B. (1970). Steroid secretion by the human testis. *The Human Testis*, 407 (E. Rosemburg and C. A. Paulsen, editors) (New York: Plenum Press)
28. Mahoudeau, J. A., Bardin, E. W. and Lipsett, M. B. (1971). Metabolic clearance rate and origin of plasma dihydrotestosterone in man and its conversion of the 5α-androstanediol. *J. Clin. Invest.*, **50,** 1338
29. Longcope, C., Kato, T. and Horton, R. (1969). Conversion of blood androgen to estrogens in normal adult men and women. *J. Clin. Invest.*, **48,** 2191
30. Perlman, P. L. (1950). The functional significance of testis cholesterol in the rat. Histochemical observations on testes following hypophysectomy and experimental cryptorchidism. *Endocrinology*, **46,** 347
31. Niemi, M. and Ikonen, M. (1962). Cytochemistry of oxidative enzyme systems in the Leydig cells of the rat testis and their functional significance. *Endocrinology*, **70,** 167
32. Catt, K. J., Dufau, M. L. and Tsuhara, T. (1972). Radioligand-receptor assay of LH and HCG. *J. Clin. Endocrinol. Metab.*, **34,** 123
33. Kuehl, F. A., Jr., Patanelli, D. J., Tarnoff, J. and Hames, J. L. (1970). Testicular adenyl cyclase: stimulation by the pituitary gonadotrophins. *Biol. Reprod.*, **2,** 154
34. Johnson, B. H. and Ewing, L. L. (1971). FSH and the regulation of testosterone secretion in rabbit testes. *Science, N.Y.*, **73,** 635
35. Bartke, A. (1971). Effect of prolactin on spermatogenesis in hypophysectomised mice. *J. Endocrinol.*, **49,** 317
36. Hafiez, A. A., Lloyd, C. W. and Bartke, A. (1972). The role of prolactin in the regulation of testis function. The effects of prolactin and LH on the plasma levels of testosterone and androstanedione in hypophysectomized rats. *J. Endocrinol.*, **52,** 327
37. Naftolin, F., Ryan, K. J. and Petro, Z. (1972). Aromatization of androstanediol by the anterior hypothalamus of adult male and female rats. *Endocrinology*, **90,** 295
38. Davidson, J. M. (1969). Feedback of gonadotrophin secretion. *Frontiers in Neuroendocrinology*. 343 (W. F. Ganong and L. Martini, editors) (New York: Oxford University Press)
39. Martini, L. (1970). Les mecanismes de 'retro' action courte'. *Colloques Nationaux du Centre National de la Recherche Scientifique.* No. 927 (Paris: Neuroendocrinologie)

40. Schneider, H. P. G. and McCann, S. M. (1969). Stimulation of release of LH-releasing factor from hypothalmic tissue by dopamine *in vitro*. *J. Reprod. Fert.*, **18,** 178
41. Katongole, C. B., Naftolin, F. and Short, R. V. (1971). Relationship between blood levels of LH and testosterone in bulls and the effects of sexual stimulation. *J. Endocrinol.*, **50,** 457
42. Rowe, P. H., Racey, P. A., Lincoln, G. A., Lelane, J., Stephenson, M. J., Shenton, J. C. and Glover, T. D. (1973). Temporal variations of testosterone levels in the peripheral blood plasma of men. *J. Endocr.* (in press)
43. Haltmeyer, G. C. and Eik-Nes, K. D. (1969). Plasma levels of testosterone in male rabbits following copulation. *J. Reprod. Fert.*, **19,** 273
44. Katongole, C. B. (1971). Androgens in domestic animals. *Ph.D. Thesis, University of Cambridge.*
45. Millar, R. P. and Glover, T. D. (1973). Regulation of seasonal sexual activity in an ascrotal mammal, the rock hyrax (*Procavia capensis*). *J. Reprod. Fert. Suppl.* (in press)
46. Brüggemann, J., Adam, A. and Karg, H. (1969). ICSH-Bestimmungen in Hypophysen von Rehbocken (*Capreolus capreolus*) and Hirschen (*Cervus elaphus*) unter Berucksichtigung des Saisoneinflusses. *Acta Endocrinol, Copenhagen*, **48,** 569
47. Short, R. V. and Mann, T. (1966). The sexual cycle of a seasonally breeding mammal, the roebuck (*Capreolus capreolus*). *J. Reprod. Fert.*, **12,** 337
48. Fawcett, D. W. and Burgos, M. H. (1956). Observations on the cytomorphosis of the germinal and interstitial cells of the human testis. Ciba Found. Colloq. Ageing (G. E. W. Wolstenholme and E. C. P. Millar, editors) (Boston: Little Brown Co.)
49. Fawcett, D. W. and Ito, S. (1958). Observations on the cytoplasmic membrane of testicular cells, examined by phase contrast electron microscopy. *J. Biophys. Biochem. Cytol.*, **4,** 135
50. Andre, J. (1962). Contribution à la connaissance du chonidriome: étude de ses modifications ultrastructurales pendant la spermatogénèse. *J. Ultrastruct. Res., Suppl.* **3,** 185
51. Nicander, L. and Ploën, L. (1969). Fine structure of spermatogonia and primary spermatocytes in rabbits. *Z. Zellforsch.*, **99,** 221
52. Roosen-Runge, E. C. (1951). Quantitative studies on spermatogenesis in the albino rat. II. The duration of spermatogenesis and some effects of colchicine. *Amer. J. Anat.*, **88,** 163
53. Leblond, C. P. and Clermont, Y. (1952). Definition of the stages of the cycle of the seminiferous epithelium in the rat. *Ann. N.Y. Acad. Sci.*, **55,** 548
54. Oakberg, E. F. (1956). A description of spermiogenesis in the mouse and its use in analysis of the cycle of the seminiferous epithelium. *Amer. J. Anat.*, **99,** 391
55. Roosen-Runge, E. C. and Giesel, L. O. (1950). Quantitative studies on spermatogenesis in the albino rat. *Amer. J. Anat.*, **87,** 1
56. Cleland, K. W. (1951). The spermatogenic cycle of the guinea-pig. *Austr. J. Sci. Res. Ser. B.*, **4,** 344
57. Hochereau, M. T. (1963). Étude comparée de la vague spermatogénetique chez le taureau et chezle rat. *Ann. Biol. Anim. Biochim. Biophys.*, **3,** 5
58. Heller, C. G. and Clermont, Y. (1964). Kinetics of the germinal epithelium in man. *Recent Prog. Horm. Res.*, **20,** 545
59. Hochereau-de Reviers, M. T. (1970). Etudes des divisions spermatogoniales et due renouvellement de la spermatogonie souche chez le taureau. *Doctoral Thesis* (Paris: Faculté des Sciences)
60. Monesi, V. (1962). Autoradiographic study of DNA synthesis and the cell cycle in spermatogonia and spermatocytes of mouse testis using tritriated thymidine. *J. Cell. Biol.*, **14,** 1
61. Burgos, M. H. and Fawcett, D. W. (1955). Studies on the fine structure of the mammalian testis. I. Differentiation of spermatids in the cat (*Felis domestica*). *J. Biophys. Biochem. Cytol.*, **1,** 287
62. Fawcett, D. W. and Phillips, D. M. (1967). Further observations on mammalian spermatogenesis (hamster, guinea-pig and chinchilla). *J. Cell. Biol.*, **35,** 152
63. Nagano, T. (1968). Fine structural relation between the Sertoli cell and the differentiating spermatid in the human testis. *Z. Zellforsch.*, **89,** 39
64. Fawcett. D. W. and Phillips, D. M. (1969). Observations on the release of spermatozoa and on changes in the head during passage through the epididymis. *J. Reprod. Fert. Suppl.*, **6,** 405

65. Courot, M., Hochereau-de Reviers, M. T. and Ortavant, R. (1970). Spermatogenesis. *The Testis*, Vol. 1, 339 (A. D. Johnson, W. R. Gomes and N. L. Vandemark, editors) (New York: Academic Press)
66. Clermont, Y. (1972). Kinetics of spermatogenesis in mammals: seminiferous epithelium cycle and spermatogonial renewal. *Physiol. Rev.*, **52,** 198
67. Monesi, V. (1971). Chromosome activities during meiosis, and spermiogenesis. *J. Reprod. Fert. Suppl.*, **13,** 1
68. Bishop, D. W. (1969). Testicular enzymes as fingerprints in the study of spermatogenesis. *Reproduction and Sexual Behaviour.*, 261 (M. Diamond, editor) (Bloomington: Indiana University Press)
69. Randolph, R. W., Lostroh, A. J., Grattarola, R., Squire, P. G. and Li, C. H. (1959). Effect of ovine interstitial cell-stimulating hormone on spermatogenesis in the hypophysectomized mouse. *Endocrinology*, **65,** 433
70. Woods, M. C. and Simpson, M. E. (1961). Pituitary control of the testis of the hypophysectomized rat. *Endocrinology*, **69,** 91
71. Lostroh, A. J. (1963). Effect of follicle stimulating hormone and interstitial cell stimulating hormone on spermatogenesis in Long-Evans rats hypophysectomized for six months. *Acta Endocrinol. Copenhagen*, **43,** 592
72. Berswordt-Wallrabe, R. von and Neumann, F. (1968). Successful reinitiation and restoration of spermatogenesis in hypophysectomized rats with pregnant mare's serum after a long-term regression period. *Experientia*, **24,** 499
73. Courot, M. (1971). Establissement de la spermatogenese chez l'agneau (*Ovis aries*). Etude expérimentale de son controle gonadotrope; importance des cellules de la lignée sertolienne. *Doctoral Thesis* (University of Paris)
74. Steinberger, E. (1971). Hormonal control of mammalian spermatogenesis. *Physiol. Rev.*, **51,** 1
75. Heller, C. G. and Nelson, W. O. (1948). Classification of male hypogonadism and a discussion of the pathologic physiology, diagnosis and treatment. *J. Clin. Endocrinol. Metab.*, **8,** 345
76. Paulsen, C. A. (1968). Effect of human chorionic gonadotrophin and human menopausal gonadotrophin therapy on testicular function. *Gonadotropins*, 491 (E. Rosemberg, editor) (Los Altos, Calif. Geron-X)
77. MacLeod, J., Pazianos, A. and Ray, B. S. (1964). Restoration of human spermatogenesis by menopausal gonadotrophins. *Lancet*, **1,** 3196
78. Mancini, R. E., Seiguer, A. C. and Lloret, A. P. (1969). Effect of gonadotropins on the recovery of spermatogenesis in hypophysectomized patients. *J. Clin. Endocrinol. Metab.*, **29,** 467
79. Clermont, Y. and Harvey, S. C. (1967). Effects of hormones on spermatogenesis in the rat. *Ciba Fdn. Colloq. Endocrinol.*, **16,** 173
80. Courot, M. (1967). Endocrine control of the supporting germ cells of the impuberal testis. *J. Reprod. Fert. Suppl.*, **2,** 89
81. Moore, C. R. and Price, D. (1937). Some effects of synthetically prepared male hormone (androsterone) in rat. *Endocrinology*, **21,** 313
82. Ludwig, D. J. (1950). The effect of androgen on spermatogenesis. *Endocrinology*, **46,** 453
83. Steinberger, E. and Nelson, W. O. (1955). Effect of hypophysectomy, cryptorchidism, estrogen and androgen upon the level of hyaluronidase in the rat testis. *Endocrinology*, **56,** 429
84. Boccabella, A. V. (1963). Reinitiation and restoration of spermatogenesis with testosterone propionate and other hormones after a long-term post-hypophysectomy regression period. *Endocrinology*, **72,** 787
85. Steinberger, E. and Duckett, G. E. (1965). Effect of estrogen or testosterone initiation and maintenance of spermatogenesis in the rat. *Endocrinology*, **76,** 1184
86. Kalra, S. P. and Prasad, M. R. N. (1967). Effect of FSH and testosterone propionate on spermatogenesis in immature rats treated with clomiphene. *Endocrinology*, **81,** 965
87. Heller, C. G., Nelson, W. O., Hill, I. B., Henderson, E., Maddock, W. O., Junck, E. C., Paulsen, C. A. and Mortimore, G. E. (1950). Improvement in spermatogenesis following depression of the human testis with testosterone. *Fert. Steril.*, **1,** 415
88. Lacy, D. and Pettitt, J. A. (1970). Sites of hormone production in the mammalian testis and their significance in the control of male fertility. *Brit. Med. Bull.*, **26,** 87

89. Johnsen, S. G. (1969). Two types of Sertoli cells in man. *Acta Endocrinol., Copenhagen*, **61,** 111
90. Steinberger, A. and Steinberger, E. (1965). Differentiation of rat seminiferous epithelium in organ culture. *J. Reprod. Fert.*, **9,** 243
91. Johnsen, S. G. (1970). The stage of spermatogenesis involved in the testicular-hypophyseal feed-back mechanism in man. *Acta Endocrinol., Copenhagen*, **64,** 193
92. Mancini, R. E., Castro, A. and Seiguer, A. C. (1967). Histologic localization of FSH and LH in the ram testis. *J. Histochem. Cytochem.*, **15,** 516
93. Kretser, D. M. de., Burger, H. G., Fortune, D., Hudson, B., Long, A. R., Paulsen, C. A. and Taft, H. P. (1972). Hormonal, histological and chromosomal studies in adult males with testicular disorders. *J. Clin. Endocrinol. Metab.*, **35,** 392
94. Lacy, D. (1967). The seminiferous tubule in mammals. *Endeavour*, **26,** 101
95. Setchell, B. P. (1970). Testicular blood supply, lymphatic drainage and secretion of fluid. *The Testis*, Vol. 1, 101 (A. D. Johnson, W. R. Gomes and N. L. Vandemark, editors) (New York: Academic Press)
96. Voglmayr, J. K., Waites, G. M. H. and Setchell, B. P. (1966). Studies on spermatozoa and fluid collected directly from the testis of the conscious ram. *Nature, London*, **210,** 861
97. Tuck, R. R., Waites, G. M. H., Young, J. A. and Setchell, B. P. (1969). Composition of fluid secreted by the seminiferous tubules of the rat collected by catheterisation and micropuncture techniques. *Austr. J. Exp. Biol. Med. Sci.*, **47,** 32
98. Setchell, B. P. and Waites, G. M. H. (1971). Exocrine secretion of the testis and spermatogenesis. *J. Reprod. Fert. Suppl.*, **13,** 15
99. Roosen-Runge, E. C. (1961). The rete testis in the albino rat: its structure, development and morphological significance. *Acta Anat.*, **45,** 1
100. Setchell, B. P., Voglmayr, J. K. and Waites, G. M. H. (1969). A blood-testis barrier restricting passage from blood into rete testis fluid but not into lymph. *J. Physiol. London*, **200,** 73
101. Setchell, B. P. (1970). Testicular blood supply, lymphatic drainage and secretion of fluid. *The Testis, Vol.* 1, 101 (A. D. Johnson, W. R. Gomes and N. L. Vandemark, editors) (New York: Academic Press)
102. Setchell, B. P. (1967). The blood-testicular barrier in sheep. *J. Physiol. London*, **189,** 63P
103. Kormano, M. (1967). Dye permeability and alkaline phosphatase activity of testicular capillaries in the post-natal rat. *Histochemie*, **9,** 327
104. Clermont, Y. (1958). Contractile elements in the limiting membrane of the semi niferous tubules of the rat. *Exp. Cell Res.*, **15,** 438
105. Lacy, D. and Rotblat, J. (1960). Study of normal and irradiated boundary tissue of the seminiferous tubules of the rat. *Exp. Cell Res.*, **21,** 49
106. Ross, M. H. (1967). The fine structure and development of the peritubular contractile cell component in the seminiferous tubules of the mouse. *Amer. J. Anat.*, **121,** 523
107. Fawcett, D. W., Heidger, P. M. and Leak, L. V. (1969). Lymph vascular system of the interstitial tissue of the testis as revealed by electron microscopy. *J. Reprod. Fert.*, **19,** 109
108. Dym, M. and Fawcett, D. W. (1970). The blood-testis barrier in the rat and the physiological compartmentation of the seminiferous epithelium. *Biol. Reprod.*, **3,** 308
109. Nicander, L. (1967). An electron microscopical study of cell contacts in the seminiferous tubules of some mammals. *Z. Zellforsch.*, **83,** 375
110. Fawcett, D. W., Ito, S. and Slautterback, D. (1959). The occurrence of intercellular bridges in groups of cells exhibiting synchronous differentiation. *J. Biophys. Biochem. Cytol.*, **5,** 453
111. Flickinger, C. and Fawcett, D. W. (1967). The junctional specializations of Sertoli cells in the seminiferous epithelium. *Anat. Rec.*, **158,** 207
112. Bloom, G. and Nicander, L. (1961). On the ultrastructure and development of the protoplasmic droplet of spermatozoa. *Z. Zellforsch.*, **55,** 833
113. Nicander, L. and Bane, A. (1962). Fine structure of boar spermatozoa. *Z. Zellforsch.*, **57,** 390
114. Bedford, J. M. (1963). Morphological changes in rabbit spermatozoa during passage through the epididymis. *J. Reprod. Fert.*, **5,** 169
115. Fawcett, D. W. and Hollenberg, R. D. (1963). Changes in the acrosome of guinea-pig spermatozoa during passage through the epididymis. *Z. Zellforsch.*, **60,** 276

116. Bedford, J. M., Calvin, H. and Cooper, G. W. (1972). The maturation of spermatozoa in the human epididymis. *J. Reprod. Fert.* Suppl. **18**, p. 199
117. Bedford, J. M. (1965). Changes in the fine structure of the rabbit sperm head during passage through the epididymis. *J. Anat.*, **99**, 891
118. Lindahl, P. E. and Kihlstrom, J. E. (1952). Alterations in specific gravity during ripening of bull spermatozoa. *J. Dairy Sci.*, **35**, 393
119. Lavon, U., Volcani, R., Amir, D. and Danon, D. (1966). The specific gravity bull spermatozoa from different parts of the reproductive tract. *J. Reprod. Fert.*, **12**, 597
120. Lindahl, P. E., Kihlstrom, J. E. and Ström, B. (1953). Further studies on the light reflection in bull spermatozoa. *Ark. Zool.*, **4**, 303
121. Glover, T. D. (1961). Disintegrated spermatozoa from the epididymis. *Nature, London*, **190**, 185
122. Amann, R. P. and Almquist, J. O. (1962). Reproductive capacity of dairy bulls. VIII. Morphology of epididymal sperm. *J. Dairy Sci.*, **45**, 1516
123. Gledhill, B. L., Gledhill, M. P., Rigler, R. and Ringertz, N. R. (1966). Changes in deoxyribonucleoprotein during spermiogenesis in the bull. *Exp. Cell Res.*, **41**, 652
124. White, I. G. and Wales, R. G. (1961). Comparison of epididymal and ejaculated semen of the ram. *J. Reprod. Fert.*, **2**, 225
125. Ortavant, R. (1953). Existence d'une phase critique dans la maturation épididymaire des spermatozides de bélier et de taureau. *Compt. Rend. Soc. Biol. Paris*, **147**, 1552
126. Glover, T. D. (1962). The reaction of rabbit spermatozoa to nigrosin eosin following ligation of the epididymis. *Int. J. Fert.*, **7**, 1
127. Bedford, J. M. (1963). Changes in the electrophoretic properties of rabbit spermatozoa during passage through the epididymis. *Nature, London*, **200**, 1178
128. Bedford, J. M. (1965). Non-specific tail-tail agglutination of mammalian spermatozoa. *Exp. Cell Res.*, **38**, 654
129. Cooper, G. W. and Bedford, J. M. (1971). Acquisition of surface charge by the plasma membrane of the mammalian spermatozoa during epididymal maturation. *Anat. Rec.*, **169**, 300
130. Bedford, J. M. (1966). Development of the fertilizing ability of spermatozoa in the epididymis of the rabbit. *J. Exp. Zool.*, **163**, 319
131. Fulka, J. and Koefoed-Johnson, H. H. (1966). The influence of epididymal passage in rabbits on different spermatozoan characteristics including fertilising capacity. *Ann. Rep. Roy. Vet. Agric. Coll., Sterility Inst., Copenhagen*, 213
132. Orgebin-Crist, M. C. (1967). Maturation of spermatozoa in the rabbit epididymis. Fertilizing ability and embryonic mortality in does inseminated with epididymal spermatozoa. *Ann. Biol. Anim. Biochim. Biophys.*, **7**, 373
133. Orgebin-Crist, M. C. (1965). Passage of spermatozoa labelled with Thymidine-^{3}H through the ductus epididymidis of the rabbit. *J. Reprod. Fert.*, **10**, 241
134. Gaddum, P. and Glover, T. D. (1965). Some reactions of rabbit spermatozoa to ligation of the epididymis. *J. Reprod. Fert.*, **9**, 119
135. Gaddum, P. (1969). Sperm maturation in the male reproductive tract: development of motility. *Anat. Rec.*, **161**, 471
136. Bedford, J. M. (1967). Effect of duct ligation on the fertilizing ability of spermatozoa from different regions of the rabbit epididymis. *J. Exp. Zool.*, **166**, 271
137. Paufler, S. K. and Foote, R. H. (1968). Morphology, motility and fertility of spermatozoa recovered from different areas of ligated rabbit epididymides. *J. Reprod. Fert.*, **17**, 125
138. Cross, B. A. and Glover, T. D. (1958). The hypothalamus and seminal emission. *J. Endocrinol.*, **16**, 385
139. Kuntz, A. and Morris, R. H. (1946). Components and distribution of the spermatic nerves and the nerves of the vas deferens. *J. Comp. Neurol.*, **85**, 33
140. Norberg, K. A., Risley, P. C. and Ungerstedt, U. (1966). Andrenergic innervation of the male reproductive ducts in some mammals. I. The distribution of andrenergic nerves. *Z. Zellforsch.*, **76**, 278
141. El-Badawi, A. and Schenk, E. A. (1967). The distribution of cholinergic and adrenergic nerves in the mammalian epididymis. *Amer. J. Anat.*, **121**, 1
142. Koefoed-Johnsen, H. H. (1961). On the function of the epididymis. I. The rate of epididymal passage of spermatozoa. *Ann. Rep. Roy. Vet. Agric. Coll., Sterility Inst., Copenhagen*, 57

143. Lino, B. F., Braden, A. W. H. and Turnbull, K. E. (1967). Fate of unejaculated spermatozoa. *Nature, London*, **213,** 594
144. Amann, R. P. and Almquist, J. O. (1962). Reproductive capacity of dairy bulls. VI. Effects of unilateral vasectomy and ejaculation frequency on sperm reserves; aspects of epididymal physiology. *J. Reprod. Fert.*, **3,** 260
145. Roussel, J. D., Stallcup, O. T. and Austin, C. R. (1967). Selective phagocytosis of spermatozoa in the epididymis of bulls, rabbits and monkeys. *Fert. Steri.*, **18,** of spermatozoa in the epididymis of bulls, rabbits and monkeys. *Fert. Steril.*, **18,**
146. Phadke, A. M. and Phadke, G. M. (1961). Occurrence of macrophages in the semen and in the epididymis in cases of male infertility. *J. Reprod. Fert.*, **2,** 400
147. Phadke, A. M. (1964). Fate of spermatozoa in cases of obstructive azoospermia and after ligation of the vas deferens in man. *J. Reprod. Fert.*, **7,** 1
148. Hammond, J. and Asdell, S. A. (1926). The vitality of the spermatozoa in the male and female reproductive tracts. *Brit. J. Exp. Biol.*, **4,** 155
149. Tesh, J. M. and Glover, T. D. (1969). Ageing of rabbit spermatozoa in the male tract and its effect on fertility. *J. Reprod. Fert.*, **20,** 287
150. Glover, T. D. (1958). Experimental induction of seminal degeneration in rabbits. *Stud. Fert.*, **10,** 80
151. Glover, T. D. (1960). Spermatozoa from the isolated cauda epididymidis of rabbits and some effects of artificial cryptorchidism. *J. Reprod. Fert.*, **1,** 121
152. Cummins, J. M. and Glover, T. D. (1970). Artificial cryptorchidism and fertility in the rabbit. *J. Reprod. Fert.*, **23,** 423
153. MacLeod, J. and Hotchkiss, R. S. (1941). The effect of hyperpyrexia upon spermatozoa counts in men. *Endocrinology*, **28,** 780
154. Weisel, G. F. (1949). The seminal vesicles of *Gillichthys*, a marine teleost. *Copeia*, **1949,** 101
155. Glover, T. D. and Nicander, L. (1971). Some aspects of structure and function in the mammalian epididymis. *J. Reprod. Fert. Suppl.*, **13,** 39
156. Glover, T. D. and Sale, J. B. (1968). Reproductive system of the male rock hyrax (*Procavia* and *Heterohyrax*). *J. Zool., London*, **156,** 351
157. Nicander, L. and Glover, T. D. (1973). Regional histology and fine structure of the epididymal duct in the golden hamster (*Mesocricetus auratus*). *J. Anat.*, **114,** 347
158. Friend, D. S. and Farquhar, M. G. (1967). Functions of coated vesicles during protein absorption in the rat vas deferens. *J. Cell Biol.*, **35,** 357
159. Crabo, B. (1965). Studies on the composition of epididymal content in bulls and boars. *Acta Vet. Scand.*, **6,** *Suppl.* 5
160. Burgos, M. H. (1964). Uptake of colloidal particles by cells of the caput epididymidis. *Anat. Rec.*, **148,** 587
161. Reid, B. L. and Cleland, K. W. (1957). The structure and function of the epididymis. I. The histology of the rat epididymis. *Austr. J. Zool.*, **5,** 223
162. Martan, J. (1969). Epididymal histology and physiology. *Biol. Reprod., Suppl.* **1,** 134
163. Hamilton, D. W. (1972). The mammalian epididymis. *Reproductive Biology*, 268 (H. Balin and S. Glasser, editors) (Amsterdam: Excerpta Medica)
164. Suzuki, F. (1973). Effect of castration on the epididymal epithelium of the golden hamster. (in preparation)
165. Cross, B. A. (1959). Hypothalamic influences on sperm transport in the male and female genital tract. Recent progress in the endocrinology of reproduction, 167 (C. Lloyd, editor) (New York: Academic Press)
166. Holstein, A. F. (1967). Muskulatur und Motilität des Nebenhodens beim Kaninchen. *Z. Zellforsch.*, **57,** 692
167. Kormano, M. (1968). Microvascular structure of the rat epididymis. *Ann. Med. Exp. Fenn.*, **46,** 113
168. Kormano, M. (1968). Penetration of intravenous trypan blue into the rat testis and epididymis. *Acta histochem.* (*Jena*), 133
169. Kormano, M. and Penttila, A. (1968). Distribution of endogenous and administered 5-hydroxytryptamine in the rat testis and epididymis. *Ann. Med. Exp. Fenn.*, **46,** 468
170. Kormano, M. and Penttila, A. (1968). The distribution and content of intravenously injected dopamine in the testis and epididymis of the rat. *Pharmacology*, **1,** 199
171. Scott, T. W., Wales, R. G., Wallace, J. C. and White, I. G. (1963). Composition of ram

epididymal and testicular fluid and the biosynthesis of glycerylphosphorylcholine by the rabbit epididymis. *J. Reprod. Fert.*, **6,** 49

172. Crabo, B. and Gustafsson, B. (1964). Distribution of sodium and potassium and its relation to sperm concentration in the epididymal plasma of the bull. *J. Reprod. Fert.*, **7,** 337
173. Gustafsson, B. (1966). Luminal contents of the bovine epididymis under conditions of reduced spermatogenesis, luminal blockage and certain sperm abnormalities. *Acta Vet. Scand.*, Suppl. **17**
174. Wales, R. G., Wallace, J. C. and White, I. G. (1966). Composition of bull epididymal testicular fluid. *J. Reprod. Fert.*, **12,** 139
175. Levine, N. and Marsh, D. J. (1971). Micropuncture studies of the electrochemical aspects of fluid and electrolyte transport in individual seminiferous tubules, the epididymis and vas deferens in rats. *J. Physiol., London*, **213,** 557
176. White, I. G. and Wales, R. G. (1961). Comparison of epididymal and ejaculated semen of the ram. *J. Reprod. Fert.*, **2,** 225
177. Mann, T. (1964). *Biochemistry of Semen and of the Male Reproductive Tract* (London: Methuen)
178. Martan, J. and Risley, P. C. (1962). Holocine secretory cells of the rat epididymis. *Anat. Rec.*, **146,** 173
179. Linnetz, L. J. and Amann, R. P. (1968). The male rabbit. II. Histochemistry of the epididymis and ampulla as influenced by sperm output. *J. Reprod. Fert.*, **16,** 343
180. Setchell, B. P., Hinks, B. T., Voglmayr, J. K. and Scott, T. W. (1967). Amino acids in ram testicular fluid and semen and their metabolism by spermatozoa. *Biochem. J.*, **105,** 1061
181. Setchell, B. P., Voglmayr, J. K., Scott, T. W. and Waites, G. M. H. (1969). Characteristics of testicular spermatozoa and the fluid which transports them into the epididymis. *Biol. Reprod.* Suppl. **1,** 40
182. Risley, P. (1963). Physiology of the male accessory organs. *Mechanisms concerned with conception*, 73 (C. G. Hartman, editor) (New York: Macmillan)
183. Montagna, W. and Hamilton, J. B. (1952). Histochemical studies of human testes. II. The distribution of glycogen and other periodic acid—Schiff reactive substances. *Anat. Rec.*, **112,** 237
184. Nicander, L. (1957). On the regional histology and cytochemistry of the ductus epididymidis in rabbits. *Acta Morph. Neerl. Scand.*, **1,** 99
185. Nicander, L. (1957). Studies on the regional histology and cytochemistry of the ductus epididymidis in stallions, rams and bulls. *Acta Morph. Neerl. Scand.*, **1,** 337
186. Cavazos, L. F. (1958). Effects of testosterone propionate on histochemical reactions of epithelium of rat ductus epididymidis. *Anat. Rec.*, **132,** 209
187. El Gohary, M., Cavazos, L. F. and Manning, J. P. (1962). Effects of testosterone on histochemical reactions of epithelium of hamster ductus epididymidis and seminal vesicle. *Anat. Rec.*, **144,** 229
188. Maneely, R. E. (1959). Epididymal structure and function. A historical and critical review. *Acta Zool. Stockholm*, **40,** 1
189. Nicander, L. (1970). On the morphological evidence of secretion and absorption in the epididymis. *Morphological aspects of Andrology*, Vol. 1, 121 (A. F. Holstein and E. Horstmann, editors) (Berlin: Grosse Verlag)
190. Fouquet, J. -P. and Gouha, S. (1969). Histochemical studies on the enzymes of glycogen metabolism in hamster epididymis. *Histochemie*, **17,** 89
191. Allen, J. M. (1961). The histochemistry of glucose-6-phosphatase in the epididymis of mouse. *J. Histochem. Cytochem.*, **9,** 681
192. Hayashi, M. (1964). Distribution of β-glucuronide activity in rat tissue employing the naphthal AS-B1 glucuronide hexagonium pararosanilin method. *J. Histochem. Cytochem.*, **12,** 659
193. Allen, J. M. and Slater, J. J. (1961). A cytochemical analysis of the lactic dehydrogenase diphosphopyridine nucleotide—diaphorase in the epididymis of the mouse. *J. Histochem. Cytochem.*, **9,** 221
194. Blackshaw, A. W. and Samisoni, J. I. (1967). Histochemical localisation of some dehydrogenase enzymes in the bull testis and epididymis. *J. Dairy Sci.*, **50,** 747
195. Niemi, M. (1965). The fine structure and histochemistry of the epithelial cells of the rat vas deferens. *Acta Anat.*, **60,** 207

196. Allen, J. M. and Slater, J. J. (1957). A chemical and histochemical study of alkaline phosphatase and aliesterase in the epididymis of normal and castrate mice. *Anat. Rec.*, **129,** 255

197. McGadey, J., Baillie, A. H. and Ferguson, M. M. (1966). Histochemical utilisation of hydroxysteroids by the epididymis. *Histochemie*, **7,** 212

198. Moniem, K. A. (1972). Histochemical localisation of 3β-hydroxysteroid dehydrogenase in the rat epididymis. *J. Reprod. Fert.*, **28,** 461

199. Hamilton, D. W., Jones, A. L. and Fawcett, D. W. (1968). Sterol biosynthesis from (1-^{14}C) acetate in the epididymis and vas deferens of the mouse. *J. Reprod. Fert.*, **18,** 156

200. Hamilton, D. W., Jones, A. L. and Fawcett, D. W. (1969). Cholesterol biosynthesis in the mouse epididymis and ductus deferens: A biochemical and morphological study. *Biol. Reprod.*, **1,** 167

201. Hamilton, D. W. and Fawcett, D. W. (1970). *In vitro* synthesis of cholesterol and testosterone from acetate by rat epididymis and vas deferens. *Proc. Soc. Exp. Biol. (N.Y.)*, **133,** 693

202. Moniem, K. A. and Glover, T. D. (1972). Comparative histochemical localisation of lysosomal enzymes in mammalian epididymides. *J. Anat.*, **111,** 437

203. Marquis, N. R. and Fritz, I. B. (1965). Effects of testosterone on the distribution of carnitine, acetylcarnitine and carnitine acetyltransferase in tissues of the reproductive system of the male rat. *J. Biol. Chem.*, **240,** 2197

204. Graham, E. F. and Pace, M. M. (1967). Some biochemical changes in spermatozoa due to freezing. *Cryobiology*, **4,** 75

205. Jones, R. and Glover, T. D. (1973). The effects of castration on the composition of epididymal plasma. *J. Reprod. Fert.*, **34,** 405

206. Orgebin-Crist, M. C., Dyson, A. L. M. B. and Davies J. (1973). Hormonal regulation of epididymal sperm maturation. *Excerpta Medica* (in press)

207. Lubicz-Nawrocki, C. M. and Glover, T. D. (1973). The influence of the testis on the survival of spermatozoa in the epididymis of the golden hamster (*Mesocricetus auratus*). *J. Reprod. Fert.* (in press)

208a. Niemi, M. and Tuohimaa, P. (1971). The mitogenic activity of testosterone in the accessory sex glands of the rat in relation to its conversion to dihydrotestosterone. *Basic Actions of Sex Steroids on Target Organs*, 258 (P. O. Hubinot, F. Leroy, P. Galand and S. Karger, editors) (Basel: S. Karger)

208b. Back, D. J. (1973). The passage of ^{3}H-testosterone in the cauda epididymidis of the rat. *J. Reprod. Fert.*, **35,** 586

209. Blaquier, J. A. (1971). Selective uptake and metabolism of androgens by rat epididymis. The presence of a cytoplasmic receptor. *Biochem. Biophys. Res. Commun.*, **45,** 1076

210. Hansson, V. and Tveter, K. J. (1971). Uptake and binding *in vivo* of ^{3}H-labelled androgen in the rat epididymis and ductus deferens. *Acta Endocrinol. Copenhagen*, **66,** 745

211. Danzo, B. J., Orgebin-Crist, M.-C. and Strott, A. C. (1972). Rabbit epididymal receptor for 5α-dihydrotestosterone. *IV International Congress of Endocrinology*

212. Lubicz-Nawrocki, C. M. (1973). The effect of testosterone and its metabolites on the viability of hamster epididymal spermatozoa. *Biol. Reprod.* (in press)

213. Williams-Ashman, H. G. (1970). Biochemistry of testicular androgen action. *The Androgens of the Testis* 117 (K. B. Eik-Nes, editor) (New York: Marcel Dekker Inc.)

214 Liao, S. (1968). Evidence for a discriminatory action of androgenic steroids on the synthesis of nucleolar ribonucleic acids in prostatic nuclei. *Amer. Zool.*, **3,** 223

215. Moniem, K. A. (1972). Some histochemical features of the mammalian epididymis. *Ph.D. Thesis, University of Liverpool.*

216. Risley, P. L. (1958). The contractile behaviour *in vivo* of the ductus epididymis and vasa efferentia of the rat. *Anat. Rec.*, **130,** 471

217. Cross, B. A. (1959). Hypothalamic influences on sperm transport in the male and female genital tracts. *Recent Progress in the Endocrinology of Reproduction* 167 (C. Lloyd, editor) (New York: Academic Press)

218. Davis, J. R., Langford, G. A. and Kirby, P. J. (1970). The testicular capsule. *The Testis.* Vol. 1, 218 (A. D. Johnson, W. R. Gomes and N. L. Vandemark, editors) (New York: Academic Press)

219. Roosen-Runge, E. C. (1951). Motions of the seminiferous tubules of the rat and dog. *Anat. Rec.*, **109,** 413

220. Hovatta, O. (1972). Contractility and structure of adult rat seminiferous tubules in organ culture. *Z. Zellforsch.*, **130,** 171
221. Niemi, M. and Kormano, M. (1965). Contractility of the seminiferous tubule of the post-natal rat testis, and its response to oxytocin. *Ann. Med. Exp. Fenn.*, **43,** 40
222. Macmillan, E. W. (1954). Observations on the isolated vaso-epididymal loop and on the effects of experimental subcapital epididymal obstructions. *Stud. Fert.*, **6,** 57
223. Reid, B. L. (1959). The structure and function of the epididymis. *Austr. J. Zool.*, **7,** 22
224. Skinner, J. D. and Rowson, L. E. A. (1968). Some effects of unilateral cryptorchism and vasectomy on sexual development of the pubescent ram and bull. *J. Endocrinol.*, **42,** 311
225. Mann, T., Short, R. V., Walton, A., Archer, R. K. and Miller, W. C. (1957). The 'tail-end' sample of stallion semen. *J. Agric. Sci.*, **49,** 301
226. Bruchovsky, N. and Wilson, J. D. (1968a). Conversion of testosterone to 5α DHT by rat prostate *in vivo* and *in vitro*. *J. Biol. Chem.*, **243,** 2012
227. Bruchovsky, N. and Wilson, J. D. (1968b). The intranuclear binding of testosterone and 5α DHT by rat prostate. *J. Biol. Chem.*, **243,** 5953
228. Mainwaring, W. I. P. (1969). The binding of [1,2-^{3}H] testosterone within nuclei of the rat prostate. *J. Endocrinol.*, **44,** 323
229. Anderson, K. M. and Liao, S. (1968). Selective retention of 5α DHT by prostate nuclei. *Nature, London*, **219,** 277
230. Gloyna, R. E. and Wilson, J. D. (1969). A comparative study of the conversion of testosterone to 17β-hydroxy-5α-androstan-3-one (dihydrotestosterone) by prostate and epididymis. *J. Clin. Endocrinol. Metabol.*, **29,** 970
231. Maule, J. P. (1962). *The Semen of Animals and Artificial Insemination.* (Farnham Royal: Commonwealth Agricultural Bureaux)
232. Bishop, M. W. H., Campbell, R. C., Hancock, J. L. and Walton, A. (1954). Semen characteristics and fertility in the bull. *J. Agric. Sci.*, **44,** 227
233. Harvey, C. (1945). A method for estimating the fertility of spermatozoa. *Nature, London*, **155,** 368
234. MacLeod, J. (1965). Seminal cytology in the presence of varicocele. *Fert. Steril.*, **16,** 735
235. Eliasson, R. (1971). Standards of investigation of human semen *Andrologie*, **3,** 49
236. Gledhill, B. L., Darzynkiewicz, Z. and Ringertz, N. R. (1971). Changes in deoxyribonucleoprotein during spermiogenesis in the bull: increased [^{3}H] actinomycin D binding of nuclear chromatin of morphologically abnormal spermatozoa. *J. Reprod. Fert.*, **26,** 25
237. Nelson, L. (1967). Sperm motility. *Fertilization.* Vol. 1, 27 (C. B. Metz and A. Monroy, editors) (New York: Academic Press)
238. Segal, S. T. and Laurence, K. A. (1962). Automatic analysis of particulate matter in human semen. *Ann. N.Y. Acad. Sci.*, **99,** 271
239. Gordon, D. L., Moore, D. J., Thorslund, T. and Paulsen, C. A. (1965). The determination of size and concentration of human sperm with an electronic particle counter. *J. Lab. Clin. Med.*, **65,** 506
240. Freund, M. (1963). Effect of frequency of emission on semen output and estimate of daily sperm production in man. *J. Reprod. Fert.*, **6,** 269
241. Freund, M. (1966). Standards for the rating of human sperm morphology. *Int. J. Fert.*, **11,** 97
242. Shakkebaek, N. E. and Beatty, R. A. (1970). Studies on mieotic chromosomes and spermatozoan heads in mice treated with LSD. *J. Reprod. Fert.*, **22,** 141
243. MacLeod, J. (1965). The semen examination. *Clin. Obstet. Gynecol.*, **8,** 115
244. Moench, G. L. (1931). Sperm morphology in relation to fertility. *Amer. J. Obstet. Gynecol.*, **22,** 199
245. Flechon, J-E. (1972). Cytochemie ultrastructurale de l'épididyme de Lapin: absorption et sécrétion. *J. Microscopie*, **14,** 47a
246. Gaddum, P. and Blandau, R. J. (1970). Proteolytic reaction of mammalian spermatozoa on gelatin membranes. *Science, N.Y.*, **170,** 749
247. Campbell, R. C., Dott, H. M. and Glover, T. D. (1956). Nigrosin-eosin as a stain for differentiating live and dead spermatozoa. *J. Agric. Sci.*, **48,** 1
248. Eliasson, R. and Treichl, L. (1971). Supravital staining of human spermatozoa. *Fert. Steril.*, **22,** 134
249. Eliasson, R. (1970). Oxygen consumption of human semen. *Biol. Reprod.*, **3,** 369

250. Eliasson, R. (1971). Oxygen consumption of human spermatozoa in seminal plasma and a Ringer solution. *J. Reprod. Fert.*, **27,** 385
251. Karjalanien, K. and Niemi, M. (1969). Oxygen consumption of human spermatozoa and seminal plasma as measured by a micro-diver technique. *Scand. J. Clin. Lab. Invest.*, **23,** *Suppl.* 108, 79
252. Peterson, R. N. and Freund, M. (1970). ATP synthesis and oxidative metabolism in human spermatozoa. *Biol. Reprod.*, **3,** 47
253. Mann, T. (1971). Biochemical appraisal of human semen. *Fertility Disturbances in Men and Women* (C. R. Joel, editor) (Basel:Karger)
254. Eliasson, R. (1968). Biochemical analyses of human semen in the study of the physiology and pathophysiology of male accessory genital glands. *Fert. Steril.*, **19,** 344
255. Hirschhäuser, C. and Kionke, M. (1971). Demonstration of muramidase (lysozyme) in human seminal plasma. *Life Sci.*, **10,** 333
256. Polge, C., Smith, A. U. and Parkes, A. S. (1949). Revival of spermatozoa after vitrification and dehydration at low temperatures. *Nature, London*, **164,** 666
257. Parkes, A. S. (1968). The biology of spermatozoa and artificial insemination. *Marshall's Physiology of Reproduction.* Vol. 1, 161 (A. S. Parkes, editor) (London: Longmans Green)
258. Bunge, R. G. (1960). Further observations on freezing human spermatozoa. *J. Urol.*, **83,** 192
259. Sherman, J. K. (1963). Improved methods of preservation of human spermatozoa by freezing and freeze-drying. *Fert. Steril.*, **14,** 49
260. Freund, M. and Wiederman, J. (1966). Factors affecting the dilution, freezing, and storage of human semen. *J. Reprod. Fert.*, **11,** 1
261. Sherman, J. K. (1964). Research on frozen semen. *Fert. Steril.*, **15,** 485
262. Richardson, D. W. and Sadlier, R. M. F. S. (1967). The toxicity of various non-electrolytes to human spermatozoa and their protective effects during freezing. *J. Reprod. Fert.*, **14,** 439
263. Lovelock, J. E. and Bishop, M. W. H. (1959). Prevention of freezing damage to living cells by dimethylsulphoxide. *Nature, London*, **183,** 1394
264. Zimmerman, S. J., Maude, M. B. and Moldawer, M. (1964). Freezing and storage of human semen in fifty healthy medical students. A comparative study of glycerol and dimethylsulphoxide as a preservative. *Fert. Steril.*, **15,** 505
265. Ackerman, D. R. and Sod-Moriah, U. A. (1968). DNA content of human spermatozoa after storage at low temperatures. *J. Reprod. Fert.*, **17,** 1
266. Bunge, R. G. and Sherman, J. K. (1953). Fertilizing capacity of frozen human spermatozoa. *Nature, London*, **172,** 767
267. Murphy, D. P. (1964). Donor insemination. *Fert. Steril.*, **15,** 528
268. Murphy, D. P. and Torrano, E. F. (1965). Donor insemination: study of 112 women. *Fert. Steril.*, **17,** 273
269. Fjallbrant, B. and Ackerman, D. R. (1969). Cervical mucus penetration *in vitro* by fresh and frozen-preserved human semen specimens. *J. Reprod. Fert.*, **20,** 515
270. Ackerman, D. R. (1970). Hyaluronidase in human semen and sperm suspensions subjected to temperature shock and to freezing. *J. Reprod. Fert.*, **23,** 521
271. Healey, P. (1969). Effect of freezing on the ultrastructure of the spermatozoon of some domestic animals. *J. Reprod. Fert.*, **18,** 21
272. Roussel, J. D. and Austin, C. R. (1967). Preservation of primate spermatozoa by freezing. *J. Reprod. Fert.*, **13,** 333
273. Norman, C., Goldberg, E., Porterfield, I. B. and Johnson, C. E. (1960). Prolonged survival of human spermatozoa in chemically defined media at room temperature. *Nature, London*, **188,** 760
274. Amelar, R. D. and Hotchkiss, R. S. (1965). The split ejaculate. *Fert. Steril.*, **16,** 46
275. Harvey, C. and Jackson, M. H. (1955). A method of concentrating human semen. *J. Clin. Pathol.*, **8,** 341
276. Harvey, C. (1957). The use of partitioned ejaculates in investigating the role of accessory secretions in human semen. *Stud. Fert.*, **8,** 3
277. Behrman, S. J. (1959). Artificial insemination. *Fert. Steril.*, **10,** 248
278. Behrman, S. J. (1961). Artificial insemination. *Int. J. Fert.*, **6,** 291

10
Regulation of Ovarian Steroidogenesis: Gonadotrophins, Enzymes, Prostaglandins, Cyclic-AMP, Luteolysins

J. HAMMERSTEIN
Abteilung für Gynäkologische Endokrinologie, Klinikum Steglitz, Freie Universität Berlin

---

10.1 GENERAL ASPECTS

10.1.1 Species differences

> 'This "evolutionary monotony"...... is in distinct contrast to the staggering variety of ways in which the problem of reproduction has been solved'. Nalbandov[1]

It would be far beyond the scope of this survey to deal with all the interspecies variability in the regulation of ovarian steroidogenesis. Even if the subject were limited to primates and common laboratory and domestic mammals, the variability of their respective regulatory principles is perplexing.
The usual classification

1. Species with menstrual cycles (human, haplorrhine primates)
2. Species with oestrous cycles
   (a) with spontaneous ovulations and long cycles (sheep, pig, horse, dog, goat, guinea-pig etc.),
   (b) with spontaneous ovulations and short cycles (rat, mouse, golden and Chinese hamster etc.),
   (c) with ovulations induced by mating (rabbit, ferret, cat etc.)

is of limited value in this respect since the regulatory mechanism of the ovarian function may differ considerably even among species of the same category. For instance, formation and maintenance of corpora lutea of the cycle is totally dependent on hypophyseal support in the sheep but more or less independent in the pig. The guinea-pig, which is the only common rodent

Pathways of ovarian steroidogenesis

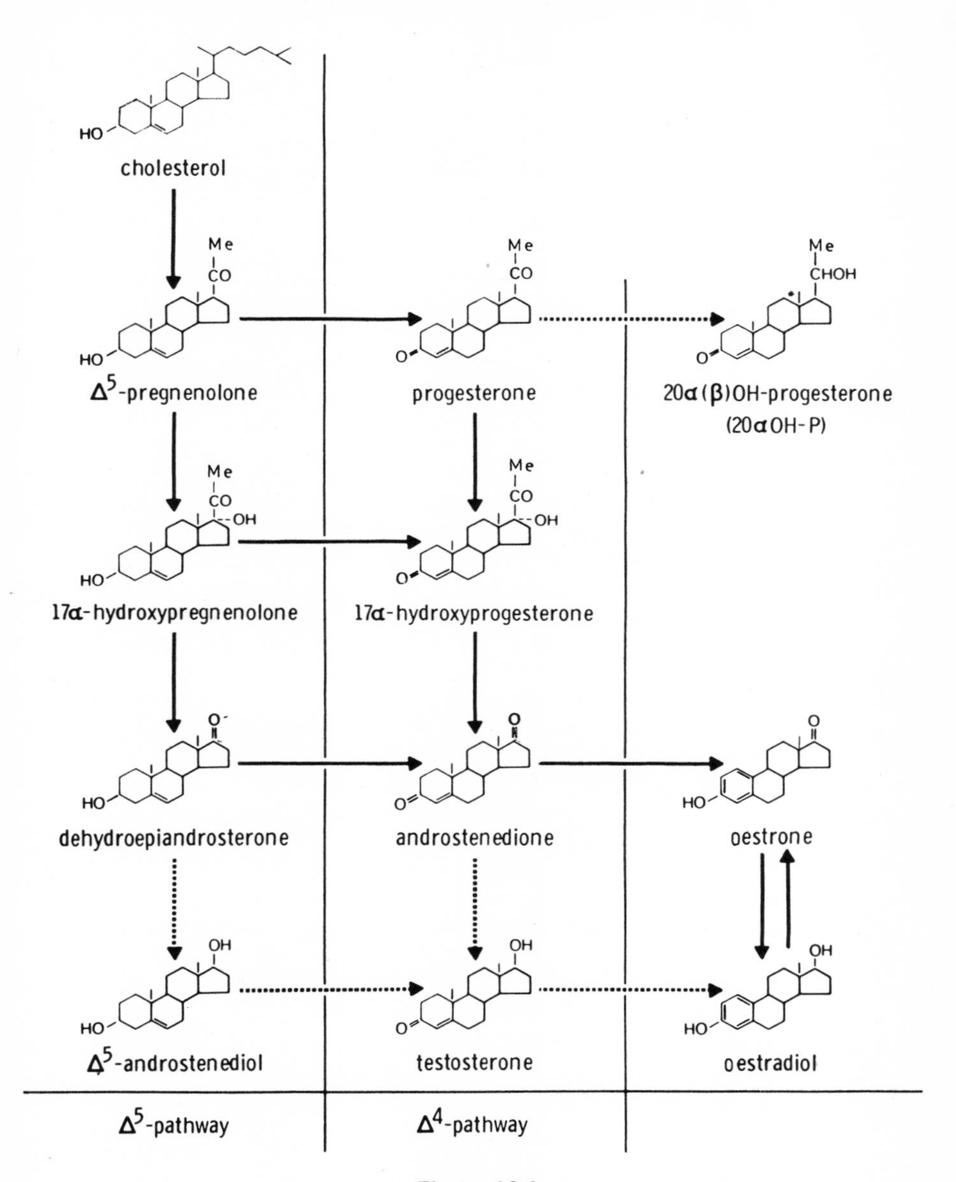

Figure 10.1

within the group of species with spontaneous long oestrous cycles, takes an intermediate position since its luteal function becomes autonomous from day 3 or 4 of the cycle onwards and persists for a period of time which is longer than normal if the pituitary is removed.

In view of these species differences, one has to be very cautious with generalisations and final conclusions of any kind. Attention will be focused in this review mainly on some more recent achievements and subjects of actual interest with the priority given to the situation in the human. The control of ovarian steroidogenesis throughout pregnancy can be dealt with only marginally.

10.1.2 Ovarian steroidogenesis

Cholesterol, independently of whether it originates from the circulation or is synthesised by the ovary itself, is the key precursor for all steroidal hormones synthesised by this gland. Once Δ^5-pregnenolone is formed from cholesterol under direct gonadotrophic control, all further transformations follow either the Δ^4- or the Δ^5-pathway (Figure 10.1). The route which is predominantly or exclusively operative varies both among species and tissue compartments. In the human ovary, for example, the Δ^4-pathway is the one which predominates in luteal tissue and granulosa cells whereas in the thecal and interstitial cell the Δ^5-route is preferred.

The hormonal end products depend on the enzymatic equipment of the respective tissue compartment in quantitative rather than qualitative terms and are obviously not to be influenced qualitatively by gonadotrophins. As the only well-investigated exception from this general rule, prolactin has been found to selectively inhibit the 20α-hydroxysteroid dehydrogenase activity of rat corpora lutea without influencing the total progestin formation[2]. Thus, with decreasing prolactin levels more progesterone will be inactivated to its 20α-OH-analogue which process appears to be of fundamental importance for the regulation of the cycle in this species.

Whereas theca cells are believed to be the main producers of the oestrogens, and granulosa cells are thought to make predominantly progestins, the interstitial tissue is not so clearly defined in this respect. In the human, this ovarian compartment forms, besides other steroids, androgens which may gain pathophysiological relevance. In the rabbit the production of 20α OH-P by the interstitium is worth mentioning because of the participation of this steroid in bringing about ovulation.

It is well established that corpora lutea of all species produce progesterone as the principal physiological product. However, only rarely are progesterone and its 20β (α) OH-P analogue the final steroids on the steroidogenic pathway synthesised by luteal cells, as is the case in the cow. Human corpora lutea, have been demonstrated to form, in addition to the progestins, 17α- hydroxyprogesterone, androstenedione, oestradiol and oestrone in appreciable amounts from precursors as small as acetate[3]. The majority of mammalian species take an intermediate position between the two extremes; their corpora lutea form 17α-hydroxyprogesterone and androstenedione in addition to progestins, but no oestrogens[4].

10.1.3 Gonadotrophins

Both the gametogenic and incretory function of the ovaries are under hypophyseal control. Consequently, these endocrine glands quickly become atrophic and inactive if the function of the pituitary ceases. In the human and the guinea-pig, pregnancy can be considered an exception from this rule, since in these species the corpora lutea of pregnancy may continue to elaborate sex hormones even in the complete absence of pituitary function.

Follicle stimulating hormone (FSH) and luteinising hormone (LH) are the two hypophyseal gonadotrophins of fundamental importance for the regulation of the cycle in all non-pregnant mammals. In species like the rat and the mouse, prolactin is another hypophyseal hormone, which appears to be essential for the ovarian, especially the luteal function. The term luteotropic (or luteotrophic) hormone (LTH) should no longer be used synonymously with prolactin since it has become obvious in recent years that a luteotropic complex rather than a single luteotropic hormone is usually responsible for the maintenance of the corpus luteum among the mammalian species.

The trophoblast of primates makes chorionic gonadotrophin (xCG), also a proteohormone, which plays a decisive role in prolonging luteal life span throughout the first trimester of pregnancy. It is equal in function with LH, but differs in structure. Equidae produce a special gonadotrophin in the endometrial cups of the uterus. This pregnant mare's serum gonadotrophin (PMS) has both FSH and LH properties. It is thought to be responsible for the formation of a new set of so-called accessory corpora lutea between the 40th and 80th day of pregnancy.

All gonadotrophins are glucoproteins except prolactin which lacks the carbohydrate moiety. Very recently the amino acid sequence of various gonadotrophins has been assessed; only the primary structure of FSH is not yet fully clarified. The secondary and tertiary (conformational) structure of the gonadotrophins is still under investigation.

FSH, LH and hCG have the composition of two subunits in common. While the α-moieties of these three gonadotrophins are identical and interchangeable, the β-subunits vary considerably both between the hormones and to some extent also between the species. Thus the β-subunits are the determinants of biological activity and immunogenicity. Nonetheless, they are almost inactive in the dissociated state but regain their biological properties after recombination with an α-unit of any origin. Prolactin differs from the other gonadotrophins in that it has a single-chain structure made up of 198 amino acids.

10.2 CONTROL OF OVARIAN STEROIDOGENESIS THROUGHOUT THE MENSTRUAL (OESTROUS) CYCLE: *in vivo* OBSERVATIONS

10.2.1 Gonadotrophic control of the follicular phase

10.2.1.1 Biological importance of FSH

As early as 1942, Greep *et al.*[5] described the biological activity of highly purified FSH as only morphogenic: 'In hypophysectomised immature female rats thylakentrin (=FSH) brought about growth of Graafian follicles without accompanying luteinisation, effected no growth of the uterus, or cornification of the vaginal epithelium, and failed to stimulate the ovarian interstitial cells'.

Since this statement was made, much controversy has been going on as to whether FSH in addition to its morphogenic potency is also steroidogenic in action. In support of Greep's view Eshkol and Lunefeld[6] were only able to stimulate follicular development and ovarian weight gain in infantile intact mice whereas the weight of the uterus was not influenced when using urinary gonadotrophin preparations of human origin which had been immunologically rendered free of LH activity by Donini. Petrusz *et al.*[7] on the other hand, stated that human urinary FSH *per se* is capable of inducing oestrogen synthesis in hypophysectomised female rats. Rosemberg and Joshi[8] also arrived at the conclusion that Donini's FSH preparation was steroidogenic when used at high doses. Whether this is due to some residual LH contamination and whether such high FSH dosages have any biological meaning still remains open for discussion.

Using ovine FSH with an activity 30–35 times higher than that of NIH-FSH-S1 and an LH contamination below 1%, Lostroh and Johnson[9] were unable to influence steroidogenesis in the ovaries of hypophysectomised rats. According to these authors, FSH is the only hypophyseal hormone to stimulate the development of the follicles up to the antrum stage. FSH, however, fails to maintain follicular development beyond medium size since, even in the presence of high concentrations of FSH, autolysis cannot be prevented. The latter was thought to be due to the lack of oestrogens which are supposed to stimulate the mitotic activity of the granulosa cells.

As long as no more *in vivo* data are available, it appears justifiable to stick to Greep's old concept and to consider FSH a morphogenic gonadotrophin whose main function is to stimulate the follicular development up to the Graafian stage. In support of this view, highly purified human pituitary FSH has recently been shown to be unable to stimulate oestrogen production in an amenorrhoic woman lacking LH basal secretion, whereas human menopausal gonadotrophin (HMG) at a much lower dosage did so, due to its LH content[10]. Since FSH also provides the morphological preconditions for the LH-controlled steroidogenesis it is quite understandable that LH or hCG treatment in hypogonadotrophic amenorrhoic women fails to stimulate steroidogenesis unless pretreatment with FSH has been carried out.

10.2.1.2 The steroidogenic action of LH

Ample experimental as well as clinical evidence points so clearly to LH as being the steroidogenic gonadotrophin in the follicular phase that no further comment is needed. As already mentioned, LH can express its action only in the presence of follicles which have reached some degree of maturity under the influence of FSH. Thus, FSH and LH should be considered synergists with regard to the elaboration of ovarian hormones during the follicular phase.

Detailed information on the precise regulatory mechanisms is, however, lacking. In this respect recent experiments of Moor, who kept whole sheep follicles in culture for several days, are of special interest since their outcome tells us how little we actually know about these processes[11]. Thus, follicles of the appropriate stage of maturation, once they have been activated to produce oestrogens by endogenous gonadotrophins, continue to make oestrogens over a couple of days *in vitro* with no necessity to add gonadotrophins. Only the one or two largest follicles of an ovary and only those originating from day 14 and 15 but not from day 16 of the oestrous cycle were found capable of synthesising oestrogens in appreciable amounts. Similarly, Channing when studying the steroidogenic potency of granulosa cells in culture, found out that the progesterone production of these cells goes on for days spontaneously once the initial 'activation' by gonadotrophins has taken place.

10.2.1.3 Local effects of steroid hormones inside the ovary

The regulatory mechanism which guarantees that, in the human, only a single follicle reaches the state of full maturation in the course of the follicular phase is still unrevealed. In this respect a local action of the steroidal hormones formed in the follicles has been considered by Hoffmann[13,14]. By means of intra-ovarian injections of 50 µg oestradiol in the form of microcrystals at laparotomy on day 7–9 of the cycle, this author was able to postpone ovulation for at least 7 days as judged from endometrium biopsies taken between days 21 and 24 of the cycle and from the delayed onset of menstruation. On the other hand, intra-ovarian application of oestradiol microcrystals on day 11 or 12 failed to influence the timing of the cycle. Similarly, microcrystals of progesterone (500–1000 µg) when injected into the ovary during the early follicular phase were shown to delay ovulation. This effect was, however, limited to those ovaries not bearing the corpus luteum of the foregoing cycle. It should be mentioned that the amounts of steroids administered in these experiments were too low to cause systemic effects. On the basis of these findings, a local action of follicular and luteal steroids on follicular development appears probable. This action might be brought about by interference with the gonadotrophin action at the ovarian cellular level.

10.2.2 Regulation of steroidogenesis at the preiovulatory phase

10.2.2.1 In the human

The interplay between gonadotrophins and steroids in bringing about ovulation is still incompletely understood. This is especially true for the quantitative relationship between hypophyseal and ovarian hormones at this stage of the cycle. In the human, a rise in plasma oestrogen and 17α-hydroxyprogesterone levels usually precedes the ovulatory LH surge by 1–3 days. The latter reaches its maximum within 40 h prior to ovulation. Since oestrogen levels often start to increase preovulatorily a few days before the first LH rise can be detected, it would appear that LH secretion is already at its optimum with regard to the stimulation of ovarian steroidogenesis prior to its pre-ovulatory rise, i.e. at very low levels. In other words, at this stage of the cycle, increase of oestrogen production by the ripening follicle is a consequence of follicular maturation under the continuous influence of FSH rather than one of increasing LH levels. In this context one should, however, be aware of the possibility that the plasma gonadotrophin levels do not necessarily reflect hypophyseal gonadotrophin secretion because of 'peripheral consumption', even if this is unlikely to occur to a great extent. Still more obscure is the regulatory mechanism which causes the plasma oestrogen levels to decrease sharply before ovulation, that is at very high LH and FSH levels. Also at this stage of the cycle it is conceivable that steroids formed by the follicle are involved in a sort of a intra-ovarian regulation. Oestrogens do not appear to play a role in this respect. For instance, in a recent report it was shown that ovulation was induced by means of HMG/hCG treatment in the absence of oestrogens in a woman with primary amenorrhoea due to 17β-desmolase deficiency[15]. Progesterone, on the other hand, might be of importance since a slight preovulatory increase of plasma progesterone has clearly been demonstrated in women who underwent laparotomy at ovulation time. This progesterone is formed more or less exclusively by the granulosa cells inside the follicle where it accumulates due to difficulties in leaving the non–vascularised follicle. It is quite possible that this intrafollicular progesterone exerts local effects upon the surrounding theca cells to diminish steroidogenesis.

10.2.2.2 In sub-primate mammalia

In agreement with the above findings in the human MER-25, an oestrogen antagonist, failed to seriously interfere with ovulation when given 12 and 6 h before hCG to immature rats primed with PMS[16]. On the other hand, inducement of ovulation by LH in PMS-primed immature rats was abolished if cyanoketone, an inhibitor of the 3β-hydroxy-steroid dehydrogenase, was administered up to 6 h after LH[17]. Since the transformation of pregnenolone to progesterone is catalysed by this enzyme, progesterone as well as other steroids further down the biosynthetic pathway might be involved in the mechanism of ovulation under the influence of LH in this species. Whereas additional evidence is available indicating that the progestins are integrated in the process of ovulation under the influence of LH, Ferrin *et al.* failed to suppress ovulation in the rat by administering antibodies against progesterone in the morning of pro-oestrous[18]. Even though these authors provided convincing data for an efficient inactivation of circulating progesterone by means of the antisera, one wonders whether the antibodies used were able to enter the follicles and thus to interfere with the progesterone action inside the follicle.

The regulatory mechanism leading to ovulation, already poorly understood, was recently found to be further complicated by prostaglandins coming into play. It can only be mentioned here that indomethacin, an inhibitor of prostaglandin synthesis, was shown to block the induction of ovulation by endogenous as well as exogenous LH or hCG in rats and rabbits[19,20]. Interestingly, the luteinising and steroidogenic action of LH was not abolished under the *in vivo* conditions chosen nor did indomethacin interfere with these two processes when injected directly into the follicle. Apart from the retention of the ovum, the process of luteinisation of the unruptured follicle was hardly distinguishable from normal, both in the morphological and steroidogenic respect. The dissociation of the ovulatory from the luteinising and the steroidogenic properties of LH by blockage of the intraovarian prostaglandin synthesis represents one of the most intriguing recent findings in the whole field and needs further investigation.

10.2.3 Maintenance of luteal function

10.2.3.1 Interspecies variation of luteotropic complexes

According to Bartosik and Romanoff[21] a luteotrophic hormone may be defined 'as a pituitary hormone which acts on the corpus luteum of hypophysectomised, non-pregnant mammals to maintain the histological integrity and functional capacity (progesterone secretion) for a duration which is equivalent to that observed in the intact animal'. It should be added that hormones of non–hypophyseal origin may also belong to this group, e.g. the oestrogens in the rabbit.

With regard to the nature of the luteotropic principles among the various species, the physiological situation in the intact cyclic female is not at all revealing because of the complete lack of correlation between blood and tissue levels of the respective luteotropins on the one hand, and the amount of hormones formed by the corpora lutea on the other hand. When considering the growing body of analytical data now available for most of the common species, one could, in fact, become doubtful whether any regulatory correlation exists between the hypophyseal and the ovarian hormones at this stage of the cycle. With more refined experimental measures such as hysterectomy, hypophysectomy, hypophyseal stalk transection, use of highly purified proteohormones, application of hormone antisera etc. it becomes quite clear, however, that the luteal function is not autonomous and has no 'inherent life span'—whatever this means—but is dependent on one or more luteotropic hormones in the majority of mammals. The wide variety of luteotropic complexes among the species which is tentatively put together in Table 10.1 precludes any generalisation as to how the corpus luteum of the cycle is maintained in mammals.

Research data are often controversial in this field mainly because of the large number of experimental approaches in use, as mentioned above. Apart from the degree of purity and the dosage of the suspected luteotropins administered in such investigations, the timing of the experimental design needs careful consideration in order to minimise the confusion. How complicated the situation can become has recently been shown in the guinea-pig, in which administration of oestrogens in pharmacological dosages late in the cycle has been proved to be luteotropic and antiluteolytic; this effect was not abolished by hysterectomy or hypophysectomy. Physiological amounts of oestrogens given within the first 5 days after ovulation, however, proved luteolytic. This action could be fully overcome by hysterectomy and partially

Table 10.1 Factors necessary for normal maintenance of the cyclic luteal function after ovulation

Species	*Pituitary*	*Prolactin*	*LH*	*FSH*	*Oestrogens (intra-ovarian action)*
Human	—	—	(+)	—	(luteolytic)
Cattle		(+)	+	—	—
Sheep	+	+	+	—	—
Pig	—	—	—	—	(+)
Rabbit (pseudopregnant)	+	—	(+)	(+)	+
Hamster (pseudopregnant)	+	+	—	+	(+)
Guinea pig	—	—	—	—	
Rat (pseudopregnant)	+	+	+	—	+

Data in brackets: either necessity not fully proved or luteotropic action only indirect

by hypophysectomy[22]. Until very recently, hypophyseal support was not felt essential for maintenance of the human corpus luteum both to full maturity and to full length as will be discussed in more detail in a separate section. No clear-cut results in this respect have been obtained in the hypophysectomised Rhesus monkey[23] either. A distinct shortening of the luteal phase after administration of hCG antiserum to intact animals, however, clearly indicates that LH is luteotropic in this primate[24]. Also in the cyclic pig, hypophysectomy, or hypophyseal stalk-transection does not appear to interfere basically with the normal luteal life span and function. In the guinea-pig, both hypophysectomy and hypophyseal stalk-transection, if performed 1–4 days after ovulation, even results in a prolongation of luteal life span in spite of the presence of the uterus with its well-known luteolytic potential[22]. The presumptive autonomy of the corpus luteum, once it is formed in the above species, is still a matter of discussion and has recently been called into question[21]. In this context, the possibility should be considered that the gonadotrophic actions outlast hypophysectomy due to a long half-life and/or high receptor affinity of these hormones. This possibility will be discussed in more detail in the next section.

For a long time in the past prolactin has been thought to be the most important luteotropic hormone. It is notably luteotropic in rodents such as hamsters, mice and rats, but only in addition to other luteotropic principles. In the latter species, prolactin has even been found luteolytic toward the end of the luteal life span[25]. In sheep, its participation in the control of the corpus luteum has been somewhat controversial; but recent work[26] clearly showed that prolactin (together with LH) is essential for both the morphological integrity and the progesterone secretion of the corpus luteum. This conclusion was derived from experiments in sheep that had undergone hysterectomy between day 9 and 12 of the cycle followed by hypophysectomy 20–50 days later. In the cow also, prolactin, when perfused through the ovaries in high concentrations, has been claimed to be luteotropic since it was able to stimulate the steroidogenesis of isolated bovine ovaries[21]. In numerous other experiments with divergent design, prolactin failed to influence steroid secretion in the cow. After hypophyseal stalk-transection in heifers, however, continuation of progesterone secretion was observed at a lower level than normal, suggesting that prolactin is actually luteotropic in this species. Thus, with the experimental and analytical methods becoming more and more sophisticated, a re-evaluation of the role of prolactin as an important luteolytic factor is to be observed. In this respect a permissive rather than a direct action of prolactin on steroidogenesis should be considered. This would explain the difficulty of demonstrating its involvement in the control of the luteal hormone production on the basis of more simple experimental designs.

LH which has been described in the previous section as essential in all mammalian species for follicular steroidogenesis, ovulation, and luteinisation is not needed any longer in the sow, guinea-pig and hamster once ovulation has taken place. In the latter species, FSH and prolactin form the luteotropic complex.

The rabbit is an exception to this in as far as the follicular oestrogen is the ultimate luteotropic principle. Thus corpora lutea have been demonstrated to regress rapidly after destruction by x-irradiation of the antral follicles which are considered to be the main source of the intraovarian oestrogens[27]. The resulting luteolysis could be overcome by the administration of oestrogens but not of LH or FSH. Nevertheless, these two gonadotrophins are also involved in the control of the corpus luteum function since they are necessary for maintenance of the follicular growth as described in the previous section. Likewise, oestrogens appear to be luteotropic in the rat: after destruction of the antral follicles by means of x-irradiation, prolactin was no longer luteotropic unless oestrogens were administered[28]. It has also been shown that in the hypophysectomised-hysterectomised rat, oestrogens are luteotropic[29]. Finally, some tenuous evidence is available that oestrogens are also luteotropic in the sow. In all three species experimental evidence is in favour of an intraovarian action of the oestrogens close to the site of production. The involvement of oestrogens in the feed-back regulatory systems between the ovary and the hypothalamus/pituitary or uterus will not be discussed here.

10.2.3.2 Luteal maintenance in women

In women with hypogonadotrophic amenorrhoea due to partial or total lack of hypophyseal function, ovulations can be induced by means of sequential administration of HMG and hCG. Corpora lutea formed in consequence of this treatment produce normal or elevated amounts of progesterone and oestrogens and proceed to normal length without further gonadotrophic support as shown by several authors. This has been taken as evidence in favour of the hypothesis that in the human, in contradistinction to the majority of the mammalian species, the corpus luteum of the menstrual cycle is autonomous throughout its whole life span. This view would appear to be further supported by the fact that in regularly menstruating women, plasma FSH and LH levels reach their lowest levels during the luteal phase and apparently fail to show any gross relationship to the luteal function (but see below).

The concept of luteal autonomy in the human has recently been called in question by Vande Wiele *et al.*[30]. After induction of ovulation in women with amenorrhoea due to hypophyseal failure, these authors observed a distinctly shortened luteal life span if an LH preparation of hypophyseal origin (HLH) instead of hCG was used. Under these conditions, plasma levels of progesterone and oestrogens remained high for only 5 or 6 postovulatory days, followed by menstruation within another 2 days. The function of these corpora lutea could be extended up to 14 days, however, by postovulatory administration of 400 i.u. HLH daily. After this time, luteal regression could not be prevented by continuation of this treatment with the menstrual flow starting on day 17 or 18 after the initial application of HLH. Plasma progesterone and oestrogens were found in the upper range of normal until slightly elevated in these therapeutic experiments and the first hormonal signs of luteal regression occurred 3 to 4 days prior to the onset of menstruation.

The difference in length of the induced luteal phase depending on whether hCG or HLH had been administered, is best explained by the metabolic difference in half-life between the two gonadotrophin preparations used[31]. While the disappearance rates follow a double exponential curve for both gonadotrophic preparations, HLH is removed faster than hCG; thus the half-life of the slower process is 4 h for LH as compared with 23 h for hCG. By calculation, the hCG dosages used in ovulation-inducing therapy should thus be sufficient to support the corpus luteum for the full cycle's length. Actually, however, hCG values were found by Kosasa *et al.*[32], to drop to zero prior to the time when nidation is thought to happen in three out of six amenorrhoic women treated sequentially with HMG and hCG, followed by conception and pregnancy.

As further evidence for LH being necessary for keeping the luteal function going, a positive correlation between serum LH levels and the life span of human corpora lutea has been assessed[33]. In this context, mention should also be made that LH is obviously taken up and possibly 'utilised' by the ovaries at both phases of the cycle as may be judged from plasma LH levels being lower in the ovarian effluent than in the peripheral circulation[34]. Interestingly, however, no difference in the LH content between the effluent of the ovary bearing the corpus luteum and that of the opposite gonad was found. It should also be mentioned that no evidence for peripheral '*utilisation*' *of LH* was detected in women rendered anovulatory by treatment with oestrogen–progesterone preparations.

No doubt luteal function is far from being at a maximum under physiological conditions. This can easily be demonstrated by administration of hCG to normal women during the luteal phase. If plasma progesterone is taken as a criterion, the stimulatory effect of hCG is dose-dependent.

In the case of conception, hCG may be detected radioimmunologically only 2–4 days after nidation is thought to have happened[32]. It appears reasonable to consider this early hCG to be responsible for the prevention of luteal breakdown after nidation at the time of menstruation. In an attempt to mimic the endogenous hCG levels in women in early pregnancy by administering increasing daily doses of hCG from 300 to 80 000 I.U. starting between the 6th and 8th postovulatory day, Geiger and Kaiser[35] were able to prolong luteal function up to 33 days with menstrual bleeding occurring 4–7 days after the last injection. The urinary excretion of oestrogens and pregnanediol thus induced was comparable with that in early pregnancy in the same women. These experimental data as well as similar results from related studies clearly indicate that hCG alone is sufficient to maintain luteal function over longer periods of time, unless one considers the possibility that a sequence of ovulations is induced by this sort of treatment rather than a prolongation of the life span of one single corpus luteum[36].

While all the above-mentioned findings are in favour of LH/hCG to be the main if not the only luteotropic principle in the human, the role of FSH remains obscure. When considering the experimental data gained from amenorrhoeic women under sequential HMG/hCG treatment for ovulation induction, one should take into account that the half-life of FSH is about as long as that of hCG. Thus, an outlasting action of FSH upon luteal function cannot simply be ruled out. In any event, FSH might only exert auxiliary effects upon luteal steroidogenesis in view of the failure of HLH to maintain luteal function to full length after pre-treatment with HMG (see above). It has also been claimed that luteal insufficiency is due to an inappropriate production of FSH in the follicular phase rather than to an abnormal LH secretion[37].

As far as prolactin is concerned, no evidence is available at present to show that this gonadotrophin plays any role in the control of luteal steroidogenesis in the human. In the course of the menstrual cycle, no major changes in the plasma levels of biologically or radioimmunologically determined prolactin have been discovered so far.

10.2.4 Luteolysis

10.2.4.1 Interspecies variations

It was not until very recently that some light has been thrown upon the regulatory mechanisms which cause luteal break-down in the non-pregnant cycling female. In the majority of mammals studied in this respect, the uterus plays a central role in maintaining normal periodicity of the oestrous cycle. In rodents like the mouse, rat, hamster, and rabbit the uterus acts upon the length of the cycle only in the pseudopregnant animal but not in the unmated one. It was often found in these species that removal of the whole uterus or parts of it was followed by a definite prolongation of luteal life-span. This led to the postulate that luteolysins were formed in the uterus. These hormones are thought to abolish the viability of the corpus luteum. How this is achieved is still an open question to which quite different answers have been given so far. Interactions of the luteolysins with the hypophyseal or follicular luteotropins at the luteal level are becoming more and more probable[38]. Since 1968, an increasing body of evidence has been accumulated in sheep, cattle, horses, guinea-pigs, rats, hamsters, and rabbits which indicates that prostaglandins (PG) may act as luteolysins and might be identical with the postulated uterine hormones.

In primates, the uterus is obviously not involved in the termination of luteal function. This might have something to do with the prompt hCG production of the primate trophoblast after nidation mentioned above, a property which is lacking in the sub-primate mammalia. Also, in the ferret, dog, and special marsupials, uterine participation in regulating the luteal life span has not been demonstrated so far. A peculiarity of these species is that pregnancy is either as long as the luteal life span or even shorter, as in the marsupials. Therefore no signal for survival is necessary from the foetus to the ovary.

The following section will be mainly concerned with the regulatory mechanisms of luteolysis in the human and the sheep as the best-documented examples for non-uterine and uterine luteolysis.

10.2.4.2 Non–uterine luteolysis in the human

In the human, luteal life span is limited to 10–14 days unless pregnancy occurs, in which case the corpus luteum preserves its capacity to produce hormones throughout the whole pregnancy[39]. Whereas all evidence is in favour of hCG being the decisive factor in preventing the demise of the corpus luteum in the case of pregnancy as explained above, we are at a loss to properly define the triggering mechanism leading to luteolysis in the infertile human menstrual cycle.

Hysterectomy obviously does not influence luteal activity as has been demonstrated by several groups on the basis of determinations of plasma progesterone or urinary pregnanediol and oestrogens. For instance, not less than 37 out of 48 preclimacteric, hysterectomised women have been recently reported to exhibit ovulatory cycles of normal length and undisturbed hormonal excretion[40]. The same appears to be true in the Rhesus monkey[23]. Likewise, a normal cyclical pattern of serum FSH, LH, oestradiol, progesterone, and 17α-hydroxyprogesterone was found in a case of congenital absence of the uterus[41]. Also, human endometrium failed to shorten the prolonged luteal life span in hysterectomised pseudopregnant hamsters when placed into the cheek pouch of these animals, whereas endometrial tissue of the hamster and the rat did so[42]. Thus all the evidence is against a uterine participation in the luteolytic processes in the human and the same is true for the Rhesus monkey[23].

If one rejects the idea of an inherent life span of the human corpus luteum, other regulatory mechanisms have to be looked for. In this respect the possibility of an intraovarian involvement of steroid hormones in the luteolytic processes deserves our attention. Contrary to comparable data obtained in the rabbit[43], Hoffmann[13] has shown that microcrystals of oestradiol, when implanted at surgery into the ovary bearing the corpus luteum, shortened the luteal life span considerably; and so did microcrystals of testosterone isobutyrate[44]. Crystalline progesterone, on the other hand, did not exert any influence upon luteal activity when administered in the same manner. Likewise, the implantation of 50 μg of crystalline oestradiol into an ovary not bearing a fresh corpus luteum was without effect. In all these experiments the amount of steroids administered locally was too low to exert systemic effects. Again, these results point to an intraovarian action of oestrogens and may be androgens which might be of importance for the control of luteal life span. This concept if further supported by observations in women with cervical cancer who underwent x-irradiation of the ovaries between day 12 and 19 and successively Wertheim hysterectomy between day 43 and 84 after their last menstrual period. In each of the 23 cases, the ovaries had an atrophic appearance with virtually no developing follicles except for one corpus luteum in one of the two ovaries which looked functional[45]. If irradiation was performed in a control group of women before day 10 or after day 20 of the cycle, no corpora lutea were detectable in the completely atrophic ovaries. If one assumes that the growing follicles of the luteal phase are the source of an increasing production of oestrogens and that these oestrogens exert inhibitory effects upon the corpus luteum, then the prolonged activity of such irradiated corpora lutea in the absence of surrounding follicles would be quite understandable. It has to be taken into account, however, that human corpora lutea themselves produce considerable amounts of oestrogens. Whether this means a sort of luteal 'self-destruct mechanism'[23] is open to doubt at present. If this concept were correct, it is difficult to see how in the case of pregnancy the luteolytic action of oestrogens is overcome.

It is now well established that oestrogens when given in extremely high doses just postovulatorily, will prevent nidation and/or pregnancy. As recently shown by Gore *et al.*[46] as well as our group, this treatment results in premature decrease of plasma progesterone levels and in early menstruation as long as the medication was started within 3 days after ovulation. If such treatment was begun around the time of nidation, oestrogens failed to be 'luteolytic'[46, 47]. Similar observations have also been made in the Rhesus monkey[23]. If in this species a slight continuous elevation of plasma oestrogen levels was established by means of implanted silastic capsules filled with oestradiol, this was found to be sufficient to cause a fall of plasma luteal progesterone levels to zero with an average 6, 4 days in advance without consistently influencing the plasma LH values[48]. Whether the outcome of these therapeutic experiments in women and monkeys has any meaning for the physiological control of luteal life span remains to be clarified.

Administration of excessive dosages of progestational agents like chlormadinone acetate, d-norgestrel, medroxyprogesterone acetate or norethisterone during the luteal phase has also been shown to decrease the endogenous plasma progesterone levels to nearly preovulatory values within a few days. However, this effect should not be termed luteolytic since it is easily overcome by additional administration of hCG[49].

Human endometria contain appreciable amounts of prostaglandins, namely PGF_2, the concentration of which being considerably higher in the luteal than in the follicular phase[50]. It was therefore tempting to assume that prostaglandins play a role in terminating human corpus luteum function. Several clinical trials have, however, failed to confirm this hypothesis. If anything, a slight acceleration of luteal breakdown has been observed to occur secondary to $PGF_{2\alpha}$[51]. Other prostaglandins have not been tried in the human as yet. In this context it is of interest to note that in the monkey infusions of $PGF_{2\alpha}$ at low concentration proved to be stimulatory, those at high concentration to be inhibitory in terms of the progesterone level in the ovarian venous blood[52]. This inhibitory effect of $PGF_{2\alpha}$ was easily overcome by hCG administration. The ovarian blood flow did not change in these *in vivo* perfusion experiments of the ovary and can therefore be ruled out as the primary site of action of the prostaglandins[38].

10.2.4.3 Uterine luteolysis in sheep

In contrast to the human, in sheep the uterus is basically involved in the termination of luteal function. Following removal of the whole uterus prior to day 13 of the 16-day oestrous cycle, the corpus luteum is maintained for virtually the full length of gestation as was first shown by Wiltbank and Casida in 1956.

After removal of the uterus horn adjacent to the ovary containing the corpus luteum, the length of the luteal phase is approximately doubled. If, however, the contralateral uterus horn is removed, the length of the cycle remains unchanged. Likewise, a prolongation of luteal function is achieved by simple surgical separation of the ovary from the uterus unless adhesions are formed[53]. These observations—among many others—suggest that the luteolytic signal is transmitted from the uterus to the ovary by a local rather than a systemic route. Nonetheless, the latter might also be operative to some extent, since the corpus luteum is maintained for a much longer time following total rather than partial hysterectomy, no matter which part of the uterus has been removed. Accordingly, prolongation of luteal function has been shown to be roughly proportional to the amount of uterus removed[54].

The luteolytic action of the ovine uterus is overcome by pregnancy. It has been demonstrated experimentally that the intrauterine presence of an embryo is essential by day 12 or 13 if the corpus luteum is to be maintained as in normal pregnancy[54]. Even if the embryo is left in the uterus only for 24 h, the luteal life span is extended from 16 to 24 days[42].

In connection with the above phenomena, transplanted embryos prevent luteolysis only if they are placed into the uterus horn adjacent to the ovary bearing the corpus luteum. Transplantation into the contralateral uterus horn is promptly followed by demise of the corpus luteum. This again points to a local transmission route for the postulated luteolysins. Inhibition of luteolysis can also be achieved by repeatedly injecting cell preparations of ovine embryos into the uterus during the critical phase of luteolysis[50]. Homogenates of 14- or 15- but not of 25-day embryos were effective in this respect[42]. The mechanism by which the embryo or preparations of it inhibits the release of uterine luteolysins is still a mystery.

Concerning the transmission of the luteolytic signal from the uterus to the ovary, neither a neural pathway nor a humoral transport via the oviduct or the lymphatic system appears to play a role[56]. Ligation of the uterine vein, however, is followed by luteal retention which strongly indicates that the luteolysin is taken up and carried to the ovary by the uterine vein[57]. This view is further supported by the elegant work of McCracken, Baird and Goding[56] who performed cross-circulation experiments in ewes, the donor animals of which were bearing utero–ovarian transplants at the neck and the recipients solely ovarian autographs. While the former were cycling regularly, the latter displayed luteal retention as expected. When utero–ovarian venous blood of day 15 of the oestrous cycle from the donor was infused into the ovarian artery of the recipient, the steady progesterone secretion characteristic for luteal retention in the recipient rapidly dropped and oestrous behaviour appeared within 24 h. Blood from donor ewes of day 2 or 10 of the oestrous cycle failed to be effective in this respect. Similarly, blood collected from uterine veins of normal sheep by day 14 of the oestrous cycle was able to suppress the ovarian progesterone secretion markedly when infused into the artery of an ovine ovary *in situ* on day 10 of the oestrous cycle. Blood from the jugular vein of the same day of the cycle or uterine venous blood of day 8 were without effect[58]. Since the utero–ovarian veins by-pass the ovary and are not of the portal type, it has for long been and still is to some extent a mystery how the luteolytic uterine signal reaches the ovary under these circumstances. As an attempt to solve this enigma, a countercurrent exchange mechanism between the utero–ovarian vein and the ovarian artery has been postulated to be the main means by which uterine luteolysin may reach the ovary without entering the general circulation[56, 59]. This hypothesis is supported by the observation that separation of the coiled ovarian artery from the closely connected utero–ovarian vein is sufficient to stop uterine luteolysis in the sheep[60]. Further evidence will be given below.

10.2.4.4 Prostaglandins and luteolysis

Shortly after $PGF_{2\alpha}$ was shown in 1968 to be luteolytic in the pseudopregnant rat, prostaglandins were also found to exert similar effects in various other species, e.g. the hamster, rabbit, and guinea-pig. In sheep $PGF_{2\alpha}$ and to a lesser degree $PGF_{1\alpha}$ and PGE_1 have been proved to cause a decrease in progesterone secretion by the autotransplanted ovary when administered into the ovarian artery[61, 62]. Not only was there a prompt drop in the progesterone levels to undetectable values within 24 h after $PGF_{2\alpha}$ had been administered, this event was also consecutively followed, as in normal cycles, by short-term rises in the secretion of oestrogens and LH, the latter in coincidence with the appearance of oestrous behaviour. Unlike the rat, luteolysis was not correlated with a stimulation of the 20α-hydroxysteroid dehydrogenase[61].

When infused directly into the ovine uterine vein over periods of 6–9 h, amounts as small as 20–40 µg $PGF_{2\alpha}$ per hour have been found sufficient to cause premature luteal break-down[59]. This is only 5–10 times the amount of $PGF_{2\alpha}$ necessary to cause luteolysis, when infused directly into the ovine ovarian artery and much less than if the same effect is to be achieved via the general circulation. This observation may also be taken to support the counter-current exchange hypothesis.

If prostaglandins, namely $PGF_{2\alpha}$, were identical with the ovine uterine luteolysin, one would expect these compounds to be present in both the uterus as well as the uterine venous blood by the end of the ovine oestrous cycle. This has been recently demonstrated. When PGF was analysed in ovine uterine venous blood throughout the cycle, a steep increase (up to 20 fold) was found during days 14–16 of the cycle, that is at the time when plasma progesterone levels had just fallen sharply[63, 64] and a new wave of follicular growth in seen. Likewise, the endometrial concentration of $PGF_{2\alpha}$ reaches a sharp peak just at the end of the ovine oestrous cycle[65]. Unexpectedly, however, $PGF_{2\alpha}$ levels both in the endometrium and the uterine venous blood were still higher on day 13 when conception had occurred[66]. This raises the question of how the conceptus overcomes the luteolytic impulses of the uterus other than by suppression of the uterine prostaglandin secretion.

An answer cannot be given at present; but a luteotropic compound produced and secreted by the embryo should be taken into consideration[59].

The regulation of the uterine production of prostaglandins is not clear either. Since progesterone when given early in the cycle is known to shorten the ovine oestrous cycle as long as a uterus is present, it has been suggested that progesterone induces prostaglandin production of the uterus. However, progesterone administration to ovarectomised ewes every other day over a period of 10 days only caused slight increase in the peripheral levels of PGF. When oestradiol was given after such 10 day's priming with progesterone, PGF levels in the peripheral blood rose dramatically, just as under physiological conditions prior to oestrous. In addition, immunisation of such sheep against oestradiol blocked any PGF release into the general circulation. These findings led Caldwell *et al.*[67] to postulate that the luteal break-down is initiated by an increase of the ovarian oestrogen secretion which in turn stimulates $PGF_{2\alpha}$ release into the uterine vein as well as sharply decreases the ovarian progesterone production. In accordance with this concept is the finding that destruction of the wave of developing follicles by x-ray irradiation in the otherwise intact sheep prevents luteal regression at the expected time[68].

If the sequence of events as proposed by Caldwell *et al.*[67] is correct one still is at a loss to explain by which means the first increase of the ovarian oestrogen production on day 13–14 of the cycle is triggered and how it is suppressed in the case of pregnancy. At least no change in the gonadotrophin and progesterone levels can be detected before day 15 of the cycle[69].

With ^3H-$PGF_{2\alpha}$ just recently becoming available, it was possible to study the above-mentioned 'countercurrent exchange hypothesis' in more detail. Surprisingly enough, the first radioactivity becomes detectable in the ovarian artery not earlier than 20–30 min after the onset of an infusion of this labelled presumptive luteolysin into the ovine utero–ovarian vein. Peak values were reached some 30–35 min after the 1 h infusion had been stopped, thus indicating a very slow penetration rate of the labelled compound through the walls of the ovarian vessels. For comparison, the radioactivity of blood drawn from the iliac artery remained extremely low throughout the whole infusion and post-infusion period[63].

Nothing definitive is known as to how the prostaglandins exert their luteolytic action at the ovarian level. The hypothesis of Pharris[38] that prostaglandins act by reducing the ovarian blood flow drastically was based on experiments with very high concentrations of $PGF_{2\alpha}$. At physiological PGF_2 levels, a clear-cut influence upon this parameter has not yet been detected. It should be mentioned, however, that even at still lower levels, near the threshold of effectiveness, the corpora lutea may undergo regression by ischaemia in some areas of the gland. This might point to the possibility that $PGF_{2\alpha}$ acts by diverting the intraovarian blood flow into arterio–venous shunts and thus leading to local ischaemia of the corpus luteum[59].

An interference with the luteotropic complex at the ovarian level as the primary site of action of the prostaglandins in bringing about luteolysis is also under serious discussion[38]. In intact as well as in hypophysectomised sheep the luteolytic effect of endogenous prostaglandins at the respective phase of the oestrous cycle can be overcome almost indefinitely by continuous LH infusion[70]. However, when in this species LH was infused together with

$PGF_{2\alpha}$ into the artery of an ovary autotransplanted to the neck, LH failed to counteract the effect of the prostaglandin and luteolysis resulted[71]. In other species such as rat, rabbit, hamster, and monkey, evidence also is accumulating that $PGF_{2\alpha}$ and the respective luteotropic complex (see Table 10.1) are antagonists in terms of luteolysis at the ovarian level[38, 52, 72].

10.3 REGULATION OF STEROIDOGENESIS AT THE CELLULAR LEVEL

10.3.1 Limitations to *in vitro* techniques

Under suitable *in vitro* conditions, steroid formation in appropriate ovarian tissue preparations goes on for several hours without the need for gonadotrophins to be added to the incubation medium. Nonetheless, addition of LH, hCG and cAMP often causes a definite further increase in steroidogenesis in such an incubation system. The extent to which this happens varies considerably among tissues of both the same kind and same functional state. This might be due, among other things, to the amount of endogenous gonadotrophins being more or less fixed at the specific receptors located in the cellular membranes. Thus, the amount of gonadotrophins pre-existing in the ovaries is one of the unknowns in experiments of this sort and may contribute considerably to the great variability and limited reproducibility of the quantitative results often observed.

In the vast majority of *in vitro* experiments designed to clarify the influence of gonadotrophins and their transmitters upon the ovarian steroidogenesis, intact-cell preparations, namely slices, rather than homogenates or subcellular fractions have been used. As in the case of the adrenals[73], the integrity of the ovarian cells and their subcellular structures is a prerequisite for the steroidogenesis to proceed normally in response to appropriate stimuli. For instance, disintegration of bovine luteal cells by homogenisation has been shown to abolish the progesterone increase in response to LH stimulation although this can be reliably elicited if luteal slices are used[74]. Studies with broken-cell preparations are predominantly suited to investigations of single enzymatic steps on a qualitative rather than a quantitative basis. The use of homogenates or subcellular fractions for investigating the action of gonadotrophins *in vitro* may thus be misleading both qualitatively and quantitatively and should therefore be regarded with reservation.

Most data in this field have been collected from short-term incubations of a few minutes to 2 h duration, under which conditions only acute responses to *tropic* stimuli can be visualised. From what has been said above on the possibility of outlasting *in vivo* effects of gonadotrophins for periods as long as a whole luteal phase, it is quite clear that the elucidation of long-term *tropic* effects cannot be the objective of such short-term investigations, unless special experimental measures or treatments *in vivo* precede the *in vitro* studies. For the clarification of permanent and/or permissive hormone actions upon ovarian steroidogenesis, single cell and tissue cultures as pioneered by Channing[12] are of special value.

According to general belief, LH as well as cyclic-AMP (cAMP) trigger a cascade of intracellular reactions which finally result in a stimulated transformation of cholesterol into Δ^5-pregnenolone, as will be discussed in more detail below. Once the process of overall steroidogenesis has been set going by the gonadotrophin stimulus, all steps further down the steroid pathway are no longer under direct gonadotrophic control but follow the primary regulatory impulse more or less automatically, with the cellular equipment of enzymes and co-factors being rate-limiting. The one well-known exception to this rule, i.e. the inhibitory action of prolactin on the 20α-hydroxysteroid dehydrogenase activity of rat corpora lutea, will be referred to in the next section. In any event, one has to be very cautious with regard to reports on the influence of gonadotrophins on single enzymatic steps of steroid transformation following Δ^5-pregnenolone formation.

Although the biosynthetic products of steroidogenesis are necessarily piling up in the usual *in vitro* systems and might thereby impair the process of steroidogenesis, this effect appears to be only of minor importance. In human corpora lutea at least, the profile of newly synthesised steroids is surprisingly constant, no matter what the physiological conditions of the tissue *in vivo* and the duration of incubation had been. Also, neither LH nor cAMP have been found to be able to basically change this pattern[39].

10.3.2 Gonadotrophic control of cellular steroidogenesis

10.3.2.1 General outline

On the basis of *in vivo*- and *in vitro*-findings, LH must be considered the decisive gonadotrophin for the control of steroid biosynthesis in the vast majority of species and ovarian tissue compartments. With regard to the stimulation of *follicles* and *interstitial tissue* to produce steroid hormones, the reviewer is not aware of any report indicating a hypophyseal hormone other than LH to be of primary importance. Even for those species in which the corpora lutea are not believed to be under direct LH control (see Table 10.1), this obviously holds true. For instance, both the preovulatory progesterone rise in the interstitial tissue of the hamster and the oestrogen production by the follicles of the rabbit are increased *in vitro* only by LH[75,76].

Also the *corpora lutea* of the majority of species studied in this respect are stimulated *in vitro* by LH, but not by other gonadotrophins, to increase steroid formation. Even in rats and mice, in which the corpora lutea have been thought for almost two decades to be regulated solely by prolactin, LH and not prolactin has been found to stimulate progestin synthesis *in vitro*. According to Armstrong *et al.*[77] this response can, however, be reversed by pretreatment of immature, hyperstimulated rats *in vivo* with LH 2 h or more before the ovaries are excised for the *in vitro* studies. Under these circumstances, the formation of progesterone was markedly reduced while the *de novo* formation of progesterone from labelled acetate was increased. For many reasons, these findings are best interpreted by a rapid precursor depletion (presumably cholesterol acetate) secondary to LH. As the only exception known, isolated rabbit corpora lutea fail to respond to LH *in vitro* but do so to oestrogens, which is consistent with the *in vivo* findings mentioned above[78]. In this species, LH only indirectly influences luteal steroidogenesis through its stimulatory ability upon the oestrogen production of the follicles.

While among the common mammalia a direct or indirect involvement of LH in the control of the ovarian steroidogenesis is well established, the role of prolactin which is anyway limited to some species not including primates, is far from being clearly defined. All *in vitro* attempts to acutely stimulate ovarian steroid formation by prolactin have failed so far[4]. This negative statement by no means implies that prolactin is without direct influence on the biosynthetic processes within the ovary under appropriate conditions. The evidence available is, however, in favour of prolactin playing a permissive rather than a direct steroidogenic role. To give an example, intact rabbits in which the interstitial cell cholesterol has been depleted by LH, regain their ability to respond to acute LH stimulation more rapidly if the cholesterol stores have been replenished by chronic pretreatment with prolactin[79]. Similarly in cyclic rats, LH has been found to stimulate *in vitro* the progesterone production of corpora lutea only if originating from animals pretreated with prolactin[80]. In fact, a great body of data which will be discussed below in some detail has been accumulated to show that prolactin controls cholesterol storage and turnover in species such as rabbits and rats, thereby participating in steroid biosynthesis in a decisive way.

Whereas gonadotrophins, as outlined above, do not stimulate isolated enzymatic steps along the biosynthetic pathway after Δ^5-pregnenolone, *in vivo* pretreatment of rats with prolactin has been shown to cause a fourfold increase of the *in vitro* production of progesterone by corpora lutea 3 days of age. Since in these and many other experiments the synthesis of 20α-OH-P was decreased to the same extent, 'total progestin' formation was not altered by prolactin[80], thus indicating that prolactin inhibits the 20α-hydroxysteroid dehydrogenase in this species. Similar changes can be induced in pseudopregnant rats by hypophysectomy but can be completely prevented by prolactin treatment[81].

Little is known as to whether FSH in addition to its morphogenic action upon follicular development exerts permissive effects on the steroid biosynthesis comparable with that of prolactin. The biochemical basis of its participation in the maintenance of the corpus luteum function in the hamster as established *in vivo* remains to be clarified. Although cultured porcine and monkey granulosa cells originating from small follicles have been reported to respond to FSH or FSH and LH respectively with an increased progesterone secretion rate[12], this is obviously due to the trophic expression of FSH upon highly immature cells rather than to a direct steroidogenic action.

10.3.2.2 Gonadotrophin–receptor interactions at the ovarian level

The field of gonadotrophin–receptor interactions has become of increasing interest in recent years. In the ovaries, as in the testes and adrenals, binding of the circulating gonadotrophins by specific receptors is now generally believed to be the initial step in the sequence of events at the cellular level leading finally to an increased ovarian steroidogenesis. There is convincing evidence in favour of these receptors being located in the outer layer of the plasma membrane in such a way that the gonadotrophins need not enter the cell in order to exert their specific stimulatory effects.

From several investigations with ^{125}I-labelled gonadotrophins it has become apparent that LH and hCG may become specifically bound to luteal cells of a variety of species including man[82]. Since only LH and hCG but not FSH or gonadotrophin subunits compete with ^{125}I-LH for the receptor sites, this phenomenon may form the basis for a highly specific cell-free radioligand receptor assay for LH and hCG as a substitute for the common biological tests.

In rats and mice, ^{125}I-LH is specifically bound not only to luteal but also to thecal cells. Again LH, hCG as well as anti-LH but not FSH may interfere with this binding process when injected prior to or simultaneously with the labelled hormone. If ovaries after such an *in vivo* pretreatment were subjected to ultrasonic cell disintegration and differential centrifugation, the ^{125}I-linked radioactivity was found predominantly in the plasma membrane fraction, thus supporting the above view[83]. In comparable studies, however, the bulk of radioactivity was found in subcellular fractions being distinctly distinguishable from the plasma membrane fraction[84]. Contrary to the membrane-bound ^{125}I-gonadotrophin, this intracellular by bound radio-activity could not be displaced by LH or hCG and appears thus to be unspecifically fixed[85]. In the light of these recent findings one will possibly have to re-evaluate the concept of the gonadotrophins being unable to enter the cell, whatever the physiological importance of this phenomenon might be.

^{125}I-hCG has been shown to bind also to porcine granulosa cells cultured *in vitro*[86]. The binding capacity of granulosa cells originating from large follicles was 10 to 1000 times greater than that of cells from middle sized or small follicles[87]. Competitive displacement was achieved with hCG and LH of human, porcine and bovine origin, while FSH had only limited effect and the α- and β-subunits of hCG were inert in this respect. In this experimental model, the binding was found to be directly proportional to the number of cultured cells over a wide range of concentrations[86]. Thus, after 10 min of incubation at 37°C about 1000 molecules of ^{125}I-hCG were assessed to be bound to one cell, When free rabbit ovarian cell suspensions were studied, the respective figure was 360 molecules of ^{125}I-LH per cell, only 1.5% of which, however, were specifically bound[84]. After *in vivo* administration of the labelled gonadotrophin, the amount of bound molecules was less than one tenth that found in the *in vitro* study[84].

Little is known about FSH-receptor interactions at the ovarian cell level. On the basis of two different experimental approaches, some evidence is available to indicate that FSH is bound to the granulosa cells of moderately large and large follicles and here again predominantly to the basal cell layer[83, 88].

As far as prolactin is concerned, this hormone appears to bind mainly to all types of corpora lutea, at least in the rat[88]. It should finally be mentioned that in the cytoplasm of rabbit corpora lutea an oestrogen receptor has been isolated in concentrations comparable with those in the endometrium[89, 90]. It is tempting to assume that the existence of such a receptor is closely linked to the luteotropic property of oestrogens in this species.

10.3.2.3 Role of cyclic-AMP as second messenger of the gonadotrophic stimulus

As in the case of other polypeptide and proteohormones like insulin, glucagon parathormone, vasopressin, TSH, and ACTH, cAMP serves also for LH as an intracellular mediator of the hormonal signal. Thus, Sutherland's second messenger concept has likewise been found valid for primates including man[12, 39], large domestic animals like cattle[91, 92], and pig[93, 94] as well as small laboratory animals such as rats[95], mice[96], and rabbits[97, 98]. Proof that cAMP plays such an essential role in the regulation of ovarian steroidogenesis is based on the following findings:

1. Addition of cAMP to various *in vitro* systems such as whole ovaries, whole follicles, slices of corpora lutea and granulosa cell cultures (but not broken cell preparations) mimics the respective cellular response to LH in the species mentioned above. Nucleotides different in structure from cyclic 3′,5′-AMP have not been found effective in this regard[91]. Whether this applies also to cyclic guanosine 3′,5′-monophosphate (cGMP), the only cyclic nucleotide other than cAMP known to occur in nature, remains to be established. Cycloheximide as well as puromycin inhibit the stimulatory action on steroidogenesis of both LH and cAMP, thus indicating that the protein formation involved in these regulatory processes does not precede cAMP stimulation but is a consequence of it[92, 95].
2. LH and cAMP are not additive with respect to the stimulation of steroid hormone formation, as has been demonstrated in slices of both cow corpora lutea[91] and whole rabbit ovaries[97] as well as in whole follicles[99].
3. cAMP is piling up within the ovarian cell secondary to LH or hCG stimulation as shown in slices of bovine corpora lutea[92], homogenates of ovarian interstitial tissue from pseudopregnant rabbits[98], whole mouse ovaries[96], cultured porcine granulosa cells[94], and whole rabbit Graafian follicles[99].
4. The intracellular accumulation of cAMP secondary to LH is a matter of a few minutes and considerably precedes the increase of steroid formation *in vitro*[92, 94].

Theoretically, an increase of intracellular cAMP may come about in two ways, either by a stimulation of the adenylate cyclase which catalyses the formation of cAMP from ATP or by an inhibition of the cyclic 3′,5′-nucleotide phosphodiesterase by which cAMP is inactivated to 5′AMP:

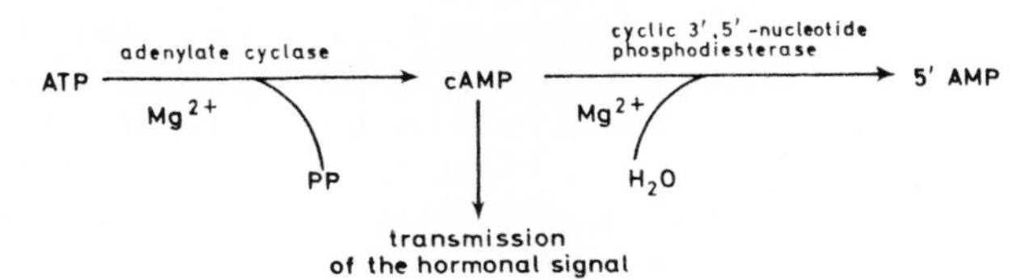

All evidence is in favour of the first of these two possibilities being operative. Thus, if inhibitors of the phosphodiesterase like theophylline were added to *in vitro* systems of homogenates of bovine corpora lutea or rabbit ovarian interstitial tissue or of porcine granulosa cell cultures, the stimulatory effect of LH on cAMP formation was found to be potentiated rather than inhibited. The latter response would be expected if LH acted by suppressing phosphodiesterase activity[93, 94, 98, 100].

While phosphodiesterase is accumulating in the 100 000 g supernatant at differential centrifugation of luteal homogenates[101], adenylate cyclase remains in the plasma membrane fraction. This latter enzyme must therefore be located in close neighbourhood of the gonadotrophin receptors. A *receptor-adenylate cyclase* complex may even exist which might undergo conformational changes secondary to gonadotrophin–receptor interactions and thereby activate the catalytic properties of the adenylate cyclase[98]. Although this issue is under intensive research at present, nothing definite is known as to the nature of the suggested receptor–enzyme complex, nor has the mechanism of adenylate cyclase activation been clarified. It also remains to be investigated as to whether Ca^{2+} is an essential participant in these membrane-located processes in analogy to its postulated role in the adrenals. Finally, it is totally unknown how the hormonal signal is passed on to the ultimate site of its regulatory action on steroidogenesis. A cAMP-dependent protein kinase which has recently been detected in the ovary might be involved in this process[102].

10.3.3 Involvement of prostaglandins

10.3.3.1 Influence of exogenous prostaglandins on steroid biosynthesis in vitro

Whereas $PGF_{2\alpha}$ has been found to be unequivocally luteolytic *in vivo* in a great variety of subprimate species, the respective data obtained *in vitro* are rather confusing. In some species such as cow, mouse, and pig experimental observations point to the paradox that prostaglandins are luteolytic *in vivo* but luteotropic *in vitro*. This was at first thought to be true also for the rabbit and the rat, but now that more information is available it is no longer tenable. In primates in which prostaglandins obviously are not an obligatory mediator of luteolysis, stimulatory rather than inhibitory effects of prostaglandins on the ovarian steroidogenesis *in vitro* have been observed.

In the rabbit Bedwani and Horton[103] had to revoke their preliminary observation that PGE_2 was effective in stimulating the formation of 20α OH-P (but not that of progesterone) by chopped ovaries in the presence of hCG and PMS. Later on it was also found by others that both PGE_2 and $PGF_{2\alpha}$ failed to stimulate *in vitro* steroid formation in this species. Sellner and Wickersham[104], on the contrary, have demonstrated trophic effects of $PGF_{2\alpha}$ upon progesterone biosynthesis when using sliced ovarian tissue. However, in bisected corpora lutea from pregnant rabbits cultured over periods of 6 h, $PGF_{2\alpha}$ in the absence of additional gonadotrophins inhibited progesterone production by 50%[105]. On the basis of this intriguing finding it is tempting to assume that the observed luteolytic effect of prostaglandins in this species is of long-term rather than short-term character.

The rat is another example of the diversity of results to be obtained depending on the experimental design. *In vitro* studies with minced ovaries from pseudopregnant rats led Pharriss *et al.* to postulate that $PGF_{2\alpha}$ enhances the ovarian formation of progesterone and 20α-OH-P, the latter to a lesser extent than the former[106]. When ovaries from pregnant rats were used, just the opposite, i.e. an increase of 20α OH-P but not of progesterone was found[107]. In contradistinction, $PGF_{2\alpha}$ failed to exert any effect at a dose range between 1 and 10 000 ng/ml on progesterone biosynthesis by slices of 'heavily luteinised' rat ovaries, nor were $PGF_{1\alpha}$, 17-keto-$PGF_{2\alpha}$ or PGE_1 effective in this respect[108]. Whereas no change in the incorporation rate of labelled acetate into progesterone under the influence of prostaglandins was seen when this experimental model was used, a decreased accumulation of radioactivity in the progesterone fraction from acetate-1-^{14}C occurred under the influence of prostaglandins both in the above-mentioned short-term incubations of Pharriss *et al.*[106] and in long-term organ cultures of rat corpora lutea[109].

As opposed to the rat and the rabbit, in the cow, slices as well as homogenates of corpora lutea can be stimulated by prostaglandins with regard to the formation of progesterone both from labelled acetate and unlabelled precursors[110, 111]. In these experiments, the four prostaglandins tested were found to be of the following order of activity $E_2 > E_1 > F_{2a} > A_1$. On the molecular basis, PGE_2 was calculated to be half as active as LH[110, 111].

In the human, too, formation of progesterone and 20αOH-P by slices of corpora lutea of the menstrual cycle has been found to be stimulated by PGE_2 but not by $PGF_{2\alpha}$ at comparable concentrations[112]. Luteal tissue originating from midpregnancy, however, did respond to some extent also to $PGF_{2\alpha}$[113]. Recent work of Channing[114] with luteinised monkey granulosa cells kept in culture over periods of 2–3 weeks is of special interest in this respect. In the absence of exogenous gonadotrophins, prostaglandins were found capable of stimulating progesterone biosynthesis from day 4 of the culture on up to day 20. The order of effectiveness was $PGE_2 = PGE_1 \gg PGA_1 > PGF_{2\alpha}$. At maximal LH stimulation, however, both $PGF_{2\alpha}$ and PGE_2 at a dosage level of 10 μg/ml inhibited instead of stimulated progesterone formation in this system, indicating that prostaglandins may interfere directly with the steroidogenic action of LH as mentioned above.

In summary, the data available do not allow far-reaching generalisations concerning the influence of prostaglandins on the ovarian steroid production *in vitro*. Too many poorly understood variables such as species, tissue and cell pecularities, sorts of *in vivo* pretreatment, variation in the experimental design including timing, differences in the biological profiles of the prostaglandins used etc. add to the confusion. One should also be aware of the fact that almost all effects of prostaglandins observed *in vitro* have been elicited at *pharmacological* (1 μg ml^{-1} medium) rather than *physiological* (low ng ml^{-1} range) levels. It is true that Channing[114] recently demonstrated a slight stimulatory effect upon progesterone secretion of luteinised cells kept in culture already at a more or less physiological PGE_1 or PGE_2 milieu (10 ng ml^{-1}), yet maximal response was not achieved below 10 μg ml^{-1}. Since the amount of prostaglandins entering the cells *in vitro* is unknown, any statement on physiological or pharmacological levels of prostaglandins should be treated with some reservation. In any case, an early destruction of the prostaglandins *in vitro* does not appear to occur as was clearly shown for PGE_1 in incubations with rabbit ovarian tissue[103].

10.3.3.2 Incidence of prostaglandins in the ovary and its stimulation

Although the incidence of prostaglandin-like material in ovine and bovine ovaries has already been demonstrated in 1937 by Von Euler and

Hammarström, it was only very recently that the biological importance of this finding was fully recognised. Thus, Chasalow and Pharriss[105] were the first to demonstrate the *de novo* synthesis of PG-like compounds from labelled arachidonic acid by rat ovarian homogenates. This production could be abolished by pretreatment of the animals with anti-LH serum. Addition of LH *in vivo* or *in vitro* to the animals so treated was able to overcome this inhibitory effect. The first report of a trophic hormone-sensitive prostaglandin synthesis system was recently confirmed by Demers *et al.*[109] who observed that LH addition to long-term organ cultures of rat corpora lutea was capable of increasing the synthesis of PGF. Interestingly, PGE_2 increased the production of PGF by the luteal cells at a 50-100 fold rate. It was the idea of these authors that activation of the cholesterol esterase activity was the starting point for PG-production facilitated by the liberation of the appropriate fatty acids.

In superfusion experiments of monkey ovaries, again, a stimulatory effect of LH, but not of FSH or prolactin, on $PGF_{2\alpha}$ formation was seen[116]. However, when monkey granulosa cells in culture were exposed to eicosa-5, 8,11,14-tetraynoic acid, an inhibitor of prostaglandin biosynthesis, it lasted 6–8 days until the LH-stimulated morphological luteinisation and progesterone secretion started to slightly decrease, thereby calling in question the acute role of prostaglandins in the regulation of steroidogenesis and its *in vitro* formation under the conditions chosen in this species. PGF-like material has also been demonstrated to be present in whole ovaries, interstitial tissue and corpora lutea of the pseudopregnant rabbit. But in contrast to the finding in the rat, LH was not capable of stimulating prostaglandin synthesis nor were FSH or prolactin active in this respect[103, 116]. The Graafian follicle of the rabbit, on the other hand, has been shown to produce increasing amounts of PGF and PGE under the influence of hCG[117].

10.3.3.3 Interrelationship between prostaglandins and cAMP

In 1970, Kuehl *et al.* advanced the hypothesis that PGE is interposed between LH and cAMP in the regulation of steroidogenesis and thereby functions as a second messenger instead of cAMP[96]. Accordingly, cAMP should act as an obligatory 'third messenger' in passing on the hormonal information within the cell. This concept is based on data from short-term *in vitro* experiments with whole mouse ovaries in which LH, PGE_1 and PGE_2 have been found to increase cAMP accumulation. Addition to the medium of 7-oxa-13-prostynoic acid, an inhibitor of prostaglandin action but not of its synthesis, abolished this stimulatory effect. Whereas this finding in fact points to an obligatory place of PGE between LH and cAMP in this regulatory process, a similar effect was not achieved in cultured porcine granulosa cells by the same inhibitor and comparable dose levels[94]. Also in bovine luteal cells excessively high doses have been found necessary to elicit an effect of this sort[118]. In any event, since in the presence of the prostaglandin synthetase inhibitor fluoroindomethacin, the LH-induced increase of cAMP remained uninfluenced both in mouse ovaries and bovine luteal cell preparations, a direct LH-stimulation of the initial synthesis of prostaglandins can be excluded with certainty[118]. How PGE can then be mobilised by LH is totally unknown at present.

Apart from these recent limitations of the original hypothesis of Kuehl *et al.* additional points concerning the interpretation of the above data are open to criticism. Firstly, the high inhibitor concentration necessary to exert the blocking effect under discussion has been shown to bring about cellular necrosis when acting together with LH or PGE_2 upon monkey granulosa cells kept in long-term cultures[119]. Thus, the above findings might be due to an unspecific toxic effect of the inhibitor rather than to a specific action, although a competitive antagonism of 7-oxa-13-prostynoic acid against PGE has been demonstrated. Secondly, LH and PGE_1 or PGE_2 at maximum stimulatory levels have been found additive in terms of increased cAMP in all *in vitro* systems tested so far[94, 111, 118]. Whereas in the case of incubated whole mouse ovaries it might be 'reasonable to assume that the increment of cAMP induced by prostaglandins represent the sum of the contributions by all cell types capable of this response'[96] which would include reactions in addition to those triggered by LH, the same obviously does not apply to single cell-type cultures as used by Channing. Finally, the magnitude of the cAMP increase in response to LH as compared with the response to prostaglandins differs considerably. Thus, in the whole mouse ovary preparation much more cAMP builds up secondary to PGE than to LH, while the reverse is true in single cell preparations from porcine granulosa cells or bovine luteal cells mentioned above.

As intriguing as Kuehl's hypothesis is, much more information is needed before the precise position of PGE in the regulation of steroidogenesis can adequately be described. In this context, also a place following cAMP should be considered since the synthesis of PGE by mouse ovaries was stimulated by cAMP *in vitro*[118].

Whereas all experimental data quoted so far are indicative of PGE being the decisive prostaglandin with regard to the regulation of ovarian steroidogenesis (maybe via cAMP), it has recently been shown by Goldberg *et al.*[120] that in rat uterus preparations cGMP formation is under the control of PGF_2, as cAMP is supposed to be under the control of PGE. Should the same turn out to be true for the ovary and cGMP be found to oppose the stimulatory effects of cAMP on steroidogenesis, this then would indeed lend support to the concept of a *dual regulatory system* of the ovarian steroidogenesis as now suggested by Kuehl *et al.*[118]. The paradox that PGF is luteolytic *in vivo* in many species, while PGE appears to be luteotropic and steroidogenic *in vitro* would then be better understood. At present, all such considerations are purely speculative but they will stimulate scientific activity in this field beyond doubt.

10.3.4 Ultimate site of action of the gonadotrophins upon ovarian steroidogenesis

From numerous *in vitro* experiments with ovarian tissues of various species it has become obvious that the *de novo* synthesis of steroids from labelled acetate may increase as a consequence of LH of hCG being added to the medium[77, 121]. Whether this is the result of a primary action of LH upon a very early stage of the formation of cholesterol as suggested by Savard *et al.*[121] or whether this happens secondary to a LH-induced depletion of the intracellular cholesterol stores along a negative intracellular feed-back regulatory mechanism as postulated by Armstrong[122], has been a matter of controversy in the past. In any event, the contribution of cholesterol formed *de novo* within the ovary to the overall steroidogenesis must be very limited and has been calculated on the basis of acute *in vitro* experiments not to exceed 16%[123]. Beyond any doubt, the bulk of the ovarian cholesterol available for further transformation into steroids originates from circulating cholesterol. This has convincingly been shown both in pseudopregnant rabbits and PMS/hCG treated immature rats after *in vivo* labelling of the ovarian cholesterol pool by intravenous administration of tritiated or ^{14}C-cholesterol[124, 125]. Owing to rapid equilibration of this labelled sterol among the respective body pools, progestins appreciably labelled were soon found in both the blood and the ovarian tissue, with the specific activities of tissue progestins and cholesterol being similar. Under these experimental conditions, LH either given *in vivo* or added to the medium *in vitro* was able to increase the formation of labelled progestins without changing the specific activity. This would be an unexpected finding if LH were to stimulate primarily the *de novo* synthesis of cholesterol within the ovary. In accordance with these findings, effective inhibition of the cholesterol synthesis by means of AY 9944 did not interfere with the stimulatory effect of LH on the *in vitro* synthesis of progesterone and its 20α-OH analogue by ovarian interstitial tissue of mature rabbits as well as by corpora lutea of cows[122, 126]. It is clear from these and several other data that LH acts mainly if not exclusively upon the biosynthesis of progesterone somewhere along the pathway between cholesterol and Δ^5-pregnenolone[127, 128, 129].

It has been hypothesised by Behrman and Armstrong[130] that both the cholesteryl synthetase and the esterase which regulate storage and depletion of the cholesterol esters within the ovarian cell are under direct gonadotrophic control and might be the main site of gonadotrophic action in the regulation of steroidogenesis. The starting point for this concept has been the observation that the cholesterol esterase activity in the 100 000 g supernatant of homogenised luteinised rat ovaries was stimulated to some extent by *in vivo* LH-pretreatment. The implications of these results with respect to the intracellular regulatory mechanisms of cholesterol turnover might be limited, however, since the measured increase of the specific enzyme activity was much too low to fully account for the high de-esterification rate of stored cholesterol esters observed secondary to LH[131]. More recently, Behrman *et al.*[132], on the basis of *in vivo* and *in vitro* findings in the pseudopregnant rat treated with anti-LH serum, suggested 'that the action of LH and prolactin may thus be complementary in that prolactin maintains adequate levels of both cholesteryl esterase and synthetase, and LH controls the activity of these preformed enzymes'. $PGF_{2\alpha}$ has been found to oppose the postulated tropic effects of prolactin in that the cholesterol synthetase is blocked by the prostaglandin which leads to a depressed ovarian cholesterol ester turnover and a reduced availability of cholesterol ester for transformation into progesterone[133].

Despite an impressive body of information on this subject having accumulated during recent years[131], this concept is far from being generally recognised and might be restricted to species such as rats and mice, in which prolactin is integrated in the regulation of steroidogenesis. For example, one might argue in analogy to the above described controversy concerning the suspected effect of LH upon early stages of cholesterol biosynthesis, that the changes in intracellular storage and depletion of cholesterol esters secondary to the action of LH and/or prolactin are only the consequence of the primary steroidogenic action of these gonadotrophins at quite another site within the cell. To obviate this sort of argumentation, Behrman *et al.*[134] designed experiments using superovulated immature rats which had been treated with aminoglutethimide phosphate, notably an inhibitor of the cholesterol side-chain cleavage enzyme, as well as with LH. As expected, the progesterone content of the ovaries was markedly decreased and rendered resistant to LH by aminoglutethimide, while at the same time the concentration of esterified cholesterol was considerably increased. Of special importance was the finding that the well-known cholesterol ester-depleting effect of LH—although at an elevated level—was fully preserved under these experimental conditions. The thus demonstrated separation of the steroidogenic from the cholesterol ester-depleting action of LH would, in fact, indicate that making cholesterol esters available for steroidogenesis is a direct effect of LH and not an indirect one triggered by the consumption of cholesterol for the biosynthesis of steroid hormones. As intriguing as the outcome of these experiments has been, it would be, nonetheless, inconclusive should it be proved true that the primary action of LH/cAMP is directed towards the synthesis of a carrier protein as discussed below. Then, the increased production of such a cholesterol carrier protein secondary to LH stimulation which is not influenced by aminoglutethimide could also be the cause of intracellular cholesterol ester depletion.

The cholesterol side-chain cleavage enzyme system has most often been suspected to be the final locus of gonadotrophin action[127, 128] but this suggestion has never been directly proved. Similar to the situation in the adrenals[135], the activity of this enzyme complex in slices of bovine corpora lutea and in mitochondria from rabbit ovarian interstitial tissue was not stimulated by LH or cAMP although the production of progesterone was increased under the chosen conditions[4, 136]. This would imply that the enzyme system under discussion is not rate-limiting with regard to the formation of progesterone.

Since similarities in the control of steroidogenesis between the adrenals and ovaries do exist in almost all respects, the assumption might also be justified that ACTH and LH share a common mechanism in stimulating steroid formation in the respective endocrine glands. With this premise in mind, findings in the study of adrenals during recent years which point to a protein being involved under the control of ACTH or cAMP in the regulation of steroidogenesis, should be considered an important hint as to how LH might exert its activity upon the transformation of cholesterol into Δ^5-pregnenolone. As recently reviewed by Gill[135], synthesis of this adrenal protein under the influence of ACTH and cAMP is inhibited *in vitro* by both puromycin and cycloheximide but not by actinomycin D. cAMP is therefore supposed to regulate the synthesis of this protein at the level of translation of preformed, relatively stable messenger RNA. Its rapid turnover with a calculated half-life of 10 min also fulfils the prerequisite for its taking an important position in an acute regulatory process such as stimulation of steroidogenesis. Very recently Ungar *et al.*[137] reported the isolation of such a protein from adrenal mitochondria which—without having the characteristics of an enzyme—was found able to stimulate the cholesterol side-chain cleavage activity many-fold. This adrenal activator protein was also thought to have a carrier function concerning the transport of 'cholesterol from the cell cytoplasm to the mitochondria where the complex can serve as the active substrate for the side-chain cleavage by the P-450 system'. However, no evidence has so far been presented that this protein is under direct control of ACTH and/or cAMP. In the ovary also, the transport of cholesterol to the mitochondria as well as the involvement of a protein with a short turnover has been suggested to be the direct site of action of LH, cAMP, and the respective transmitters[4,95,136]. In any event, future development in this special area of investigation will be looked forward to with great fascination. Somehow or other a rapidly formed protein will be found to be operative as part of the cAMP-controlled chain of reactions leading finally to a stimulated transformation of cholesterol to Δ^5-pregnenolone.

References

1. Nalbandov, A. V. (1973). *J. Reprod. Fert.*, **34,** 1
2. Lindner, H. R. and Shelesnyak, M. C. (1967). *Acta Endocr.* (*Kbh.*), **56,** 27
3. Hammerstein, J., Rice, B. F. and Savard, K. (1964). *J. Clin. Endocr.*, **24,** 597
4. Savard, K. (1973). *Biol. Reprod.*, **8,** 183
5. Greep, R. O., Dyke van, H. B. and Chow, B. F. (1942). *Endocrinology*, **30,** 635
6. Eshkol, A. and Lunenfeld, B. (1967). *Acta Endocr.* (*Kbh.*), **54,** 91
7. Petrusz, P., Robyn, C. and Diczfalusy, E. (1970). *Acta Endocr.* (*Kbh.*), **63,** 454
8. Rosemberg, E., Joshi, S. R. (1968). *Gonadotropins 1968*, 91 (E. Rosemberg, editor) (Los Altos, Calif. Geron-X)
9. Lostroh, A. J. and Johnson, R. E. (1966). *Endocrinology*, **79,** 991
10. Jewelewicz, R., Warren, M., Dyrenfurth, I. and Vande Wiele, R. L. (1971). *J. Clin. Endocr. Metab.*, **32,** 688
11. Moor, R. M. (1973). *J. Reprod. Fert.*, **32,** 545
12. Channing, C. P. (1970). *Rec. Progr. Horm. Res.*, **26,** 589
13. Hoffmann, F. (1960). *Geburtsh. Frauenhk.*, **20,** 1153
14. Hoffmann, F., Kayser, W. and Bergk, K. H. (1970). *Geburtsh. Frauenhk.*, **30,** 347
15. Tillinger, K. G., Johannisson, E., Lisboa, B. P., Reaside, J. I. and Diczfalusy, E. (1970). *Acta Obstet. Gynec. Scand.*, **49,** 35
16. Lipner, H. and Wendelken, L. (1971). *Proc. Soc. Exp. Biol. Med.*, **136,** 1141
17. Lipner, H. and Greep, R. O. (1971). *Endocrinology*, **88,** 602
18. Ferin, M., Tempone, A., Zimmering, P. and Vande Wiele, R. L. (1969). *Endocrinology*, **85,** 1070
19. Armstrong, D. T., Moon, Y. S. and Grinwich, D. L. (1973). *Advances in the biosciences*, **9,** 90
20. O'Grady, J. P., Caldwell, B. V., Auletta, F. J. and Speroff, L. (1972). *Prostaglandins*, **1,** 97
21. Bartosik, D. B. and Romanoff, E. B. (1969). *The Gonads*, 211 (New York: Appleton-Century-Crofts)
22. Illingworth, D. V. and Perry, J. S. (1973). *J. Reprod. Fert.*, **33,** 457
23. Knobil, E. (1973). *Biol. Reprod.*, **8,** 246
24. Moudgal, N. R., Macdonald, G. J. and Greep, R. O. (1971). *J. Clin. Endocr. Metab.*, **32,** 579
25. Hilliard, J. (1973). *Biol. Reprod.*, **8,** 203
26. Denamur, R., Martinet, J. and Short, R. V. (1973). *J. Reprod. Fert.*, **32,** 207
27. Keyes, P. L. and Nalbandov, A. V. (1967). *Endocrinology*, **80,** 938
28. Hixon, J. E. and Armstrong, D. T. (1971). *4th Ann. Meet. Study Reprod.*, Abst. 17
29. Takayama, M. and Greenwald, G. S. (1973). *Endocrinology*, **92,** 1405

30. Vande Wiele, R. L., Bogumil, J., Dyrenfurth, I., Ferin, M., Jewelewicz, R., Warren, M., Riskallah, T. and Mikhail, G. (1970). *Rec. Progr. Horm. Res.*, **26,** 63
31. Yen, S. S. C., Llerena, O., Little, B. and Pearson, O. H. (1968). *J. Clin. Endocr. Metab.*, **28,** 1763
32. Kosasa, T., Levesque, L., Goldstein, D. P. and Taymor, M. L. (1973). *J. Clin. Endocr. Metab.*, **36,** 622
33. Vandekerckhove, D. and Dhont, M. (1972). *Ann. d'Endocr.*, **33,** 205
34. Llerena, L. A., Guevara, A., Lobotsky, J., Lloyd, C. W., Weisz, J., Pupkin, M., Zanartu, J. and Puga, J. (1969). *J. Clin. Endocr. Metab.*, **29,** 1083
35. Geiger, W. and Kaiser, R. (1971). *Acta Endocr.*, **67,** 331
36. Short, R. V. (1967). *Ann. Rev. Physiol.*, **29,** 373
37. Yen, S. S. C., Vela, P. and Rankin, J. (1970). *J. Clin. Endocr. Metab.*, **30,** 435
38. Pharriss, B. B., Tillson, S. A. and Erickson, R. R. (1972). *Rec. Progr. Horm. Res.*, **28,** 51
39. Le Maire, W. J., Askari, H. and Savard, K. (1971). *Steroids*, **17,** 65
40. Beavis, E. L. G., Brown, J. B. and Smith, M. A. (1969). *J. Obstet. Gynaec. Brit. Cwlth.*, **76,** 969
41. Coyotupa, J., Buster, J., Parlow, A. F. and Dignam, W. J. (1973). *J. Clin. Endocr. Metab.*, **36,** 395
42. Caldwell, B. V., Rowson, L. E. A., Moor, R. M. and Hay, M. F. (1969). *J. Reprod. Fert.*, **8,** 59
43. Hammond, J. and Robson, J. M. (1951). *Endocrinology*, **49,** 384
44. Hoffmann, F. and Meger, C. (1965). *Geburtsh. Frauenhk.*, **25,** 1132
45. Rivera, A. and Sherman, A. I. (1969). *Amer. J. Obstet. Gynec.*, **103,** 986
46. Gore, B. Z., Caldwell, B. V. and Speroff, L. (1973). *J. Clin. Endocr. Metab.*, **36,** 615
47. Bacic, M., De Casparis, A. W. and Diszfalusy, E. (1970). *Amer. J. Obstet. Gynec.*, **107,** 531
48. Karsch, F. J., Krey, L. C., Weick, R. F., Dierschke, J. and Knobil, E. (1973). *Endocrinology*, **92,** 1148
49. Johansson, E. D. B. (1971). *Acta Endocr.* (*Kbh.*), **68,** 779
50. Pickles, V. R., Hall, W. J., Best, F. A. and Smith, G. N. (1965). *J. Obstet. Gynaec. Brit. Cwlth.*, **72,** 185
51. Lehmann, F., Peters, F., Breckwoldt, M. and Bettendorf, G. (1972). *Prostaglandins*, **1,** 269
52. Auletta, F. J., Speroff, L. and Caldwell, B. V. (1973). *J. Clin. Endocr. Metab.*, **36,** 405
53. Inskeep, E. K. and Butcher, R. L. (1966). *J. Anim. Sci.*, **25,** 1164
54. Moor, R. M. and Rowson, L. E. A. (1964). *Nature* (*Lond.*), **201,** 522
55. Rowson, L. E. A. and Moor, R. M. (1967). *J. Reprod. Fert.*, **13,** 511
56. McCracken, J. A., Baird, D. T. and Goding, J. R. (1971). *Rec. Progr. Horm. Res.*, **27,** 537
57. Baird, D. T. and Land, R. B. (1973). *J. Reprod. Fert.*, **33,** 393
58. Caldwell, B. V. and Moor, R. M. (1971) *J. Reprod. Fert.*, **26,** 133
59. Goding, J. R., Baird, D. T., Cumming, I. A. and McCracken, J. A. (1971). *Karolinska Symp. Res. Meth. Reprod. Endocr. 4th; Perfusion Techniques*, 169
60. Barretts, S., Blockey, M., Brown, J. M., Cumming, I. A., Goding, J. R. and Mole, B. J. (1971). *J. Reprod. Fert.*, **24,** 136
61. McCracken, J. A., Glew, M. E. and Scaramuzzi, R. J. (1970). *J. Clin. Endocr. Metab.*, **30,** 544
62. Aldridge, R. R., Barrett, S., Brown, J. B., Funder, J. W., Goding, J. R., Kaltenbach, C. C. and Mole, B. J. (1970). *J. Reprod. Fert.*, **21,** 360
63. McCracken, J. A., Carlson, J. C., Glew, M. E., Goding, J. R., Baird, D. T., Green, K. and Samuelsson, B. (1972). *Nature* (*Lond.*), **238,** 129
64. Thorburn, G. D., Cox, R. I., Currie, W. B., Restall, B. J. and Schneider, W. (1972). *J. Endocr.*, **53,** 325
65. Wilson, L., Cenedella, R. J., Butcher, R. L. and Inskeep, E. K. (1972). *J. Anim. Sci.*, **54**(1), 93
66. Wilson, L., Butcher, R. L. and Inskeep, E. K. (1972). *Prostaglandins*, **1,** 479
67. Caldwell, B. V., Tillson, S. A., Brock, W. A. and Speroff, L. (1972). *Prostaglandins*, **1,** 217
68. Karsch, F. J., Noveroske, J. W., Roche, J. F., Norton, H. W. and Nalbandov, A. V. (1970). *Endocrinology*, **87,** 1228

69. Bjersing, L., Hay, M. R., Kann, G., Moor, R. M., Naftolin, F., Scaramuzzi, R. J., Short, R. V. and Younglai, E. V. (1972). *J. Endocr.*, **52,** 465
70. Nalbandov, A. V. (1970). *Biol. Reprod.*, **2,** 7
71. Cerini, M. E. D., Chamley, W. A., Findlay, J. K. and Goding, J. R. (1972). *Prostaglandins*, **2,** 433
72. Gutknecht, G. D., Duncan, G. W. and Wyngarden, I. J. (1972). *Proc. Soc. Exp. Biol. Med.*, **139,** 406
73. Garren, L. D. (1968). *Vitamins and Hormones*, **26,** 119
74. Mason, N. R. and Savard, K. (1964). *Endocrinology*, **75,** 215
75. Mills, T. M., Davies, P. J. A. and Savard, K. (1971). *Endocrinology*, **88,** 857
76. Norman, R. L. and Greenwald, G. S. (1971). *Endocrinology*, **89,** 598
77. Armstrong, D. T., O'Brien, J. and Greep, R. O. (1964). *Endocrinology*, **75,** 488
78. Dorrington, J. H. and Kilpatrick, R. (1966). *J. Endocr.*, **35,** 53
79. Hilliard, J., Spies, H. G., Lucas, L. and Sawyer, C. H. (1968). *Endocrinology*, **82,** 122
80. Armstrong, D. T., Miller, L. S. and Knudsen, K. A. (1969). *Endocrinology*, **85,** 393
81. Armstrong, D. T., Knudsen, K. A. and Miller, L. S. (1970). *Endocrinology*, **86,** 634
82. Lee, C. Y., Coulam, C. B., Jiang, N. S. and Ryan, R. J. (1973). *J. Clin. Endocr. Metab.*, **36,** 148
83. Rajaniemi, H. and Van-Perttula, T. (1972). *Endocrinology*, **90,** 1
84. Coulson, P., Liu, T. C. and Gorski, J. (1972). *Gonadotropins*, 227, (Saxena, Beling, Gandy, editors) (New York: Wiley-Interscience)
85. Rao, Ch. V. and Saxena, B. B. (1973). *Biochim. Biophys. Acta*, **313,** 312
86. Kammerman, S., Canfield, R. E., Kolen, J. and Channing, C. P. (1972). *Endocrinology*, **91,** 65
87. Channing, C. P. and Kammerman, S. (1973). *Endocrinology*, 92, 531
88. Midgley, A. R. (1970). *Gonadotropins*, 248, (Saxena, Beling, Gandy, editors) (New York: Wiley-Interscience)
89. Lee, G., Keyes, P. and Jacobson, H. I. (1971). *Science*, **173,** 1032
90. Scott, R. S. and Rennie, P. I. C. (1971). *Endocrinology*, **89,** 297
91. Marsh, J. M. and Savard, K. (1966). *Steroids*, **8,** 133
92. Marsh, J. M., Butcher, R. W., Savard, K. and Sutherland, E. W. (1966). *J. Biol. Chem.*, **241,** 5436
93. Channing, C. P. and Seymour, J. F. (1970). *Endocrinology*, **87,** 165
94. Kolena, J. and Channing, C. P. (1972). *Endocrinology*, **90,** 1543
95. Hermier, C., Combarnous, Y. and Justisz, M. (1971). *Biochim. Biophys. Acta*, **244,** 625
96. Kuehl, F. A. Jr., Humes, J. L., Tarnoff, J., Cirillo, V. J. and Ham, E. A. (1970). *Science*, **169,** 883
97. Dorrington, J. H. and Kilpatrick, R. (1967). *Biochem. J.*, **104,** 725
98. Dorrington, J. H. and Bagget, B. (1969). *Endocrinology*, **84,** 989
99. Marsh, J. M., Mills, T. M. and LeMaire, W. J. (1972). *Biochim. Biophys. Acta*, **273,** 389
100. Marsh, J. M. (1970). *J. Biol. Chem.*, **245,** 1596
101. Stansfield, D. A., Horne, J. R. and Wilkinson, G. H. (1971). *Biochim. Biophys. Acta*, **227,** 413
102. Kuo, J. F., Krueger, B. K., Sanes, J. R. and Greengard, P. (1970). *Biochim. Biophys. Acta*, **212,** 79
103. Bedwani, J. R. and Horton, E. W. (1971). *Brit. J. Pharmacol.*, **43,** 794
104. Sellner, R. G. and Wickersham, E.W . (1972). *Biol. Reprod.*, **7,** 107
105. O'Grady, J. P., Kohorn, E. I., Glass, R. H., Caldwell, B. V., Brock, W. A. and Speroff, L. (1972). *J. Reprod. Fert.*, **30,** 153
106. Pharriss, B. B., Wyngarden, L. J. and Gutknecht, G. D. (1968). *Gonadotropins*, 121 (E. Rosemberg, editor) (Los Altos, Calif.: Geron-X)
107. Behrman, H., Yoshinaga, K. and Greep, R. O. (1971). *Ann. N.Y. Acad. Sci.*, **180,** 426
108. Wilks, J. W., Forbes, K. K. and Norland, J. F. (1973). *Prostaglandins*, **3,** 427
109. Demers, L. M., Behrman, H. R. and Greep, R. O. (1973). *Advances in the Biosciences*, **9,** 701
110. Speroff, L. and Ramwell, P. W. (1970). *J. Clin. Endocr. Metab.*, **30,** 345
111. Marsh, J. M. (1971). *Ann. N.Y. Acad. Sci.*, **180,** 416
112. Marsh, J. M. and Lemaire, W. J. (1972). *Excerpta Med. Int. Congr. Ser.*, **256,** abst. 476
113. Puri, C. P., Hingorani, V. and Laumas, K. R. (1973). *Advances in the Biosciences*, **9,** 657

114. Channing, C. P. (1972). *Prostaglandins*, **2,** 331
115. Chasalow, F. I. and Pharriss, B. B. (1972). *Prostaglandins*, **1,** 107
116. Wilks, J. W., Forbes, K. K. and Norland, J. F. (1972). *J. Reprod. Med.*, **9,** 271
117. LeMaire, W. J., Yang, N. S., Behrman, H. R. and Marsh, J. M. (1973). *Prostaglandins*, **3,** 367
118. Kuehl, F. A., Cirillo, V. J., Ham, E. A. and Humes, J. L. (1973). *Advances in the Biosciences.*, **9,** 155
119. Channing, C. P. (1972). *Prostaglandins*, **2,** 351
120. Goldberg, N. D., Haddox, M. K., Hartle, D. K., Hadden, J. W. (1973). *Pharmacology and the future of man*, 146. (Proc. 5th Int. Congr. Pharmacology) (Basel: Karger)
121. Savard, K., Marsh, J. M., Rice, B. F. (1965). *Rec. Progr. Horm. Res.*, **21,** 285
122. Armstrong, D. T. (1967). *Nature* (*Lond.*), **213,** 633
123. Savard, K., LeMaire, W. and Kumari, L. (1969). *The Gonads*, 119 (K. W. McKerns, editor) (Amsterdam: North-Holland)
124. Solod, E. A., Armstrong, D. T. and Greep, R. O. (1966). *Steroids*, **7,** 607
125. Major, P. W., Armstrong, D. T. and Greep, R. O. (1967). *Endocrinology*, **81,** 19
126. Armstrong, D. T., Lee, P. T. and Miller, L. S. (1970). *Biol. Reprod.*, **2,** 29
127. Hall, P. F. and Koritz, S. B. (1965). *Biochem.*, **4,** 1037
128. Channing, C. P. and Villee, C. A. (1966). *Biochim. Biophys. Acta*, **127,** 1
129. Armstrong, D. T. (1968). *Rec. Progr. Horm. Res.*, **24,** 255
130. Behrman, H. R. and Armstrong, D. T. (1969). *Endocrinology*, **85,** 474
131. Flint, A. P. F. and Armstrong, D. T. (1972). *Gonadotropins*, 261 (B. B. Saxena, C. G. Beling and H. M. Gandy, editors) (New York: Wiley Interscience)
132. Behrman, H. R., Moudgal, N. R. and Greep, R. O. (1972). *J. Endocr.*, **52,** 419
133. Behrman, H. R., MacDonald, G. J. and Greep, R. O. (1971). *Lipids*, **6,** 791
134. Behrman, H. R., Armstrong, D. T. and Greep, R. O. (1970). *Canad. Biochem.*, **48,** 881
135. Gill, G. N. (1972). *Metabolism*, **21,** 559
136. Joackanicz, T. M. and Armstrong, D. T. (1968). *Endocrinology*, **83,** 769
137. Ungar, F., Kan, K. W. and McCoy, K. E. (1973). *Ann. N.Y. Acad. Sci.*, **212,** 276

Index